W9-CPE-604

Introduction to Continuum Mechanics

REVISED EDITION IN SI/METRIC UNITS

PUES-17

Pergamon Unified Engineering Series

TITLES IN THE PERGAMON UNIFIED ENGINEERING SERIES

Vol. 1. W. H. DAVENPORT/D. ROSENTHAL—*Engineering: Its Role and Function in Human Society*
Vol. 2. M. TRIBUS—*Rational Descriptions, Decisions and Designs*
Vol. 3. W. H. DAVENPORT—*The One Culture*
Vol. 4. W. A. WOOD—*The Study of Metal Structures and Their Mechanical Properties*
Vol. 5. M. SMYTH—*Linear Engineering Systems: Tools and Techniques*
Vol. 6. L. M. MAXWELL/M. B. REED—*The Theory of Graphs: A Basis for Network Theory*
Vol. 7. W. R. SPILLERS—*Automated Structural Analysis: An Introduction*
Vol. 8. J. J. AZAR—*Matrix Structural Analysis*
Vol. 9. S. SEELY—*An Introduction to Engineering Systems*
Vol. 10. D. T. THOMAS—*Engineering Electromagnetics*
Vol. 11. D. ROSENTHAL—*Resistance and Deformation of Solid Media*
Vol. 12. S. J. BRITVEC—*The Stability of Elastic Systems*
Vol. 13. M. NOTON—*Modern Control Engineering*
Vol. 14. B. MORRILL—*An Introduction to Equilibrium Thermodynamics*
Vol. 15. R. PARKMAN—*The Cybernetic Society*
Vol. 16. ADEL SAADA—*Elasticity: Theory and Applications*
Vol. 17. W. M. LAI/D. RUBIN/E. KREMPL—*Introduction to Continuum Mechanics*
Vol. 18. D. K. ANAND—*Introduction to Control Systems*

Introduction to Continuum Mechanics

Revised Edition in SI/Metric Units

W. Michael Lai,
Professor of Mechanics,
Rensselaer Polytechnic Institute, Troy, N.Y.

David Rubin,
Senior Research Engineer,
Weidlinger Associates, N.Y., N.Y.

and

Erhard Krempl,
Professor of Mechanics,
Rensselaer Polytechnic Institute, Troy, N.Y.

Pergamon Press

Oxford · New York · Toronto · Sydney · Paris · Frankfurt

U.K.	Pergamon Press Ltd., Headington Hill Hall, Oxford OX3 0BW, England
U.S.A.	Pergamon Press Inc., Maxwell House, Fairview Park, Elmsford, New York 10523, U.S.A.
CANADA	Pergamon of Canada, Suite 104, 150 Consumers Road, Willowdale, Ontario M2J 1P9, Canada
AUSTRALIA	Pergamon Press (Aust.) Pty. Ltd., P.O. Box 544, Potts Point, N.S.W. 2011, Australia
FRANCE	Pergamon Press SARL, 24 rue des Ecoles, 75240 Paris, Cedex 05, France
FEDERAL REPUBLIC OF GERMANY	Pergamon Press GmbH, 6242 Kronberg-Taunus, Pferdstrasse 1, Federal Republic of Germany

Copyright © 1978 Pergamon Press Inc.

All Rights Reserved. No part of this publication may be reproduced, stored in a retrieval system or transmitted in any form or by any means: electronic, electrostatic, magnetic tape, mechanical, photocopying, recording or otherwise, without permission in writing from the publishers.

First Edition 1974

Reprinted 1977

Revised Edition in SI/Metric Units 1978

Reprinted 1979

British Library Cataloguing in Publication Data

Lai, Wei Michael
Introduction to continuum mechanics—Revised ed. in SI-metric units.—(Pergamon unified engineering series: vol. 17).
1. Continuum mechanics
I. Title II. Rubin, David, b. 1942
III. Krempl, Erhard
531 QA808.2 77-30616

ISBN 0-08-022698-1 (Hardcover)
ISBN 0-08-022699-X (Flexicover)

Printed in Great Britain by Biddles Ltd, Guildford, Surrey

Pergamon
Unified Engineering
Series

GENERAL EDITORS

Thomas F. Irvine, Jr.
State University of New York at Stony Brook
James P. Hartnett
University of Illinois at Chicago Circle

EDITORS

William F. Hughes
Carnegie-Mellon University
Arthur T. Murphy
Widener College
Daniel Rosenthal
University of California, Los Angeles

SECTIONS

Continuous Media Section
Engineering Design Section
Engineering Systems Section
Humanities and Social Sciences Section
Information Dynamics Section
Materials Engineering Section
Engineering Laboratory Section

Contents

Preface

This text is prepared for the purpose of introducing the concept of continuum mechanics to beginners in the field. Special attention and care have been given to the presentation of the subject matter so that it is within the grasp of those readers who have had a good background in calculus, some differential equations, and some rigid body mechanics. For pedagogical reasons the coverage of the subject matter is far from being extensive, only enough to provide for a better understanding of later courses in the various branches of continuum mechanics and related fields. The major portion of the material has been successfully class-tested at Rensselaer Polytechnic Institute for undergraduate students. However, the authors believe the text may also be suitable for a beginning graduate course in continuum mechanics.

We take the liberty to say a few words about the second chapter. This chapter introduces second-order tensors as linear transformations of vectors in a three-dimensional space. From our teaching experience, the concept of linear transformation is the most effective way of introducing the subject. It is a self-contained chapter so that prior knowledge of linear transformations, though helpful, is not required of the students. The third- and higher-order tensors are introduced through the generalization of the transformation laws for the second-order tensor. Indicial notation is employed whenever it economizes the writing of equations. Matrices are also used in order to facilitate computations. An appendix on matrices is included at the end of the text for those who are not familiar with matrices.

Also, let us say a few words about the presentation of the basic principles of continuum physics. Both the differential and integral formulation of the principles are presented. The differential formulations are given in Chapters 3, 4, and 6, at places where quantities needed in the formulation are defined, while the integral formulations are given later in Chapter 7. This is done for a pedagogical reason: the integral formulations as presented require slightly more mathematical sophistication on the part of a beginner and may be either postponed or omitted without affecting the main part of the text.

This text would never have been completed without the constant encouragement and advice from Professor F. F. Ling, Chairman of the Mechanics Division at RPI, to whom the authors wish to express their heartfelt thanks. Gratefully acknowledged is the financial support of the Ford Foundation under a grant which is directed by Dr. S. W. Yerazunis, Associate Dean of Engineering. The authors also wish to thank Drs. V. C. Mow and W. B. Brower, Jr. for their many useful suggestions. Special thanks are given to Dr. H. A. Scarton for painstakingly compiling a list of errata and suggestions on the preliminary edition. Finally, they are indebted to Mrs. Geri Frank who typed the entire manuscript.

Division of Mechanics
Rensselaer Polytechnic Institute

W. MICHAEL LAI,
DAVID RUBIN, and
ERHARD KREMPL

September, 1973

The Authors

W. Michael Lai (Ph.D., University of Michigan) is Professor of Mechanics at Rensselaer Polytechnic Institute, Troy, New York. He is a member of ASME, AMS, Sigma Xi, and Phi Kappa Phi.

David Rubin (Ph.D., Brown University), formerly Assistant Professor of Mechanics at Rensselaer Polytechnic Institute, Troy, New York, is senior research engineer at Weidlinger Associates, New York, New York. He is a member of ASME, Sigma Xi, Tau Beta Pi, and Chi Epsilon.

Erhard Krempl (Dr.-Ing, Technische Hochschule München) is Professor of Mechanics at Rensselaer Polytechnic Institute, Troy, New York. Before coming to RPI in 1968 he was a Research Associate at Technische Hochschule München and a Mechanics of Materials Engineer for the General Electric Company in Schenectady. Dr. Krempl is a member of ASME, ASTM, SESA, and Sigma Xi.

1

Introductions

1.1 CONTINUUM THEORY

Matter is formed of molecules which in turn consist of atoms and subatomic particles. Thus matter is not continuous. However, there are many aspects of everyday experience regarding the behaviors of materials, such as the amount of lengthening of a steel bar under the action of given forces, the rate of discharge of water in a pipe under a given head or the drag force experienced by a body moving in the air etc., which can be described and predicted with theories that pay no attention to the molecular structure of materials. The theory which aims at describing relationships between gross phenomena, neglecting the structure of material on a smaller scale, is known as the continuum theory. The continuum theory regards matter as indefinitely divisible. Thus, within this theory, one accepts the idea of an infinitesimal volume of materials referred to as a particle in the continuum, and in every neighborhood of a particle there are always materials present. Whether the continuum theory is justified or not depends on the given situation; for example, while the continuum approach adequately describes the behavior of steel in many circumstances, it does not yield results that are in accord with experimental observations in the propagation of waves of extremely small wavelength in steel. On the other hand, a rarefied gas may be adequately described by a continuum in certain circumstances. At any case, it is misleading to justify the continuum approach on the basis of the number of molecules in a given volume. After all, an infinitesimal volume in the limit contains no molecules at all. Neither is it necessary to infer that quantities occurring in continuum theory must be interpreted as certain particular statistical averages. In fact, it has been known that the same equation for a continuum can be arrived at by different hypothesis of molecular structure and definitions of gross variables. While molecular-statistical theory, whenever available, does enhance the understanding of the continuum theory, the point to be made is simply that whether the continuum theory is justified in a given situation is a matter of experimental test, not of philosophy. Suffice it to say that more than a hundred years of tests have justified such a theory in a wide variety of situations.

1.2 CONTENTS OF CONTINUUM MECHANICS

Continuum mechanics studies the response of materials to different loading conditions. Its subject matter can be divided into two main parts: (1) general principles common to all media, and (2) constitutive equations defining idealized materials. The general principles are axioms considered to be self-evident from our experience with the physical world, such as conservation of mass, principle of linear momentum, principle of moment of momentum, conservation of energy, principle of entropy, conservation of charges and magnetic flux, etc. In the present text, intended primarily for the beginners in the subject, we shall limit ourselves to situations where only the first four principles will be needed. Mathematically, there are two equivalent forms of the general principles: (1) the integral form, formulated for a finite volume of material in the continuum, and (2) the field equations for a differential volume of material (particle) at every point of the field of interest. Field equations are often derived from the integral form. They can also be derived directly from the free body of a differential volume. The latter approach seems to suit beginners better. In this text both approaches are presented, with the integral form presented toward the end of the text. Field equations are important wherever the variations of the field variables in the field is either of interest by itself or are needed to get the desired information. On the other hand, the integral forms of conservation laws lend themselves readily to certain approximate solutions.

The second major part of the theory of continuum mechanics concerns the "constitutive equations" which are used to define idealized material. Idealized materials represent certain aspects of the mechanical behavior of the natural materials. For example, for many materials under restricted conditions, the deformation caused by the application of loads disappears with the removal of the loads. This aspect of the material behavior is represented by the constitutive equation of an elastic body. Under even more restricted conditions, the state of stress at a point depends linearly on the changes of lengths and mutual angle suffered by elements at the point measured from the state where the external and internal forces vanish. The above expression defines the linearly elastic solid. Another example is supplied by the classical definition of viscosity which is based on the assumption that the state of stress depends linearly on the instantaneous rates of change of length and mutual angle. Such a constitutive equation defines the linearly viscous fluid. The mechanical behavior of real materials varies not only from material to material but also with different loading conditions for a given material. This leads to the formulation of many constitutive equations defining the many different aspects of material behavior. Within the past two decades, the theory of constitutive equations has advanced significantly toward the direction of generality and rationality. In this introductory text, we shall refrain from discussing the theory of constitutive equations generally but rather present three idealized models and study the behavior they represent by means of some solutions of simple boundary-value problems. To be more specific, the idealized materials chosen are (1) the linearly elastic solid, (2) the linearly viscous fluid including inviscid fluid, and (3) incompressible simple fluid in viscometric flows.

One important requirement which must be satisfied by all quantities used in the formulation of a physical law is that they be coordinate-invariant. In the following chapter, we discuss such quantities.

2

Tensors

As mentioned in the introduction, all laws of continuum mechanics must be formulated in terms of quantities that are independent of coordinates. It is the purpose of this chapter to introduce such mathematical entities. We shall begin by introducing a short-hand notation—the indicial notation—in Part A of this chapter, which will be followed by the concept of tensors introduced as a linear transformation in Part B.

PART A THE INDICIAL NOTATION

2A1. Summation Convention, Dummy Indices

Consider the sum

$$s = a_1x_1 + a_2x_2 + \cdots a_nx_n. \tag{A1}$$

We can write the above equation in a compact form by using a summation sign:

$$s = \sum_{i=1}^{n} a_ix_i. \tag{A2}$$

It is obvious that the following equations have exactly the same meaning as Eq. (A2)

$$s = \sum_{j=1}^{n} a_jx_j, \tag{A3}$$

$$s = \sum_{m=1}^{n} a_mx_m, \tag{A4}$$

etc.

The index i in Eq. (A2), or j in Eq. (A3), or m in Eq. (A4) is a dummy index in the sense that the sum is independent of the letter used.

We can further simplify the writing of Eq. (A1) if we adopt the following convention: Whenever an index is repeated once, it is a dummy index indicating a summation with the index running through the integral numbers $1, 2, \ldots, n$.

This convention is known as Einstein's summation convention. Using the convention, Eq. (A1) shortens to

$$s = a_i x_i. \tag{A5}$$

We also note that

$$a_i x_i = a_m x_m = a_j x_j = \ldots. \tag{A6}$$

It is emphasized that expressions such as $a_i b_i x_i$ are not defined within this convention. That is, an index should *never* be repeated more than once when the summation convention is used. Therefore, an expression of the form

$$\sum_{i=1}^{n} a_i b_i x_i$$

must retain its summation sign.

In the following we shall always take n to be 3 so that, for example,

$$a_i x_i = a_m x_m = a_1 x_1 + a_2 x_2 + a_3 x_3,$$

$$a_{ii} = a_{mm} = a_{11} + a_{22} + a_{33},$$

$$a_i \mathbf{e}_i = a_1 \mathbf{e}_1 + a_2 \mathbf{e}_2 + a_3 \mathbf{e}_3.$$

The summation convention obviously can be used to express a double sum, a triple sum, etc. For example, we can write

$$\sum_{i=1}^{3} \sum_{j=1}^{3} a_{ij} x_i x_j \tag{A7}$$

simply as

$$a_{ij} x_i x_j. \tag{A8}$$

Expanding in full, the expression (A8) gives a sum of nine terms, i.e.,

$$\begin{aligned} a_{ij} x_i x_j = {} & a_{11} x_1 x_1 + a_{12} x_1 x_2 + a_{13} x_1 x_3 + a_{21} x_2 x_1 + a_{22} x_2 x_2 \\ & + a_{23} x_2 x_3 + a_{31} x_3 x_1 + a_{32} x_3 x_2 + a_{33} x_3 x_3. \end{aligned} \tag{A9}$$

For beginners, it is probably better to perform the above expansion in two steps, first, sum over i and then sum over j (or vice versa), i.e.,

$$a_{ij} x_i x_j = a_{1j} x_1 x_j + a_{2j} x_2 x_j + a_{3j} x_3 x_j,$$

where

$$a_{1j}x_1x_j = a_{11}x_1x_1 + a_{12}x_1x_2 + a_{13}x_1x_3,$$

etc.

Similarly, the triple sum

$$\sum_{i=1}^{3}\sum_{j=1}^{3}\sum_{k=1}^{3} a_{ijk}x_ix_jx_k \tag{A10}$$

will simply be written as

$$a_{ijk}x_ix_jx_k. \tag{A11}$$

The expression (A11) represents the sum of 27 terms.

We emphasize again that expressions such as $a_{ii}x_ix_jx_j$ or $a_{ijk}x_ix_ix_jx_k$ are not defined in the summation convention, they do not represent

$$\sum_{i=1}^{3}\sum_{j=1}^{3} a_{ii}x_ix_jx_j \quad \text{or} \quad \sum_{i=1}^{3}\sum_{j=1}^{3}\sum_{k=1}^{3} a_{ijk}x_ix_ix_jx_k.$$

2A2. Free Indices

Consider the following system of three equations

$$x_1' = a_{11}x_1 + a_{12}x_2 + a_{13}x_3, \tag{A12a}$$

$$x_2' = a_{21}x_1 + a_{22}x_2 + a_{23}x_3, \tag{A12b}$$

$$x_3' = a_{31}x_1 + a_{32}x_2 + a_{33}x_3. \tag{A12c}$$

Using the summation convention, Eqs. (A12) can be written as

$$x_1' = a_{1m}x_m, \tag{A13a}$$

$$x_2' = a_{2m}x_m, \tag{A13b}$$

$$x_3' = a_{3m}x_m, \tag{A13c}$$

which can be shortened into

$$x_i' = a_{im}x_m, \qquad i = 1, 2, 3. \tag{A14}$$

An index which appears only once in each term of an equation such as the index i in Eq. (A14) is called a "free index." A free index takes on the integral number 1, 2, or 3 *one at a time*. Thus Eq. (A14) is short-hand for three equations each having a sum of three terms on its right-hand side [i.e., Eqs. (A12a, b, c)].

A further example is given by

$$\mathbf{e}_i' = Q_{im}\mathbf{e}_m, \qquad i = 1, 2, 3, \tag{A15}$$

representing

$$\mathbf{e}_1' = Q_{11}\mathbf{e}_1 + Q_{12}\mathbf{e}_2 + Q_{13}\mathbf{e}_3, \quad \text{(A15a)}$$

$$\mathbf{e}_2' = Q_{21}\mathbf{e}_1 + Q_{22}\mathbf{e}_2 + Q_{23}\mathbf{e}_3, \quad \text{(A15b)}$$

$$\mathbf{e}_3' = Q_{31}\mathbf{e}_1 + Q_{32}\mathbf{e}_2 + Q_{33}\mathbf{e}_3. \quad \text{(A15c)}$$

We note that $x_j' = a_{jm}x_m$, $j = 1, 2, 3$, is the same as Eq. (A14) and $\mathbf{e}_j' = Q_{jm}\mathbf{e}_m$, $j = 1, 2, 3$ is the same as Eq. (A15). However

$$a_i = b_j$$

is a meaningless equation. *The free index appearing in every term of an equation must be the same.* Thus the following equations are meaningful

$$a_i + k_i = c_i,$$

$$a_i + b_i c_j d_j = 0.$$

If there are two free indices appearing in an equation such as

$$T_{ij} = A_{im}A_{jm}, \qquad i = 1, 2, 3; j = 1, 2, 3 \quad \text{(A16)}$$

then the equation is a short-hand writing of 9 equations. For example, Eq. (A16) represents 9 equations; each one has a sum of 3 terms on the right-hand side. In fact

$$T_{11} = A_{1m}A_{1m} = A_{11}A_{11} + A_{12}A_{12} + A_{13}A_{13} \quad \text{(A16a)}$$

$$T_{12} = A_{1m}A_{2m} = A_{11}A_{21} + A_{12}A_{22} + A_{13}A_{23} \quad \text{(A16b)}$$

$$T_{13} = A_{1m}A_{3m} = A_{11}A_{31} + A_{12}A_{32} + A_{13}A_{33} \quad \text{(A16c)}$$

. . . .

. . . .

$$T_{33} = A_{3m}A_{3m} = A_{31}A_{31} + A_{32}A_{32} + A_{33}A_{33}. \quad \text{(A16d)}$$

Again, equations such as

$$T_{ij} = T_{ik}$$

have no meaning.

2A3. Kronecker Delta

The Kronecker delta, denoted by δ_{ij}, is defined as

$$\delta_{ij} = \begin{cases} 1 & \text{if } i = j \\ 0 & \text{if } i \neq j. \end{cases} \quad \text{(A17)}$$

That is,

$$\delta_{11} = \delta_{22} = \delta_{33} = 1,$$

$$\delta_{12} = \delta_{13} = \delta_{21} = \delta_{23} = \delta_{31} = \delta_{32} = 0.$$

In other words, the matrix

$$\begin{bmatrix} \delta_{11} & \delta_{12} & \delta_{13} \\ \delta_{21} & \delta_{22} & \delta_{23} \\ \delta_{31} & \delta_{32} & \delta_{33} \end{bmatrix}$$

is the identity matrix

$$\begin{bmatrix} 1 & 0 & 0 \\ 0 & 1 & 0 \\ 0 & 0 & 1 \end{bmatrix}.$$

We note the following

(a) $\delta_{ii} = \delta_{11} + \delta_{22} + \delta_{33} = 1 + 1 + 1 = 3,$

(b) $\delta_{1m}a_m = \delta_{11}a_1 + \delta_{12}a_2 + \delta_{13}a_3 = a_1,$ (A18)

$\delta_{2m}a_m = \delta_{21}a_1 + \delta_{22}a_2 + \delta_{23}a_3 = a_2,$

$\delta_{3m}a_m = \delta_{31}a_1 + \delta_{32}a_2 + \delta_{33}a_3 = a_3.$

Or, in general

$$\delta_{im}a_m = a_i \tag{A19}$$

(c) $\delta_{1m}T_{mj} = \delta_{11}T_{1j} + \delta_{12}T_{2j} + \delta_{13}T_{3j} = T_{1j},$

$\delta_{2m}T_{mj} = T_{2j},$

$\delta_{3m}T_{mj} = T_{3j},$

or, in general

$$\delta_{im}T_{mj} = T_{ij}. \tag{A20}$$

In particular,

$$\delta_{im}\delta_{mj} = \delta_{ij},$$

$$\delta_{im}\delta_{mj}\delta_{jn} = \delta_{in},$$

etc.

(d) If $\mathbf{e}_1, \mathbf{e}_2, \mathbf{e}_3$ are unit vectors perpendicular to each other, then

$$\mathbf{e}_i \cdot \mathbf{e}_j = \delta_{ij}. \tag{A21}$$

2A4. Permutation Symbol

The permutation symbol, denoted by ϵ_{ijk}, is defined by

$$\epsilon_{ijk} = \begin{Bmatrix} 1 \\ -1 \\ 0 \end{Bmatrix} \text{ according to whether } i, j, k \begin{Bmatrix} \text{form an even} \\ \text{form an odd} \\ \text{do not form} \end{Bmatrix} \text{ permutation of } 1, 2, 3 \tag{A22}$$

i.e.,

$$\epsilon_{123} = \epsilon_{231} = \epsilon_{312} = +1$$

$$\epsilon_{132} = \epsilon_{321} = \epsilon_{213} = -1$$

and

$$\epsilon_{111} = \epsilon_{112} = \cdots = 0.$$

We note that

$$\epsilon_{ijk} = \epsilon_{jki} = \epsilon_{kij} = -\epsilon_{jik} = -\epsilon_{ikj} = -\epsilon_{kji}. \tag{A23}$$

If $\mathbf{e}_1, \mathbf{e}_2, \mathbf{e}_3$ form a right-handed triad, then

$$\mathbf{e}_1 \times \mathbf{e}_2 = \mathbf{e}_3, \quad \mathbf{e}_2 \times \mathbf{e}_3 = \mathbf{e}_1, \text{ etc.},$$

which can be written for short as

$$\mathbf{e}_i \times \mathbf{e}_j = \epsilon_{ijk}\mathbf{e}_k. \tag{A24}$$

Now, if $\mathbf{a} = a_i\mathbf{e}_i$ and $\mathbf{b} = b_i\mathbf{e}_i$, then

$$\mathbf{a} \times \mathbf{b} = (a_i\mathbf{e}_i) \times (b_j\mathbf{e}_j) = a_ib_j(\mathbf{e}_i \times \mathbf{e}_j) = a_ib_j\epsilon_{ijk}\mathbf{e}_k,$$

i.e.,

$$\mathbf{a} \times \mathbf{b} = \epsilon_{ijk}a_ib_j\mathbf{e}_k. \tag{A25}$$

The following useful identity can be proved (*see* Problems A7)

$$\epsilon_{ijm}\epsilon_{klm} = \delta_{ik}\delta_{jl} - \delta_{il}\delta_{jk}. \tag{A26}$$

2A5. Manipulations with the Indicial Notations

(a) Substitution

If

$$a_i = U_{im}b_m \tag{i}$$

and

$$b_i = V_{im}c_m, \tag{ii}$$

then, in order to substitute the b_i's in (ii) into (i) we first change the free index in (ii) from i to m and the dummy index from m to some other letter, say n, so that

$$b_m = V_{mn}c_n. \tag{iii}$$

Now (i) and (iii) give

$$a_i = U_{im}V_{mn}c_n. \tag{iv}$$

Note (iv) represents three equations each having the sum of nine terms on its right-hand side.

(b) Multiplications

If

$$p = a_m b_m \tag{i}$$

and

$$q = c_m d_m, \tag{ii}$$

then

$$pq = a_m b_m c_n d_n. \tag{iii}$$

It is important to note that $pq \neq a_m b_m c_m d_m$. In fact the right-hand side of this expression is not even defined in the summation convention and further it is obvious that

$$pq \neq \sum_{m=1}^{3} a_m b_m c_m d_m.$$

Since the dot product of vectors is distributive, therefore if $\mathbf{a} = a_i \mathbf{e}_i$ and $\mathbf{b} = b_i \mathbf{e}_i$, then

$$\mathbf{a} \cdot \mathbf{b} = (a_i \mathbf{e}_i) \cdot (b_j \mathbf{e}_j) = a_i b_j (\mathbf{e}_i \cdot \mathbf{e}_j).$$

In particular, if $\mathbf{e}_1$, $\mathbf{e}_2$, $\mathbf{e}_3$ are unit vectors perpendicular to each other, then $\mathbf{e}_i \cdot \mathbf{e}_j = \delta_{ij}$ so that

$$\mathbf{a} \cdot \mathbf{b} = a_i b_j \delta_{ij} = a_i b_i = a_j b_j = a_1 b_1 + a_2 b_2 + a_3 b_3.$$

(c) Factoring

If

$$T_{ij} n_j - \lambda n_i = 0, \tag{i}$$

then, using the Kronecker delta, we can write

$$n_i = \delta_{ij} n_j, \tag{ii}$$

so that (i) becomes

$$T_{ij} n_j - \lambda \delta_{ij} n_j = 0.$$

Thus

$$(T_{ij} - \lambda \delta_{ij}) n_j = 0.$$

(d) Contraction

The operation of identifying two indices and so summing on them is known as contraction. For example, T_{ii} is the contraction of T_{ij},

$$T_{ii} = T_{11} + T_{22} + T_{33},$$

and T_{ijj} is a contraction of T_{ijk}

$$T_{ijj} = T_{i11} + T_{i22} + T_{i33}.$$

If

$$T_{ij} = \lambda\theta\delta_{ij} + 2\mu E_{ij},$$

Then

$$T_{ii} = \lambda\theta\delta_{ii} + 2\mu E_{ii} = 3\lambda\theta + 2\mu E_{ii}.$$

PROBLEMS

A1. Given

$$[S_{ij}] = \begin{bmatrix} 1 & 0 & 2 \\ 0 & 1 & 2 \\ 3 & 0 & 3 \end{bmatrix},$$

evaluate (a) S_{ii}, (b) $S_{ij}S_{ij}$, (c) $S_{jk}S_{jk}$, (d) $S_{mn}S_{nm}$.

A2. Determine which of these equations have an identical meaning with $a_i = Q_{ij}a'_j$.

(a) $a_l = Q_{lm}a'_m$,

(b) $a_p = Q_{qp}a'_q$,

(c) $a_m = a'_n Q_{mn}$.

A3. Given the following matrices

$$[a_i] = \begin{bmatrix} 1 \\ 0 \\ 2 \end{bmatrix}, \quad [B_{ij}] = \begin{bmatrix} 2 & 3 & 0 \\ 0 & 5 & 1 \\ 0 & 2 & 1 \end{bmatrix}, \quad [C_{ij}] = \begin{bmatrix} 0 & 3 & 1 \\ 1 & 0 & 2 \\ 2 & 4 & 3 \end{bmatrix}.$$

Demonstrate the equivalence of the following subscripted equations and the corresponding matrix equations.

(a) $D_{ji} = B_{ij}$ $\quad [D] = [B]^T$,

(b) $b_i = B_{ij}a_j$ $\quad [b] = [B]\,[a]$,

(c) $c_j = B_{ji}a_i$ $\quad [c] = [B]\,[a]$,

(d) $s = B_{ij}a_i a_j$ $\quad s = [a]^T[B]\,[a]$,

(e) $D_{ik} = B_{ij}C_{jk}$ $[D] = [B][C]$,

(f) $D_{ik} = B_{ij}C_{kj}$ $[D] = [B][C]^T$.

A4. Given that $T_{ij} = 2\mu E_{ij} + \lambda(E_{kk})\delta_{ij}$, show that (a)

$$W = \frac{1}{2} T_{ij}E_{ij} = \mu E_{ij}E_{ij} + \frac{\lambda}{2}(E_{kk})^2$$

and (b)

$$P = T_{ij}T_{ij} = 4\mu^2 E_{ij}E_{ij} + (E_{kk})^2(4\mu\lambda + 3\lambda^2).$$

A5. Given

$$[a_i] = \begin{bmatrix} 1 \\ 2 \\ 0 \end{bmatrix}, \quad [b_i] = \begin{bmatrix} 0 \\ 2 \\ 3 \end{bmatrix}, \quad [S_{ij}] = \begin{bmatrix} 0 & 1 & 2 \\ 1 & 2 & 3 \\ 4 & 0 & 1 \end{bmatrix},$$

(a) Evaluate $[T_{ij}]$ if $T_{ij} = \epsilon_{ijk}a_k$,
(b) Evaluate $[C_i]$ if $C_i = \epsilon_{ijk}S_{jk}$,
(c) Evaluate $[d_i]$ if $d_k = \epsilon_{ijk}a_i b_j$ and show that this result is the same as $d_k = (\mathbf{a} \times \mathbf{b}) \cdot \mathbf{e}_k$.

A6. (a) If $\epsilon_{ijk}T_{jk} = 0$, show that $T_{ij} = T_{ji}$.
(b) Show that $\delta_{ij}\epsilon_{ijk} = 0$.

A7. (a) Verify that

$$\epsilon_{ijm}\epsilon_{klm} = \delta_{ik}\delta_{jl} - \delta_{il}\delta_{jk}.$$

By contracting the result of part (a) show that

(b) $\epsilon_{ilm}\epsilon_{jlm} = 2\delta_{ij}$,

(c) $\epsilon_{ijk}\epsilon_{ijk} = 6$.

A8. Using the relation of Problem 7a, show that

$$\mathbf{a} \times (\mathbf{b} \times \mathbf{c}) = (\mathbf{a} \cdot \mathbf{c})\,\mathbf{b} - (\mathbf{a} \cdot \mathbf{b})\,\mathbf{c}.$$

A9. (a) If $T_{ij} = -T_{ji}$, show that $T_{ij}a_i a_j = 0$.
(b) If $T_{ij} = -T_{ji}$ and $S_{ij} = S_{ji}$, show that

$$T_{kl}S_{kl} = 0.$$

A10. Let $T_{ij} = \frac{1}{2}(S_{ij} + S_{ji})$ and $R_{ij} = \frac{1}{2}(S_{ij} - S_{ji})$, show that $S_{ij} = T_{ij} + R_{ij}$, $T_{ij} = T_{ji}$, and $R_{ij} = -R_{ji}$.

A11. Let $f(x_1, x_2, x_3)$ be an arbitrary function of x_i and $v_i(x_1, x_2, x_3)$ represent three functions of x_i. By expanding the following equations, show that they correspond to the usual formulas of differential calculus.

(a) $df = \dfrac{\partial f}{\partial x_i} dx_i$,

(b) $dv_i = \dfrac{\partial v_i}{\partial x_j} dx_j$.

A12. Let det $|A_{ij}|$ denote the determinant whose element at ith row and jth column is given by A_{ij}. Show that det $|A_{ij}| = \epsilon_{ijk} A_{i1} A_{j2} A_{k3}$.

PART B TENSORS

2B1. Tensor – A Linear Transformation

Let **T** be a transformation, which transforms any vector into another vector.† If **T** transforms $\mathbf{a}_1$ into $\mathbf{b}_1$ and $\mathbf{a}_2$ into $\mathbf{b}_2$, we write

$$\mathbf{T}\mathbf{a}_1 = \mathbf{b}_1$$

and

$$\mathbf{T}\mathbf{a}_2 = \mathbf{b}_2.$$

If **T** has the following linear properties:

$$\begin{aligned} \mathbf{T}(\mathbf{a}_1 + \mathbf{a}_2) &= \mathbf{T}\mathbf{a}_1 + \mathbf{T}\mathbf{a}_2, \\ \mathbf{T}(\alpha \mathbf{a}_1) &= \alpha \mathbf{T}\mathbf{a}_1, \end{aligned} \tag{B1}$$

where $\mathbf{a}_1$ and $\mathbf{a}_2$ are two arbitrary vectors and α an arbitrary scalar then **T** is called a **linear transformation**. It is also called a **second-order tensor** or simply **tensor**.‡

2B2. Components of a Tensor

Let $\mathbf{e}_1$, $\mathbf{e}_2$, $\mathbf{e}_3$ be unit vectors in the direction of the x_1-, x_2-, x_3-axis, respectively, of a rectangular Cartesian coordinate system. We recall that the Cartesian components of a vector **a** are given by

$$a_1 = \mathbf{e}_1 \cdot \mathbf{a},$$

$$a_2 = \mathbf{e}_2 \cdot \mathbf{a},$$

$$a_3 = \mathbf{e}_3 \cdot \mathbf{a},$$

i.e.,

$$a_i = \mathbf{e}_i \cdot \mathbf{a}.$$

Or equivalently, the vector **a** may be represented in terms of its components as

$$\mathbf{a} = a_1\mathbf{e}_1 + a_2\mathbf{e}_2 + a_3\mathbf{e}_3 = a_i\mathbf{e}_i.$$

†Throughout this text, we shall be interested only in transformation of vectors in three-dimensional Euclidean space into the same space.

‡Scalars and vectors are sometimes called the zeroth and first-order tensor, respectively. Even though they can also be defined algebraically, in terms of certain operational rules, we choose not to do so. The geometrical concept of scalars and vectors, which we assume that the students are familiar with, is quite sufficient for our purpose.

We now consider a tensor **T**. For any vector **a**, $\mathbf{b} = \mathbf{Ta}$ is a vector given by

$$\mathbf{b} = \mathbf{Ta} = a_1\mathbf{Te}_1 + a_2\mathbf{Te}_2 + a_3\mathbf{Te}_3 = a_i\mathbf{Te}_i, \tag{B2}$$

where we have used the definition of a linear transformation.

The components of **b** are given by

$$\begin{aligned} b_1 &= \mathbf{b}\cdot\mathbf{e}_1 = a_1\mathbf{e}_1\cdot\mathbf{Te}_1 + a_2\mathbf{e}_1\cdot\mathbf{Te}_2 + a_3\mathbf{e}_1\cdot\mathbf{Te}_3, \\ b_2 &= \mathbf{b}\cdot\mathbf{e}_2 = a_1\mathbf{e}_2\cdot\mathbf{Te}_1 + a_2\mathbf{e}_2\cdot\mathbf{Te}_2 + a_3\mathbf{e}_2\cdot\mathbf{Te}_3, \\ b_3 &= \mathbf{b}\cdot\mathbf{e}_3 = a_1\mathbf{e}_3\cdot\mathbf{Te}_1 + a_2\mathbf{e}_3\cdot\mathbf{Te}_2 + a_3\mathbf{e}_3\cdot\mathbf{Te}_3, \end{aligned} \tag{B3a}$$

i.e.,

$$b_i = a_j\mathbf{e}_i\cdot\mathbf{Te}_j. \tag{B3b}$$

The terms such as $\mathbf{e}_1\cdot\mathbf{Te}_1$ and $\mathbf{e}_2\cdot\mathbf{Te}_1$ are just the $\mathbf{e}_1$ and $\mathbf{e}_2$ components of $\mathbf{Te}_1$. We shall agree to write these components as $T_{11} = \mathbf{e}_1\cdot\mathbf{Te}_1$, $T_{21} = \mathbf{e}_2\cdot\mathbf{Te}_1$, $T_{12} = \mathbf{e}_1\cdot\mathbf{Te}_2$, etc. or for short

$$T_{ij} = \mathbf{e}_i\cdot\mathbf{Te}_j. \tag{B4}$$

We call T_{ij} the components of the tensor **T**. Now, in view of Eqs. (B3b) and (B4), the vector equation $\mathbf{b} = \mathbf{Ta}$ can now be written in component form as

$$\begin{aligned} b_1 &= T_{11}a_1 + T_{12}a_2 + T_{13}a_3, \\ b_2 &= T_{21}a_1 + T_{22}a_2 + T_{23}a_3, \\ b_3 &= T_{31}a_1 + T_{32}a_2 + T_{33}a_3, \end{aligned} \tag{B5a}$$

i.e.,

$$b_i = T_{ij}a_j. \tag{B5b}$$

For computation purposes, the above equations [Eqs. (B5a, b)] can be written in the following matrix form

$$\begin{bmatrix} b_1 \\ b_2 \\ b_3 \end{bmatrix} = \begin{bmatrix} T_{11} & T_{12} & T_{13} \\ T_{21} & T_{22} & T_{23} \\ T_{31} & T_{32} & T_{33} \end{bmatrix} \begin{bmatrix} a_1 \\ a_2 \\ a_3 \end{bmatrix}. \tag{B5c}$$

The matrix

$$\begin{bmatrix} T_{11} & T_{12} & T_{13} \\ T_{21} & T_{22} & T_{23} \\ T_{31} & T_{32} & T_{33} \end{bmatrix}$$

is called the **matrix of the tensor T** with respect to the set of unit vectors $\{\mathbf{e}_1, \mathbf{e}_2, \mathbf{e}_3\}$.

We note that the components in the first column of the matrix are the components of the vector $\mathbf{Te}_1$, those in the second column are the components of the vector $\mathbf{Te}_2$, and those in the third column are the components of $\mathbf{Te}_3$. That is

$$\mathbf{Te}_1 = T_{11}\mathbf{e}_1 + T_{21}\mathbf{e}_2 + T_{31}\mathbf{e}_3 = T_{j1}\mathbf{e}_j,$$
$$\mathbf{Te}_2 = T_{12}\mathbf{e}_1 + T_{22}\mathbf{e}_2 + T_{32}\mathbf{e}_3 = T_{j2}\mathbf{e}_j, \tag{B6a}$$
$$\mathbf{Te}_3 = T_{13}\mathbf{e}_1 + T_{23}\mathbf{e}_2 + T_{33}\mathbf{e}_3 = T_{j3}\mathbf{e}_j,$$

i.e.,

$$\mathbf{Te}_i = T_{ji}\mathbf{e}_j. \tag{B6b}$$

It should be obvious that the components of $\mathbf{T}$ depend on the coordinate system through the set of unit base vectors $\{\mathbf{e}_1, \mathbf{e}_2, \mathbf{e}_3\}$ just like the components of a vector change with the different orientations of $\mathbf{e}_1$, $\mathbf{e}_2$, $\mathbf{e}_3$. A vector does not depend on any coordinate system even though its components do. Similarly, a tensor does not depend on any coordinate system even though its components do.† Thus, a tensor has infinitely many matrices, one for each set of unit base vectors. We shall use the following notations for the different matrices of the same tensor. If $\{\mathbf{e}_1, \mathbf{e}_2, \mathbf{e}_3\}$, $\{\mathbf{e}_1', \mathbf{e}_2', \mathbf{e}_3'\}$ are two different bases, then with respect to $\{\mathbf{e}_1, \mathbf{e}_2, \mathbf{e}_3\}$, the matrix will be denoted by

$$[\mathbf{T}] \quad \text{or} \quad [T_{ij}],$$

and with respect to $\{\mathbf{e}_1', \mathbf{e}_2', \mathbf{e}_3'\}$, by

$$[\mathbf{T}]' \quad \text{or} \quad [T_{ij}'].$$

EXAMPLE 2.1

If $\mathbf{T}$ transforms every vector into its mirror image with respect to a fixed plane, find a matrix of $\mathbf{T}$. Also, verify that $\mathbf{T}$ is a tensor (i.e., linear transformation).

Solution. Let $\mathbf{e}_1$ be perpendicular to the plane of mirror, and $\mathbf{e}_2$, $\mathbf{e}_3$ on the plane. Then

$$\mathbf{Te}_1 = -\mathbf{e}_1,$$
$$\mathbf{Te}_2 = \mathbf{e}_2,$$
$$\mathbf{Te}_3 = \mathbf{e}_3.$$

†Any entity, such as a vector, or a tensor which is defined without reference to coordinate systems is also known as an invariant with respect to coordinates. Note however, a vector or a tensor may depend on observers (called frames). For example, velocity vector is coordinate-invariant, but not frame-invariant.

Thus [*see* Eq. (B6a)]

$$[\mathbf{T}] = \begin{bmatrix} -1 & 0 & 0 \\ 0 & 1 & 0 \\ 0 & 0 & 1 \end{bmatrix}_{\mathbf{e}_1, \mathbf{e}_2, \mathbf{e}_3}$$

Note that if a new set of base vectors is used, say $\mathbf{e}_1' = \mathbf{e}_2$, $\mathbf{e}_2' = \mathbf{e}_3$, $\mathbf{e}_3' = \mathbf{e}_1$ then

$$\mathbf{T}\mathbf{e}_1' = \mathbf{e}_1',$$

$$\mathbf{T}\mathbf{e}_2' = \mathbf{e}_2',$$

$$\mathbf{T}\mathbf{e}_3' = -\mathbf{e}_3',$$

and

$$[\mathbf{T}]' = \begin{bmatrix} 1 & 0 & 0 \\ 0 & 1 & 0 \\ 0 & 0 & -1 \end{bmatrix}_{\mathbf{e}_1', \mathbf{e}_2', \mathbf{e}_3'}.$$

The fact that $\mathbf{T}$ is a linear transformation is clearly seen from Fig. 2.1 where $\mathbf{T}(\mathbf{a}+\mathbf{b}) = \mathbf{T}\mathbf{a} + \mathbf{T}\mathbf{b}$ and $\mathbf{T}(\alpha\mathbf{a}) = \alpha\mathbf{T}\mathbf{a}$. It can, of course, be proven analytically without any difficulties.

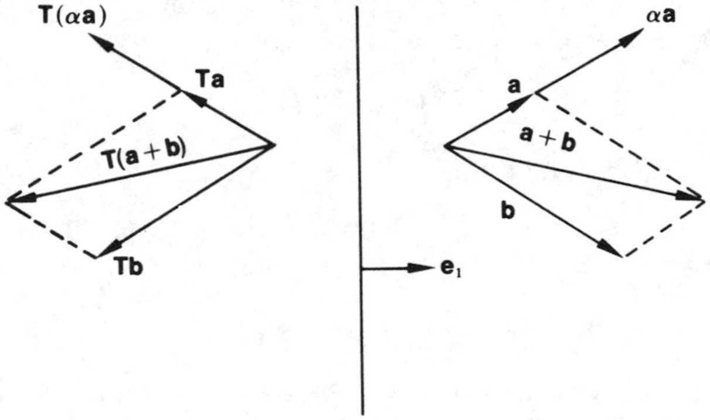

Fig. 2.1

EXAMPLE 2.2

If $\mathbf{T}$ transforms every vector into a unit vector in a fixed direction, show that $\mathbf{T}$ is not a tensor.

Solution. Let $\mathbf{n}$ be the unit vector into which all vectors transform under $\mathbf{T}$. Then for any $\mathbf{a}$ and $\mathbf{b}$ we have

$$\mathbf{T}\mathbf{a} = \mathbf{n}, \quad \mathbf{T}\mathbf{b} = \mathbf{n}, \quad \text{and } \mathbf{T}(\mathbf{a}+\mathbf{b}) = \mathbf{n}.$$

Obviously $\mathbf{T}(\mathbf{a}+\mathbf{b}) \neq \mathbf{T}\mathbf{a}+\mathbf{T}\mathbf{b}$.

2B3. Sum of Tensors

Let $\mathbf{T}$ and $\mathbf{S}$ be two tensors. The sum of $\mathbf{T}$ and $\mathbf{S}$ denoted by $(\mathbf{T}+\mathbf{S})$, is defined by:

$$(\mathbf{T}+\mathbf{S})\mathbf{a} = \mathbf{Ta}+\mathbf{Sa}, \quad \text{for any } \mathbf{a}. \tag{B7}$$

It is easily seen that $(\mathbf{T}+\mathbf{S})$, thus defined, is indeed a tensor.

The components of $(\mathbf{T}+\mathbf{S})$ are†

$$(\mathbf{T}+\mathbf{S})_{ij} = \mathbf{e}_i \cdot (\mathbf{T}+\mathbf{S})\,\mathbf{e}_j = \mathbf{e}_i \cdot \mathbf{T}\mathbf{e}_j + \mathbf{e}_i \cdot \mathbf{S}\mathbf{e}_j,$$

i.e.,

$$(\mathbf{T}+\mathbf{S})_{ij} = T_{ij}+S_{ij}. \tag{B8a}$$

In matrix notation,

$$[\mathbf{T}+\mathbf{S}] = [\mathbf{T}]+[\mathbf{S}]. \tag{B8b}$$

2B4. Dyadic Product of a and b

The dyadic product of $\mathbf{a}$ and $\mathbf{b}$ denoted by $\mathbf{ab}$ is defined to be the transformation, which transforms any $\mathbf{c}$ according to the rule:

$$(\mathbf{ab})\mathbf{c} = \mathbf{a}(\mathbf{b}\cdot\mathbf{c}). \tag{B9}‡$$

Now, for any $\mathbf{c}$, $\mathbf{d}$, α, and β, we have, from the above rule

$$\begin{aligned}(\mathbf{ab})(\alpha\mathbf{c}+\beta\mathbf{d}) &= \mathbf{a}[\mathbf{b}\cdot(\alpha\mathbf{c}+\beta\mathbf{d})]\\ &= \mathbf{a}[\alpha(\mathbf{b}\cdot\mathbf{c})+\beta(\mathbf{b}\cdot\mathbf{d})]\\ &= \alpha\mathbf{a}(\mathbf{b}\cdot\mathbf{c})+\beta\mathbf{a}(\mathbf{b}\cdot\mathbf{d})\\ &= \alpha(\mathbf{ab})\mathbf{c}+\beta(\mathbf{ab})\mathbf{d}.\end{aligned}$$

Thus $(\mathbf{ab})$ is a tensor. Its components with respect to some $\{\mathbf{e}_1, \mathbf{e}_2, \mathbf{e}_3\}$ are

$$\begin{aligned}(\mathbf{ab})_{ij} &= \mathbf{e}_i \cdot (\mathbf{ab})\mathbf{e}_j = \mathbf{e}_i \cdot [\mathbf{a}(\mathbf{b}\cdot\mathbf{e}_j)] = \mathbf{e}_i \cdot (\mathbf{a}b_j)\\ &= (\mathbf{e}_i\cdot\mathbf{a})b_j = a_ib_j, \quad \text{i.e., } (\mathbf{ab})_{ij} = a_ib_j.\end{aligned} \tag{B10a}$$

Thus

$$[\mathbf{ab}] = \begin{bmatrix} a_1b_1 & a_1b_2 & a_1b_3 \\ a_2b_1 & a_2b_2 & a_2b_3 \\ a_3b_1 & a_3b_2 & a_3b_3 \end{bmatrix} \tag{B10b}$$

†Throughout the text, $\mathbf{a}_i$ represents three vectors $\mathbf{a}_1$, $\mathbf{a}_2$, and $\mathbf{a}_3$, whereas $(\mathbf{a})_i$ represents the three components a_1, a_2, a_3 of the one vector $\mathbf{a}$. Similar notation for tensors.

‡Some authors write $(\mathbf{ab})\cdot\mathbf{c}$ for $(\mathbf{ab})\mathbf{c}$ and $\mathbf{c}\cdot(\mathbf{ab})$ for $(\mathbf{ab})^T\mathbf{c}$ (*see* definition of transpose in Section 2B7). Also, some authors write $\mathbf{a}\otimes\mathbf{b}$ for $\mathbf{ab}$.

or

$$[\mathbf{ab}] = \begin{bmatrix} a_1 \\ a_2 \\ a_3 \end{bmatrix} [b_1 \quad b_2 \quad b_3]. \tag{B10c}$$

In particular

$$[\mathbf{e}_1\mathbf{e}_1] = \begin{bmatrix} 1 & 0 & 0 \\ 0 & 0 & 0 \\ 0 & 0 & 0 \end{bmatrix}_{\mathbf{e}_i} \qquad [\mathbf{e}_1\mathbf{e}_2] = \begin{bmatrix} 0 & 1 & 0 \\ 0 & 0 & 0 \\ 0 & 0 & 0 \end{bmatrix}_{\mathbf{e}_i}$$

$$[\mathbf{e}_1\mathbf{e}_3] = \begin{bmatrix} 0 & 0 & 1 \\ 0 & 0 & 0 \\ 0 & 0 & 0 \end{bmatrix}_{\mathbf{e}_i} \qquad [\mathbf{e}_2\mathbf{e}_1] = \begin{bmatrix} 0 & 0 & 0 \\ 1 & 0 & 0 \\ 0 & 0 & 0 \end{bmatrix}_{\mathbf{e}_i}, \text{ etc.} \tag{B11}$$

Thus, it is clear that any tensor **T** can be expressed as

$$\mathbf{T} = T_{11}(\mathbf{e}_1\mathbf{e}_1) + T_{12}(\mathbf{e}_1\mathbf{e}_2) + T_{13}(\mathbf{e}_1\mathbf{e}_3) + T_{21}(\mathbf{e}_2\mathbf{e}_1) + \ldots,$$

i.e.,

$$\mathbf{T} = T_{ij}\mathbf{e}_i\mathbf{e}_j. \tag{B12}$$

2B5. Product of Two Tensors

If **T** and **S** are two tensors, then **TS** and **ST** are defined to be the transformations (easily seen to be tensors)

$$(\mathbf{TS})\mathbf{a} = \mathbf{T}(\mathbf{Sa}) \tag{B13}$$

and

$$(\mathbf{ST})\mathbf{a} = \mathbf{S}(\mathbf{Ta}). \tag{B14}$$

Thus the components of (**TS**) are

$$(\mathrm{TS})_{ij} = \mathbf{e}_i \cdot (\mathbf{TS})\mathbf{e}_j = \mathbf{e}_i \cdot \mathbf{T}(\mathbf{Se}_j) = \mathbf{e}_i \cdot \mathbf{T}S_{mj}\mathbf{e}_m = S_{mj}(\mathbf{e}_i \cdot \mathbf{Te}_m),$$

i.e.,

$$(\mathbf{TS})_{ij} = T_{im}S_{mj}. \tag{B15}$$

Similarly

$$(\mathbf{ST})_{ij} = S_{im}T_{mj}. \tag{B16}$$

Equations (B15) and (B16) are equivalent to the following matrix equations

$$[\mathbf{TS}] = [\mathbf{T}][\mathbf{S}] \tag{B17}$$

and

$$[\mathbf{ST}] = [\mathbf{S}][\mathbf{T}] \tag{B18}$$

It should be emphasized that in general $\mathbf{TS} \neq \mathbf{ST}$. That is, in general, the tensor product is *not* commutative.

If **T**, **S**, and **V** are three tensors, then

$$\big(\mathbf{T}(\mathbf{S}\mathbf{V})\big)\,\mathbf{a} = \mathbf{T}\left((\mathbf{S}\mathbf{V})\mathbf{a}\right) = \mathbf{T}\left(\mathbf{S}(\mathbf{V}\mathbf{a})\right)$$

and

$$(\mathbf{T}\mathbf{S})(\mathbf{V}\mathbf{a}) = \mathbf{T}\left(\mathbf{S}(\mathbf{V}\mathbf{a})\right),$$

i.e.,

$$\mathbf{T}(\mathbf{S}\mathbf{V}) = (\mathbf{T}\mathbf{S})\mathbf{V}. \qquad \text{(B19)}$$

Thus, the tensor product obeys associative laws.

EXAMPLE 2.3

(a) A rigid body is rotated counterclockwise through 90° about an axis. Find a matrix representing this rotation (clearly a linear transformation).

Solution. Let $\{\mathbf{e}_1, \mathbf{e}_2, \mathbf{e}_3\}$ be a set of right-handed unit base vectors with $\mathbf{e}_3$ the axis of rotation. Let **R** be the transformation. Then

$$\mathbf{R}\mathbf{e}_1 = \mathbf{e}_2,$$
$$\mathbf{R}\mathbf{e}_2 = -\mathbf{e}_1,$$
$$\mathbf{R}\mathbf{e}_3 = \mathbf{e}_3,$$

i.e., [*see* Eq. (B6a)]

$$[\mathbf{R}] = \begin{bmatrix} 0 & -1 & 0 \\ 1 & 0 & 0 \\ 0 & 0 & 1 \end{bmatrix}_{\mathbf{e}_i}.$$

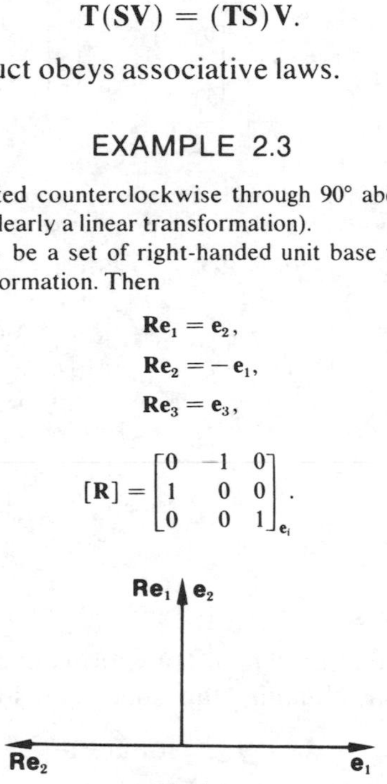

Fig. 2.2

(b) Suppose that the body of part (a) is subsequently given a 90° rotation about the $\mathbf{e}_1$-axis* by the right-hand screw rule, find the matrix representation of the resultant rotation. If the position of a point *P* was (1, 1, 0), what is its position after the two rotations?

Solution. Let **S** be the second transformation, then

$$\mathbf{S}\mathbf{e}_1 = \mathbf{e}_1,$$
$$\mathbf{S}\mathbf{e}_2 = \mathbf{e}_3,$$
$$\mathbf{S}\mathbf{e}_3 = -\mathbf{e}_2,$$

i.e.,

$$[\mathbf{S}] = \begin{bmatrix} 1 & 0 & 0 \\ 0 & 0 & -1 \\ 0 & 1 & 0 \end{bmatrix}.$$

*The notation $\mathbf{e}_1$ always refers to the original base vectors.

Since $\mathbf{S}(\mathbf{Ra}) = (\mathbf{SR})\mathbf{a}$ the resultant rotation is given by the single transformation **SR** whose components are given by the matrix

$$[\mathbf{SR}] = [\mathbf{S}][\mathbf{R}] = \begin{bmatrix} 1 & 0 & 0 \\ 0 & 0 & -1 \\ 0 & 1 & 0 \end{bmatrix} \begin{bmatrix} 0 & -1 & 0 \\ 1 & 0 & 0 \\ 0 & 0 & 1 \end{bmatrix} = \begin{bmatrix} 0 & -1 & 0 \\ 0 & 0 & -1 \\ 1 & 0 & 0 \end{bmatrix}.$$

Let the initial and final positions of P be denoted by $\mathbf{r}$ and $\mathbf{r}^*$ respectively, then

$$[\mathbf{r}^*] = [\mathbf{SR}][\mathbf{r}] = \begin{bmatrix} 0 & -1 & 0 \\ 0 & 0 & -1 \\ 1 & 0 & 0 \end{bmatrix} \begin{bmatrix} 1 \\ 1 \\ 0 \end{bmatrix} = \begin{bmatrix} -1 \\ 0 \\ 1 \end{bmatrix},$$

i.e.,

$$\mathbf{r}^* = -\mathbf{e}_1 + \mathbf{e}_3.$$

(c) Reverse the order of the rotations and find the final position of P.

Solution. Denoting the new final position of P by $\mathbf{r}^{**}$, we have

$$\mathbf{r}^{**} = \mathbf{R}(\mathbf{Sr}) = (\mathbf{RS})\mathbf{r}.$$

The components of $\mathbf{r}^{**}$ are

$$[\mathbf{r}^{**}] = [\mathbf{R}]\,[\mathbf{S}]\,[\mathbf{r}] = \begin{bmatrix} 0 & -1 & 0 \\ 1 & 0 & 0 \\ 0 & 0 & 1 \end{bmatrix} \begin{bmatrix} 1 & 0 & 0 \\ 0 & 0 & -1 \\ 0 & 1 & 0 \end{bmatrix} \begin{bmatrix} 1 \\ 1 \\ 0 \end{bmatrix}$$

$$= \begin{bmatrix} 0 & 0 & 1 \\ 1 & 0 & 0 \\ 0 & 1 & 0 \end{bmatrix} \begin{bmatrix} 1 \\ 1 \\ 0 \end{bmatrix} = \begin{bmatrix} 0 \\ 1 \\ 1 \end{bmatrix}.$$

Thus $\mathbf{r}^{**} = \mathbf{e}_2 + \mathbf{e}_3$.

2B6. Identity Tensor

The linear transformation which transforms every vector into itself is called an **identity tensor**. Denoting this special tensor by **I** we have

$$\mathbf{Ia} = \mathbf{a} \qquad \text{for any } \mathbf{a} \tag{B20}$$

and in particular

$$\mathbf{Ie}_1 = \mathbf{e}_1,$$

$$\mathbf{Ie}_2 = \mathbf{e}_2,$$

$$\mathbf{Ie}_3 = \mathbf{e}_3.$$

Thus the (Cartesian) components of the identity tensor are

$$I_{ij} = \mathbf{e}_i \cdot \mathbf{Ie}_j = \mathbf{e}_i \cdot \mathbf{e}_j = \delta_{ij}, \tag{B21a}$$

i.e.,

$$[\mathbf{I}] = \begin{bmatrix} 1 & 0 & 0 \\ 0 & 1 & 0 \\ 0 & 0 & 1 \end{bmatrix}. \tag{B21b}$$

It is obvious that the identity matrix is the matrix of **I** for *all* rectangular Cartesian coordinates.

2B7. Transpose of a Tensor

The **transpose of a tensor T**, denoted by $\mathbf{T}^T$, is defined to be that which satisfies the following identity for all vectors **a** and **b**:

$$\mathbf{a} \cdot (\mathbf{T}\mathbf{b}) = \mathbf{b} \cdot (\mathbf{T}^T\mathbf{a}). \tag{B22}$$

It can be easily seen that $\mathbf{T}^T$ is a tensor.

From the above definition, we have

$$\mathbf{e}_i \cdot (\mathbf{T}\mathbf{e}_j) = \mathbf{e}_j \cdot (\mathbf{T}^T\mathbf{e}_i).$$

Thus

$$T_{ij} = T^T_{ji}, \tag{B23}$$

i.e., the matrix of $\mathbf{T}^T$ is the transpose of the matrix of **T**. It can be easily established that

$$(\mathbf{S}\mathbf{T})^T = \mathbf{T}^T\mathbf{S}^T. \tag{B24}$$

2B8. Orthogonal Tensor

An orthogonal tensor is a linear transformation, for which the transformed vectors preserve their lengths and angles. Let **Q** denote an orthogonal tensor, then by definition, $|\mathbf{Q}\mathbf{a}| = |\mathbf{a}|$, and $\cos(\mathbf{a}, \mathbf{b}) = \cos(\mathbf{Q}\mathbf{a}, \mathbf{Q}\mathbf{b})$ for any **a** and **b**. Thus,

$$(\mathbf{Q}\mathbf{a}) \cdot (\mathbf{Q}\mathbf{b}) = \mathbf{a} \cdot \mathbf{b} \tag{B25}$$

for any **a** and **b**. But

$$(\mathbf{Q}\mathbf{a}) \cdot (\mathbf{Q}\mathbf{b}) = \mathbf{b} \cdot \mathbf{Q}^T(\mathbf{Q}\mathbf{a})$$

[*see* Eq. (B22)]. Thus

$$\mathbf{a} \cdot \mathbf{b} = \mathbf{b} \cdot \mathbf{Q}^T\mathbf{Q}\mathbf{a},$$

i.e.,

$$\mathbf{b} \cdot \mathbf{I}\mathbf{a} = \mathbf{b} \cdot \mathbf{Q}^T\mathbf{Q}\mathbf{a},$$

or

$$\mathbf{b} \cdot (\mathbf{I} - \mathbf{Q}^T\mathbf{Q})\mathbf{a} = \mathbf{0},$$

Since **a** and **b** are arbitrary, it follows that

$$\mathbf{Q}^T\mathbf{Q} = \mathbf{I}.$$

It can be shown that the transpose of an orthogonal tensor is also an orthogonal one (*see* problem B10). Thus, we also have

$$\mathbf{Q}\mathbf{Q}^T = \mathbf{I}.$$

Thus,

$$\mathbf{Q}\mathbf{Q}^T = \mathbf{Q}^T\mathbf{Q} = \mathbf{I}. \tag{B26a†}$$

In matrix notation

$$[\mathbf{Q}]\,[\mathbf{Q}]^T = [\mathbf{Q}]^T\,[\mathbf{Q}] = [\mathbf{I}] \tag{B26b}$$

and in subscript notation

$$Q_{im}Q_{jm} = Q_{mi}Q_{mj} = \delta_{ij}. \tag{B26c}$$

EXAMPLE 2.4

The tensor given in Example 2.1, being a reflection, is obviously an orthogonal tensor. Verify that $[\mathbf{T}]\,[\mathbf{T}]^T = [\mathbf{I}]$ for the $[\mathbf{T}]$ in that example. Also find the determinant of $[\mathbf{T}]$.

Solution.

$$[\mathbf{T}]\,[\mathbf{T}]^T = \begin{bmatrix} -1 & 0 & 0 \\ 0 & 1 & 0 \\ 0 & 0 & 1 \end{bmatrix} \begin{bmatrix} -1 & 0 & 0 \\ 0 & 1 & 0 \\ 0 & 0 & 1 \end{bmatrix} = \begin{bmatrix} 1 & 0 & 0 \\ 0 & 1 & 0 \\ 0 & 0 & 1 \end{bmatrix}.$$

Determinant of $[\mathbf{T}]$ is

$$\begin{vmatrix} -1 & 0 & 0 \\ 0 & 1 & 0 \\ 0 & 0 & 1 \end{vmatrix} = -1.$$

EXAMPLE 2.5

The tensor given in Example 2.3, being a rigid body rotation, is obviously an orthogonal tensor. Verify that $[\mathbf{R}]\,[\mathbf{R}]^T = [\mathbf{I}]$ for the $[\mathbf{R}]$ in that example. Also find the determinant of $[\mathbf{R}]$.

Solution.

$$[\mathbf{R}]\,[\mathbf{R}]^T = \begin{bmatrix} 0 & 1 & 0 \\ -1 & 0 & 0 \\ 0 & 0 & 1 \end{bmatrix} \begin{bmatrix} 0 & -1 & 0 \\ 1 & 0 & 0 \\ 0 & 0 & 1 \end{bmatrix} = \begin{bmatrix} 1 & 0 & 0 \\ 0 & 1 & 0 \\ 0 & 0 & 1 \end{bmatrix},$$

$$\det\,[\mathbf{R}] \equiv |\mathbf{R}| = \begin{vmatrix} 0 & 1 & 0 \\ -1 & 0 & 0 \\ 0 & 0 & 1 \end{vmatrix} = +1.$$

The determinant of the matrix of any orthogonal tensor $\mathbf{Q}$ is easily shown to be equal to either $+1$ or -1. In fact,

$$[\mathbf{Q}]\,[\mathbf{Q}]^T = [\mathbf{I}],$$

$$|[\mathbf{Q}]\,[\mathbf{Q}]^T| = |\mathbf{I}| = 1.$$

But the determinant of the product of two matrices equals the product of the individual determinant and $\det\,[\mathbf{Q}] = \det\,[\mathbf{Q}]^T$ so we have

$$|\mathbf{Q}|\,|\mathbf{Q}^T| = |\mathbf{Q}|^2 = 1.$$

Thus,

$$|\mathbf{Q}| = \pm 1.$$

The value of $+1$ corresponds to rotation and -1 corresponds to reflection.

†In other words, $\mathbf{Q}^T$ is the inverse of $\mathbf{Q}$, often denoted as $\mathbf{Q}^{-1}$.

2B9. Transformation Laws for Cartesian Components of Vectors and Tensors

Suppose $\{\mathbf{e}_1, \mathbf{e}_2, \mathbf{e}_3\}$ and $\{\mathbf{e}_1', \mathbf{e}_2', \mathbf{e}_3'\}$ are unit vectors corresponding to two rectangular Cartesian coordinate systems (*see* Fig. 2.3). It is clear that $\{\mathbf{e}_1, \mathbf{e}_2, \mathbf{e}_3\}$ can be made to coincide with $\{\mathbf{e}_1', \mathbf{e}_2', \mathbf{e}_3'\}$, through either a rigid

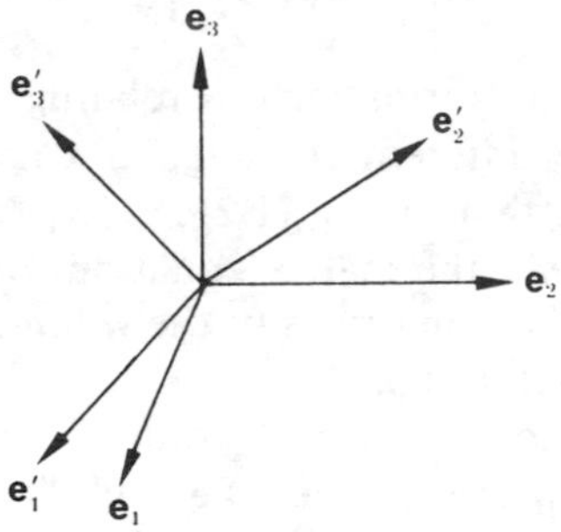

Fig. 2.3

body rotation (if both bases are same handed) or a rotation followed by a reflection (if different handed). That is, $\{\mathbf{e}_1', \mathbf{e}_2', \mathbf{e}_3'\}$ and $\{\mathbf{e}_1, \mathbf{e}_2, \mathbf{e}_3\}$ are related by an orthogonal tensor $\mathbf{Q}$ through the equations

$$\mathbf{e}_i' = \mathbf{Q}\mathbf{e}_i = Q_{mi}\mathbf{e}_m, \tag{B27a}$$

i.e.,

$$\begin{aligned}\mathbf{e}_1' &= Q_{11}\mathbf{e}_1 + Q_{21}\mathbf{e}_2 + Q_{31}\mathbf{e}_3,\\ \mathbf{e}_2' &= Q_{12}\mathbf{e}_1 + Q_{22}\mathbf{e}_2 + Q_{32}\mathbf{e}_3,\\ \mathbf{e}_3' &= Q_{13}\mathbf{e}_1 + Q_{23}\mathbf{e}_2 + Q_{33}\mathbf{e}_3,\end{aligned} \tag{B27b}$$

where

$$Q_{mi}Q_{mj} = Q_{im}Q_{jm} = \delta_{ij} \quad (\text{i.e., } \mathbf{Q}^T\mathbf{Q} = \mathbf{Q}\mathbf{Q}^T = \mathbf{I}).$$

Note that $Q_{mi} = \mathbf{e}_m \cdot \mathbf{Q}\mathbf{e}_i = \mathbf{e}_m \cdot \mathbf{e}_i' = \cos(\mathbf{e}_m, \mathbf{e}_i')$.

Now, consider any vector $\mathbf{a}$. The (Cartesian) components of $\mathbf{a}$ with respect to $\{\mathbf{e}_1, \mathbf{e}_2, \mathbf{e}_3\}$ and $\{\mathbf{e}_1', \mathbf{e}_2', \mathbf{e}_3'\}$ are $a_i = \mathbf{e}_i \cdot \mathbf{a}$ and $a_i' = \mathbf{e}_i' \cdot \mathbf{a}$ respectively. Since $a_i' = \mathbf{e}_i' \cdot \mathbf{a} = Q_{mi}\mathbf{e}_m \cdot \mathbf{a}$, we have

$$\boxed{a_i' = Q_{mi}a_m}\ . \tag{B28a}$$

In matrix form, the above equations read

$$\begin{bmatrix} a_1' \\ a_2' \\ a_3' \end{bmatrix}_{\mathbf{e}_i'} = \begin{bmatrix} Q_{11} & Q_{21} & Q_{31} \\ Q_{12} & Q_{22} & Q_{32} \\ Q_{13} & Q_{23} & Q_{33} \end{bmatrix}_{\mathbf{e}_i} \begin{bmatrix} a_1 \\ a_2 \\ a_3 \end{bmatrix}_{\mathbf{e}_i} \tag{B28b}$$

or

$$[\mathbf{a}]' = [\mathbf{Q}]^T [\mathbf{a}]. \tag{B28c}$$

Equation (B28) is the transformation law relating components of the same vector with respect to different rectangular Cartesian unit bases. It is very important to note that in Eq. (B28c), $[\mathbf{a}]' \equiv [\mathbf{a}]_{\mathbf{e}_i'}$ and $[\mathbf{a}] \equiv [\mathbf{a}]_{\mathbf{e}_i}$. That is, Eq. (B28c) is *not* the matrix equation for $\mathbf{a}' = \mathbf{Q}^T\mathbf{a}$. The distinction is that $[\mathbf{a}]$ and $[\mathbf{a}]'$ are matrices of the *same* vector, whereas $\mathbf{a}'$ is the transformed vector of $\mathbf{a}$ (through $\mathbf{Q}^T$).

Next, consider any tensor $\mathbf{T}$. The components of $\mathbf{T}$ with respect to $\{\mathbf{e}_1, \mathbf{e}_2, \mathbf{e}_3\}$ and $\{\mathbf{e}_1', \mathbf{e}_2', \mathbf{e}_3'\}$ are $T_{ij} = \mathbf{e}_i \cdot \mathbf{T}\mathbf{e}_j$ and $T_{ij}' = \mathbf{e}_i' \cdot \mathbf{T}\mathbf{e}_j'$, respectively. Since $T_{ij}' = \mathbf{e}_i' \cdot \mathbf{T}\mathbf{e}_j' = Q_{mi}\mathbf{e}_m \cdot \mathbf{T}Q_{nj}\mathbf{e}_n = Q_{mi}Q_{nj}(\mathbf{e}_m \cdot \mathbf{T}\mathbf{e}_n)$, we have

$$\boxed{T_{ij}' = Q_{mi}Q_{nj}T_{mn}}\,. \tag{B29a}$$

In matrix form Eq. (B29a) reads

$$\begin{bmatrix} T_{11}' & T_{12}' & T_{13}' \\ T_{21}' & T_{22}' & T_{23}' \\ T_{31}' & T_{32}' & T_{33}' \end{bmatrix}_{\mathbf{e}_i'} = \begin{bmatrix} Q_{11} & Q_{21} & Q_{31} \\ Q_{12} & Q_{22} & Q_{32} \\ Q_{13} & Q_{23} & Q_{33} \end{bmatrix}_{\mathbf{e}_i} \begin{bmatrix} T_{11} & T_{12} & T_{13} \\ T_{21} & T_{22} & T_{23} \\ T_{31} & T_{32} & T_{33} \end{bmatrix}_{\mathbf{e}_i} \begin{bmatrix} Q_{11} & Q_{12} & Q_{13} \\ Q_{21} & Q_{22} & Q_{23} \\ Q_{31} & Q_{32} & Q_{33} \end{bmatrix}_{\mathbf{e}_i}$$

or in short

$$[\mathbf{T}]' = [\mathbf{Q}]^T[\mathbf{T}]\,[\mathbf{Q}]\,. \tag{B29b}$$

Equivalently, we have

$$\boxed{T_{ij} = Q_{im}Q_{jn}T_{mn}'}\,. \tag{B30a}$$

Or, in matrix notation

$$[\mathbf{T}] = [\mathbf{Q}]\,[\mathbf{T}]'\,[\mathbf{Q}]^T. \tag{B30b}$$

Equation (B29a) is the transformation law relating components of the same tensor with respect to different rectangular Cartesian unit bases. Again, it is important to note that in Eqs. (B29b) and (B30b), $[\mathbf{T}]$ and $[\mathbf{T}]'$ are different matrices of the *same* tensor $\mathbf{T}$. Do not confuse Eq. (B29b) with the equation $\mathbf{T}' = \mathbf{Q}^T\mathbf{T}\mathbf{Q}$.

EXAMPLE 2.6

Given

$$[\mathbf{T}]_{\{\mathbf{e}_1,\mathbf{e}_2,\mathbf{e}_3\}} = \begin{bmatrix} 0 & 1 & 0 \\ 1 & 2 & 0 \\ 0 & 0 & 1 \end{bmatrix},$$

find $[\mathbf{T}]_{\mathbf{e}_1'\mathbf{e}_2'\mathbf{e}_3'}$ where $\{\mathbf{e}_1', \mathbf{e}_2', \mathbf{e}_3'\}$ is obtained by rotating $\{\mathbf{e}_1, \mathbf{e}_2, \mathbf{e}_3\}$ about $\mathbf{e}_3$ through 90°.

Solution. Since

$$\mathbf{e}_1' = \mathbf{e}_2,$$

$$\mathbf{e}_2' = -\mathbf{e}_1,$$

$$\mathbf{e}_3' = \mathbf{e}_3,$$

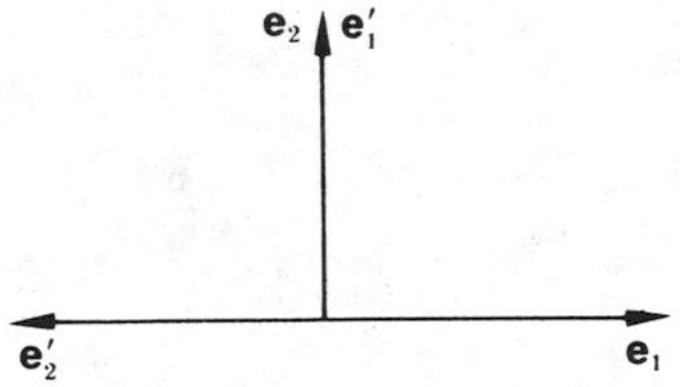

Fig. 2.4

the only nonzero elements of $\mathbf{Q}$ are

$$Q_{12} = \mathbf{e}_1 \cdot \mathbf{e}_2' = -1,$$

$$Q_{21} = \mathbf{e}_2 \cdot \mathbf{e}_1' = 1,$$

$$Q_{33} = \mathbf{e}_3 \cdot \mathbf{e}_3' = 1,$$

i.e.,

$$[\mathbf{Q}] = \begin{bmatrix} 0 & -1 & 0 \\ 1 & 0 & 0 \\ 0 & 0 & 1 \end{bmatrix}.$$

Thus

$$[\mathbf{T}]' = \begin{bmatrix} 0 & +1 & 0 \\ -1 & 0 & 0 \\ 0 & 0 & 1 \end{bmatrix}_{\mathbf{e}_i} \begin{bmatrix} 0 & 1 & 0 \\ 1 & 2 & 0 \\ 0 & 0 & 1 \end{bmatrix}_{\mathbf{e}_i} \begin{bmatrix} 0 & -1 & 0 \\ +1 & 0 & 0 \\ 0 & 0 & 1 \end{bmatrix}_{\mathbf{e}_i} = \begin{bmatrix} 2 & -1 & 0 \\ -1 & 0 & 0 \\ 0 & 0 & 1 \end{bmatrix}_{\mathbf{e}_i'},$$

i.e.,

$$T_{11}' = 2, \quad T_{12}' = -1, \quad T_{13}' = 0, \text{ etc.}$$

We note that we can also use directly the formula

$$T_{ij}' = \mathbf{e}_i' \cdot \mathbf{T}\mathbf{e}_j',$$

if the specific components of T_{ij}' are required. For example, if we need only T_{11}', then

$$T_{11}' = \mathbf{e}_1' \cdot \mathbf{T}\mathbf{e}_1' = [0, 1, 0] \begin{bmatrix} 0 & 1 & 0 \\ 1 & 2 & 0 \\ 0 & 0 & 1 \end{bmatrix} \begin{bmatrix} 0 \\ 1 \\ 0 \end{bmatrix} = [0, 1, 0] \begin{bmatrix} 1 \\ 2 \\ 0 \end{bmatrix} = 2$$

or

$$T_{11}' = \mathbf{e}_1' \cdot \mathbf{T}\mathbf{e}_1' = \mathbf{e}_2 \cdot \mathbf{T}\mathbf{e}_2 = T_{22} = 2.$$

EXAMPLE 2.7

Show that if S_{ij} are the Cartesian components of a tensor **S**, then $S_{ii} = S_{11} + S_{22} + S_{33}$ is a scalar invariant with respect to all orthogonal transformations. That is, $S_{11} + S_{22} + S_{33} = S'_{11} + S'_{22} + S'_{33}$.

Solution. The primed components are related to the unprimed components by Eq. (B29a)

$$S'_{ij} = Q_{mi} Q_{nj} S_{mn}.$$

This means

$$S'_{11} = Q_{m1} Q_{n1} S_{mn},$$

$$S'_{22} = Q_{m2} Q_{n2} S_{mn},$$

and

$$S'_{33} = Q_{m3} Q_{n3} S_{mn}.$$

Adding gives

$$S'_{ii} = S'_{11} + S'_{22} + S'_{33} = Q_{mi} Q_{ni} S_{mn}.$$

But with Eq. (B26c)

$$S'_{ii} = \delta_{mn} S_{mn} = S_{mm}.$$

The result of this example indicates that the matrices

$$\begin{bmatrix} 1 & 0 & 0 \\ 0 & 2 & 0 \\ 0 & 0 & 3 \end{bmatrix} \quad \text{and} \quad \begin{bmatrix} 1 & 0 & 0 \\ 0 & 1 & 0 \\ 0 & 0 & 3 \end{bmatrix}$$

could not be matrices of the same tensor.

Equations (B28) and (B29) state that when the components of a vector (or tensor) with respect to $\{\mathbf{e}_1, \mathbf{e}_2, \mathbf{e}_3\}$ are known, then its components with respect to any $\{\mathbf{e}'_1, \mathbf{e}'_2, \mathbf{e}'_3\}$ are uniquely determined from them. In other words, the components a_i (or T_{ij}) with respect to one set of $\{\mathbf{e}_1, \mathbf{e}_2, \mathbf{e}_3\}$ completely characterize a vector (or a tensor). Thus, it is perfectly meaningful to use a statement such as "consider a tensor T_{ij}," meaning consider the tensor **T** whose components with respect to some set of $\{\mathbf{e}_1, \mathbf{e}_2, \mathbf{e}_3\}$ are T_{ij}. In fact, an alternative way of defining a tensor is through the use of transformation laws relating components of a tensor with respect to different bases. Confining ourselves to only rectangular Cartesian coordinate systems and using unit vectors along positive coordinate directions as base vectors, we now define Cartesian components of tensors of different orders in terms of their transformation laws in the following, where the primed quantities are referred to basis $\{\mathbf{e}'_1, \mathbf{e}'_2, \mathbf{e}'_3\}$ and unprimed quantities to basis $\{\mathbf{e}_1, \mathbf{e}_2, \mathbf{e}_3\}$, the $\mathbf{e}'_i$ and $\mathbf{e}_i$ are related by $\mathbf{e}'_i = \mathbf{Q}\mathbf{e}_i$, **Q** being an orthogonal transformation

$$\begin{aligned} \alpha' &= \alpha && \textbf{zeroth-order tensor} \text{ (or } \textbf{scalar}\text{)} \\ a'_i &= Q_{mi} a_m && \textbf{first-order tensor} \text{ (or } \textbf{vector}\text{)} \\ T'_{ij} &= Q_{mi} Q_{nj} T_{mn} && \textbf{second-order tensor} \text{ (or } \textbf{tensor}\text{)} \\ T'_{ijk} &= Q_{mi} Q_{nj} Q_{rk} T_{mnr} && \textbf{third-order tensor} \end{aligned} \qquad \text{(B31)}$$

etc.

EXAMPLE 2.8

Show that (a) if A_{ijkl} and A'_{ijkl} are components, with respect to the bases $\{\mathbf{e}_i\}$ and $\{\mathbf{e}'_i\}$ respectively, of a fourth-order tensor, and if B_{ij} and B'_{ij} are components, with respect to the same bases $\{\mathbf{e}_i\}$ and $\{\mathbf{e}'_i\}$, of a second-order tensor, then $D_{ij} \equiv A_{ijkl}B_{kl}$ and $D'_{ij} \equiv A'_{ijkl}B'_{kl}$ are components of a second-order tensor; (b) if with respect to any rectangular Cartesian basis $\{\mathbf{e}_1, \mathbf{e}_2, \mathbf{e}_3\}$,

$$T_{ij} = C_{ijkl}E_{kl}, \tag{i}$$

where T_{ij} and E_{ij} are components of arbitrary second-order tensors $\mathbf{T}$ and $\mathbf{E}$, show that C_{ijkl} are the components of a fourth-order tensor. (This is known as the quotient rule.)

Solution. (a) We have

$$A'_{ijkl} = Q_{mi}Q_{nj}Q_{rk}Q_{sl}A_{mnrs}$$

and

$$B'_{ij} = Q_{mi}Q_{nj}B_{mn}.$$

Thus

$$A'_{ijkl}B'_{kl} = Q_{mi}Q_{nj}Q_{rk}Q_{sl}A_{mnrs}Q_{pk}Q_{ql}B_{pq}.$$

But

$$Q_{rk}Q_{pk} = \delta_{rp} \quad \text{and} \quad Q_{sl}Q_{ql} = \delta_{sq}.$$

Therefore

$$A'_{ijkl}B'_{kl} = \delta_{rp}\delta_{sq}Q_{mi}Q_{nj}A_{mnrs}B_{pq} = Q_{mi}Q_{nj}A_{mnpq}B_{pq}.$$

That is

$$D'_{ij} = Q_{mi}Q_{nj}D_{mn}.$$

In other words, D'_{ij} and D_{ij} are indeed related by the transformation law for a second-order tensor.

(b) Since both $\mathbf{T}$ and $\mathbf{E}$ are second-order tensors, we may replace the unprimed components by primed components in Eq. (i) to give

$$Q_{im}Q_{jn}T'_{mn} = C_{ijkl}Q_{ko}Q_{lp}E'_{op}. \tag{ii}$$

But, since Eq. (i) is valid for any basis, the primed components of $\mathbf{T}$ and $\mathbf{E}$ are also related by

$$T'_{mn} = C'_{mnop}E'_{op}. \tag{iii}$$

Substituting Eq. (iii) into Eq. (ii) we have

$$Q_{im}Q_{jn}C'_{mnop}E'_{op} = C_{ijkl}Q_{ko}Q_{lp}E'_{op}.$$

If we multiply this equation by $Q_{ir}Q_{js}$ and use Eq. (B26c), then the Q's will be brought to the right-hand side of the equation in the form

$$\delta_{rm}\delta_{sn}C'_{mnop}E'_{op} = C_{ijkl}Q_{ir}Q_{js}Q_{ko}Q_{lp}E'_{op}.$$

Bringing all the terms to the left-hand side of the equation, we have

$$(C'_{rsop} - C_{ijkl}Q_{ir}Q_{js}Q_{ko}Q_{lp})E'_{op} = 0.$$

The components of the tensor $\mathbf{E}$ are arbitrary, so that the parenthesis must identically vanish. Thus

$$C'_{rsop} = Q_{ir}Q_{js}Q_{ko}Q_{lp}C_{ijkl}$$

and C_{ijkl} does transform as the components of a fourth-order tensor.

EXAMPLE 2.9

A third-order tensor can also be defined in the following way: **M** is a third-order tensor if for any vector **a**, **Ma** is a second-order tensor and $\mathbf{M}(\alpha\mathbf{a}+\beta\mathbf{b}) = \alpha\mathbf{Ma}+\beta\mathbf{Mb}$. Defining the components of **M** by the rule

$$M_{ijk} = \mathbf{e}_i \cdot (\mathbf{M}\mathbf{e}_k)\mathbf{e}_j,$$

show that the components of **M** transform according to the law of Eq. (B31).

Solution.

$$\begin{aligned} M'_{ijk} &= \mathbf{e}'_i \cdot (\mathbf{M}\mathbf{e}'_k)\mathbf{e}'_j = Q_{mi}\mathbf{e}_m \cdot (\mathbf{M}Q_{nk}\mathbf{e}_n)Q_{pj}\mathbf{e}_p \\ &= Q_{mi}Q_{pj}Q_{nk}\{\mathbf{e}_m \cdot (\mathbf{M}\mathbf{e}_n)\mathbf{e}_p\} = Q_{mi}Q_{pj}Q_{nk}M_{mpn}. \end{aligned}$$

EXAMPLE 2.10

Using the definition given in Example 2.9, find the components of a triad (**a b c**) defined by

$$(\mathbf{a\,b\,c})\mathbf{d} = (\mathbf{a\,b})\,(\mathbf{c}\cdot\mathbf{d}).$$

Solution.

$$\begin{aligned} (\mathbf{a\,b\,c})_{ijk} &= \mathbf{e}_i \cdot \{(\mathbf{a\,b\,c})\mathbf{e}_k\}\mathbf{e}_j = \mathbf{e}_i \cdot \{(\mathbf{a\,b})\,(\mathbf{c}\cdot\mathbf{e}_k)\}\mathbf{e}_j \\ &= c_k\{\mathbf{e}_i \cdot (\mathbf{a\,b})\mathbf{e}_j\} = c_k a_i b_j = a_i b_j c_k. \end{aligned}$$

We mention here that any third-order tensor can be expressed as

$$\mathbf{M} = M_{ijk}\,\mathbf{e}_i\mathbf{e}_j\mathbf{e}_k.$$

2B10. Symmetric and Antisymmetric Tensors

A tensor **T** is said to be symmetric if $\mathbf{T} = \mathbf{T}^T$. Thus, the components of a symmetric tensor have the property,

$$T_{ij} = (\mathbf{T}^T)_{ij} = T_{ji}, \tag{B32}$$

i.e.,

$$T_{12} = T_{21}, \quad T_{13} = T_{31}, \quad \text{and} \quad T_{23} = T_{32}.$$

A tensor **T** is said to be antisymmetric if $\mathbf{T} = -\mathbf{T}^T$. Thus, the components of an antisymmetric tensor have the property

$$T_{ij} = -(\mathbf{T}^T)_{ij} = -T_{ji}, \tag{B33}$$

i.e.,

$$T_{11} = T_{22} = T_{33} = 0,$$

$$T_{12} = -T_{21}, \quad T_{13} = -T_{31}, \quad \text{and} \quad T_{23} = -T_{32}.$$

Any tensor **T** can always be decomposed into the sum of a symmetric tensor and an antisymmetric tensor. In fact,

$$\mathbf{T} = \mathbf{T}^S + \mathbf{T}^A, \tag{B34}$$

where

$$\mathbf{T}^S = \frac{\mathbf{T}+\mathbf{T}^T}{2} \quad \text{is symmetric}$$

and

$$\mathbf{T}^A = \frac{\mathbf{T}-\mathbf{T}^T}{2} \quad \text{is antisymmetric.}$$

It is not hard to prove that the decomposition is unique (*see* Problem B25).

2B11. The Dual Vector of an Antisymmetric Tensor

For any antisymmetric tensor **T**, we can always find a vector, called the **dual vector**† of **T**, denoted by $\mathbf{t}^A$, such that any vector **a** is transformed under **T** into the vector $\mathbf{t}^A \times \mathbf{a}$. That is,

$$\mathbf{Ta} = \mathbf{t}^A \times \mathbf{a}. \tag{B35}$$

In fact, the components of $\mathbf{t}^A$ can be found in the following way:

From $\mathbf{Te}_j = \mathbf{t}^A \times \mathbf{e}_j$, we have

$$T_{ij} = \mathbf{e}_i \cdot \mathbf{Te}_j = \mathbf{e}_i \cdot (\mathbf{t}^A \times \mathbf{e}_j) = \mathbf{t}^A \cdot (\mathbf{e}_j \times \mathbf{e}_i).$$

Since $\mathbf{e}_j \times \mathbf{e}_i = -\mathbf{e}_i \times \mathbf{e}_j$, T_{ij} must equal $-T_{ji}$ for $\mathbf{t}^A$ to exist. When $T_{ij} = -T_{ji}$ (i.e., **T** is antisymmetric) the nonzero components of T_{ij} are related to the components of $\mathbf{t}^A$ by the following, if $(\mathbf{e}_1, \mathbf{e}_2, \mathbf{e}_3)$ is a right-handed triad:

$$T_{32} = -T_{23} = \mathbf{e}_3 \cdot \mathbf{Te}_2 = \mathbf{t}^A \cdot (\mathbf{e}_2 \times \mathbf{e}_3) = \mathbf{t}^A \cdot \mathbf{e}_1 = t_1{}^A,$$

$$T_{13} = -T_{31} = \mathbf{e}_1 \cdot \mathbf{Te}_3 = t^A \cdot (\mathbf{e}_3 \times \mathbf{e}_1) = \mathbf{t}^A \cdot \mathbf{e}_2 = t_2{}^A,$$

and

$$T_{21} = -T_{12} = \mathbf{e}_2 \cdot \mathbf{Te}_1 = \mathbf{t}^A \cdot (\mathbf{e}_1 \times \mathbf{e}_2) = \mathbf{t}^A \cdot \mathbf{e}_3 = t_3{}^A.$$

Thus

$$\mathbf{t}^A = T_{32}\mathbf{e}_1 + T_{13}\mathbf{e}_2 + T_{21}\mathbf{e}_3 \tag{B36a}$$

or

$$\mathbf{t}^A = -(T_{23}\mathbf{e}_1 + T_{31}\mathbf{e}_2 + T_{12}\mathbf{e}_3), \tag{B36b}$$

i.e.,

$$2\mathbf{t}^A = -\epsilon_{ijk} T_{jk} \mathbf{e}_i. \tag{B36c}$$

†Also known as axial vector of **T**.

EXAMPLE 2.11

Given

$$[T_{ij}] = \begin{bmatrix} 1 & 2 & 3 \\ 4 & 2 & 1 \\ 1 & 1 & 1 \end{bmatrix},$$

decompose the tensor into a symmetric and an antisymmetric part. Also find the dual vector for the antisymmetric part. Verify $\mathbf{T}^A\mathbf{a} = \mathbf{t}^A \times \mathbf{a}$ for $\mathbf{a} = \mathbf{e}_1 + \mathbf{e}_3$.

Solution. $[\mathbf{T}] = [\mathbf{T}^S] + [\mathbf{T}^A]$, where

$$[\mathbf{T}^S] = \tfrac{1}{2}([\mathbf{T}] + [\mathbf{T}]^T) = \frac{1}{2}\begin{bmatrix} 2 & 6 & 4 \\ 6 & 4 & 2 \\ 4 & 2 & 2 \end{bmatrix} = \begin{bmatrix} 1 & 3 & 2 \\ 3 & 2 & 1 \\ 2 & 1 & 1 \end{bmatrix}.$$

$$[\mathbf{T}^A] = \tfrac{1}{2}([\mathbf{T}] - [\mathbf{T}]^T) = \frac{1}{2}\begin{bmatrix} 0 & -2 & 2 \\ 2 & 0 & 0 \\ -2 & 0 & 0 \end{bmatrix} = \begin{bmatrix} 0 & -1 & 1 \\ 1 & 0 & 0 \\ -1 & 0 & 0 \end{bmatrix},$$

i.e.,

$$[\mathbf{T}] = \begin{bmatrix} 1 & 3 & 2 \\ 3 & 2 & 1 \\ 2 & 1 & 1 \end{bmatrix} + \begin{bmatrix} 0 & -1 & 1 \\ 1 & 0 & 0 \\ -1 & 0 & 0 \end{bmatrix}.$$

The vector of $\mathbf{T}^A$ is

$$\mathbf{t}^A = -(T^A_{23}\mathbf{e}_1 + T^A_{31}\mathbf{e}_2 + T^A_{12}\mathbf{e}_3) = -(0\mathbf{e}_1 - \mathbf{e}_2 - \mathbf{e}_3) = \mathbf{e}_2 + \mathbf{e}_3.$$

Now, let $\mathbf{b} = \mathbf{T}^A\mathbf{a}$, then

$$[\mathbf{b}] = \begin{bmatrix} 0 & -1 & 1 \\ 1 & 0 & 0 \\ -1 & 0 & 0 \end{bmatrix}\begin{bmatrix} 1 \\ 0 \\ 1 \end{bmatrix} = \begin{bmatrix} 1 \\ 1 \\ -1 \end{bmatrix},$$

i.e.,

$$\mathbf{b} = \mathbf{e}_1 + \mathbf{e}_2 - \mathbf{e}_3.$$

On the other hand,

$$\mathbf{t}^A \times \mathbf{a} = (\mathbf{e}_2 + \mathbf{e}_3) \times (\mathbf{e}_1 + \mathbf{e}_3) = \mathbf{e}_1 + \mathbf{e}_2 - \mathbf{e}_3 = \mathbf{b}.$$

EXAMPLE 2.12

Given that $\mathbf{Q}$ is a rotation tensor and that $\mathbf{n}$ is a unit vector in the direction of the axis of rotation, prove that the dual vector $\mathbf{q}$ of $\mathbf{Q}^A$ is parallel to $\mathbf{n}$.

Solution. Since $\mathbf{n}$ is parallel to the axis rotation, therefore

$$\mathbf{Qn} = \mathbf{n}. \qquad \text{(i)}$$

Thus $(\mathbf{Q}^T\mathbf{Q})\mathbf{n} = \mathbf{Q}^T\mathbf{n}$. Since $\mathbf{Q}^T\mathbf{Q} = \mathbf{I}$, we have

$$\mathbf{Q}^T\mathbf{n} = \mathbf{n}. \qquad \text{(ii)}$$

From Eqs. (i) and (ii), we get

$$(\mathbf{Q} - \mathbf{Q}^T)\mathbf{n} = 0.$$

But

$$(\mathbf{Q} - \mathbf{Q}^T)\mathbf{n} = 2\mathbf{q} \times \mathbf{n},$$

where **q** is the vector of $\mathbf{Q}^A$. Thus

$$\mathbf{q} \times \mathbf{n} = 0,$$

i.e.,

$$\mathbf{q} \parallel \mathbf{n}.$$

We note that it can be shown (*see* Problem B33) that if θ denotes the right-hand rotation angle, then

$$\mathbf{q} = (\sin\theta)\mathbf{n}.$$

2B12. Eigenvalues and Eigenvectors of a Tensor T

Consider a tensor **T**. If **a** is a vector which transforms under **T** into a vector parallel to itself, i.e., $\mathbf{Ta} = \lambda\mathbf{a}$, then **a** is an **eigenvector** and λ is the corresponding **eigenvalue**.

If **a** is an eigenvector with corresponding eigenvalue λ of the linear transformation **T**, then any vector parallel to **a** is also an eigenvector with the same eigenvalue λ. In fact, for any scalar α

$$\mathbf{T}(\alpha\mathbf{a}) = \alpha\mathbf{Ta} = \alpha(\lambda\mathbf{a}) = \lambda(\alpha\mathbf{a}).$$

Thus, an eigenvector, as defined by $\mathbf{Ta} = \lambda\mathbf{a}$, has arbitrary length. For definiteness, all eigenvectors sought will be unit length.

A tensor may have eigenvectors in many different directions. In fact, since $\mathbf{Ia} = \mathbf{a}$, any vector is an eigenvector of the identity tensor **I**, with eigenvalues all equal to unity. For the tensor $\alpha\mathbf{I}$ the same is true, except that the eigenvalues are all equal to α.

Some tensors only have eigenvectors in one direction. For example, for any rotation tensor, which effects a rigid body rotation about an axis through an angle not equal to integral multiples of π, only those vectors which are parallel to the axis of rotation will remain parallel to themselves.

Let **n** be a unit eigenvector, then $\mathbf{Tn} = \lambda\mathbf{n} = \lambda\mathbf{In}$. Thus

$$(\mathbf{T} - \lambda\mathbf{I})\mathbf{n} = 0 \quad \text{with} \quad \mathbf{n} \cdot \mathbf{n} = 1.$$

Let $\mathbf{n} = \alpha_i\mathbf{e}_i$, then in component form

$$(T_{ij} - \lambda\delta_{ij})\alpha_j = 0 \quad \text{with} \quad \alpha_j\alpha_j = 1.$$

In long form, we have

$$(T_{11} - \lambda)\alpha_1 + T_{12}\alpha_2 + T_{13}\alpha_3 = 0, \tag{B37a}$$

$$T_{21}\alpha_1 + (T_{22} - \lambda)\alpha_2 + T_{23}\alpha_3 = 0, \tag{B37b}$$

$$T_{31}\alpha_1 + T_{32}\alpha_2 + (T_{33} - \lambda)\alpha_3 = 0, \tag{B37c}$$

$$\alpha_1^2 + \alpha_2^2 + \alpha_3^2 = 1. \tag{B37d}$$

Equations (B37a, b, c) are a system of linear homogeneous equations in α_1, α_2, α_3 which admits nontrivial solutions† for the α_i's only if the determinant of their coefficient vanishes. That is

$$|\mathbf{T}-\lambda\mathbf{I}| = 0, \tag{B38a}$$

i.e.,

$$\begin{vmatrix} T_{11}-\lambda & T_{12} & T_{13} \\ T_{21} & T_{22}-\lambda & T_{23} \\ T_{31} & T_{32} & T_{33}-\lambda \end{vmatrix} = 0. \tag{B38b}$$

For a given $\mathbf{T}$, T_{ij} are known, the above equation is a cubic equation in λ. It is called the characteristic equation of $\mathbf{T}$. The roots of the equation are the eigenvalues of $\mathbf{T}$. Let the roots be λ_1, λ_2, and λ_3. Corresponding to each λ_i, the components of the corresponding eigenvector are determined from the equations in Eqs. (B37). [Equations (B37a, b, c) will not be all independent when $\lambda = \lambda_i$.] The method can best be described by examples.

EXAMPLE 2.13

Show that if $T_{21} = T_{31} = 0$, then $\mathbf{e}_1$ is an eigenvector of $\mathbf{T}$ with corresponding eigenvalues equal to T_{11}.

Solution. From $\mathbf{T}\mathbf{e}_1 = T_{11}\mathbf{e}_1 + T_{21}\mathbf{e}_2 + T_{31}\mathbf{e}_3$ we have

$$\mathbf{T}\mathbf{e}_1 = T_{11}\mathbf{e}_1.$$

Thus $\mathbf{e}_1$ is an eigenvector with T_{11} as its eigenvalue. Similarly, if $T_{12} = T_{32} = 0$, then $\mathbf{e}_2$ is an eigenvector with corresponding eigenvalue T_{22}, and if $T_{13} = T_{23} = 0$, then $\mathbf{e}_3$ is an eigenvector with corresponding eigenvalue T_{33}.

EXAMPLE 2.14

If, with respect to some basis $(\mathbf{e}_1, \mathbf{e}_2, \mathbf{e}_3)$ the matrix of $\mathbf{T}$ is given by

$$[\mathbf{T}] = \begin{bmatrix} 2 & 0 & 0 \\ 0 & 3 & 4 \\ 0 & 4 & -3 \end{bmatrix}.$$

Find the eigenvalues and their corresponding eigenvectors.

Solution. The characteristic equation is

$$|\mathbf{T}-\lambda\mathbf{I}| = \begin{vmatrix} 2-\lambda & 0 & 0 \\ 0 & 3-\lambda & 4 \\ 0 & 4 & -3-\lambda \end{vmatrix} = (2-\lambda)\,(\lambda^2-25) = 0.$$

Thus, there are three distinct eigenvalues, $\lambda_1 = 2$, $\lambda_2 = 5$, and $\lambda_3 = -5$.

† $\alpha_1 = \alpha_2 = \alpha_3 = 0$ is the trivial solution for Eqs. (B37a, b, c).

Corresponding to $\lambda_1 = 2$, Eq. (B37a, b, c) give

$$0\alpha_1 = 0,$$

$$\alpha_2 + 4\alpha_3 = 0,$$

$$4\alpha_2 - 5\alpha_3 = 0,$$

and

$$\alpha_1^2 + \alpha_2^2 + \alpha_3^2 = 1.$$

Thus, $\alpha_2 = \alpha_3 = 0$ and $\alpha_1 = \pm 1$ and the eigenvector corresponding to $\lambda_1 = 2$ is $\mathbf{n}_1 = \pm \mathbf{e}_1$. We note that from the previous example, the eigenvalue 2 and the corresponding eigenvector $\pm \mathbf{e}_1$ can be written down without computation.

Corresponding to $\lambda_2 = 5$, we have

$$-3\alpha_1 = 0,$$

$$-2\alpha_2 + 4\alpha_3 = 0,$$

$$4\alpha_2 - 8\alpha_3 = 0,$$

and

$$\alpha_1^2 + \alpha_2^2 + \alpha_3^2 = 1.$$

Thus (note the second and third equations are the same),

$$\alpha_1 = 0, \quad \alpha_2 = \frac{2}{\sqrt{5}}, \quad \alpha_3 = \frac{1}{\sqrt{5}}$$

and the eigenvector corresponding to $\lambda_2 = 5$ is

$$\mathbf{n}_2 = \pm \frac{1}{\sqrt{5}} (2\mathbf{e}_2 + \mathbf{e}_3).$$

Corresponding to $\lambda_3 = -5$, similar computations give

$$\mathbf{n}_3 = \pm \frac{1}{\sqrt{5}} (-\mathbf{e}_2 + 2\mathbf{e}_3).$$

EXAMPLE 2.15

Find the eigenvalues and eigenvectors for the rotation tensor $\mathbf{R}$ corresponding to a 90° rotation about the $\mathbf{e}_3$-axis (*see* Example 2.3).

Solution. The characteristic equation is

$$\begin{vmatrix} 0-\lambda & -1 & 0 \\ 1 & 0-\lambda & 0 \\ 0 & 0 & 1-\lambda \end{vmatrix} = 0,$$

i.e.,

$$\lambda^2(1-\lambda) + (1-\lambda) = (1-\lambda)(\lambda^2+1) = 0.$$

Thus, only one eigenvalue is real, namely $\lambda_1 = 1$, the other two are complex, $\lambda_2 = +\sqrt{-1}$, $\lambda_3 = -\sqrt{-1}$. Correspondingly, there is only one real eigenvector. Only real eigenvectors are of interest to us, we shall therefore compute only the eigenvector corresponding to

$\lambda_1 = 1$. From

$$(0-1)\alpha_1 - \alpha_2 = 0,$$

$$\alpha_1 - \alpha_2 = 0,$$

$$(1-1)\alpha_3 = 0,$$

$$\alpha_1{}^2 + \alpha_2{}^2 + \alpha_3{}^2 = 1,$$

we obtain $\alpha_1 = 0$, $\alpha_2 = 0$, $\alpha_3 = \pm 1$, i.e., $\mathbf{n} = \pm \mathbf{e}_3$, which, of course, is parallel to the axis of rotation.

EXAMPLE 2.16

Find the eigenvalues and eigenvectors for the tensor

$$[\mathbf{T}] = \begin{bmatrix} 2 & 0 & 0 \\ 0 & 2 & 0 \\ 0 & 0 & 3 \end{bmatrix}.$$

Solution. The characteristic equation gives

$$(2-\lambda)^2(3-\lambda) = 0.$$

Thus $\lambda_1 = 3$, $\lambda_2 = \lambda_3 = 2$. These results are obvious in view of Example 2.13. In fact, that example also tells us that the eigenvector corresponding to $\lambda_1 = 3$ is $\pm \mathbf{e}_3$ and eigenvectors corresponding to $\lambda_2 = \lambda_3 = 2$ are $\pm \mathbf{e}_1$ and $\pm \mathbf{e}_2$. However, there are actually infinitely many eigenvectors corresponding to the double root. From Eqs. (B37a, b, c), we have, with $\lambda = 2$,

$$0\alpha_1 = 0,$$

$$0\alpha_2 = 0,$$

$$\alpha_3 = 0,$$

and

$$\alpha_1{}^2 + \alpha_2{}^2 + \alpha_3{}^2 = 1.$$

The equations are satisfied by $\alpha_3 = 0$ together with any α_1 and α_2 satisfying $\alpha_1{}^2 + \alpha_2{}^2 = 1$. Thus any unit vector perpendicular to $\mathbf{e}_3$, $\mathbf{n} = \alpha_1\mathbf{e}_1 + \alpha_2\mathbf{e}_2$, is an eigenvector.

2B13. Principal Values and Principal Directions of Real Symmetric Tensors

In the following chapters, we shall encounter several real tensors (stress tensor, strain tensor, rate of deformation tensor) which are symmetric, for which the following theorem, stated without proof, is important: "the eigenvalues of any real symmetric tensor are all real." Thus, for a real symmetric tensor, there always exist at least three real eigenvectors which we shall also call the **principal directions**. The corresponding eigenvalues are called the **principal values**.

We now prove that there always exist three principal directions which are mutually perpendicular.

Let $\mathbf{n}_1$ and $\mathbf{n}_2$ be two eigenvectors corresponding to the eigenvalues λ_1 and λ_2 respectively of a tensor $\mathbf{T}$. Then

$$\mathbf{T}\mathbf{n}_1 = \lambda_1\mathbf{n}_1 \tag{i}$$

and

$$\mathbf{T}\mathbf{n}_2 = \lambda_2\mathbf{n}_2. \tag{ii}$$

Thus

$$\lambda_1\mathbf{n}_1 \cdot \mathbf{n}_2 = \mathbf{n}_2 \cdot \mathbf{T}\mathbf{n}_1, \tag{iii}$$

$$\lambda_2\mathbf{n}_1 \cdot \mathbf{n}_2 = \mathbf{n}_1 \cdot \mathbf{T}\mathbf{n}_2. \tag{iv}$$

The definition of the transpose of $\mathbf{T}$ gives $\mathbf{n}_1 \cdot \mathbf{T}\mathbf{n}_2 = \mathbf{n}_2 \cdot \mathbf{T}^T\mathbf{n}_1$, thus for symmetric $\mathbf{T}$, $\mathbf{n}_1 \cdot \mathbf{T}\mathbf{n}_2 = \mathbf{n}_2 \cdot \mathbf{T}\mathbf{n}_1$.

From Eqs. (iii) and (iv), we have

$$(\lambda_1 - \lambda_2)(\mathbf{n}_1 \cdot \mathbf{n}_2) = 0. \tag{v}$$

It follows that if $\lambda_1 \neq \lambda_2$, then $\mathbf{n}_1 \cdot \mathbf{n}_2 = 0$, i.e., $\mathbf{n}_1$ and $\mathbf{n}_2$ are perpendicular to each other. We have thus proved that if the eigenvalues are all distinct, then the three principal directions are mutually perpendicular.

Next, let us suppose that $\mathbf{n}_1$ and $\mathbf{n}_2$ are two eigenvectors corresponding to the same eigenvalue λ. Then, by definition, $\mathbf{T}\mathbf{n}_1 = \lambda\mathbf{n}_1$ and $\mathbf{T}\mathbf{n}_2 = \lambda\mathbf{n}_2$ so that for any scalars α and β, $\mathbf{T}(\alpha\mathbf{n}_1 + \beta\mathbf{n}_2) \equiv \alpha\mathbf{T}\mathbf{n}_1 + \beta\mathbf{T}\mathbf{n}_2 = \lambda(\alpha\mathbf{n}_1 + \beta\mathbf{n}_2)$. That is, $\alpha\mathbf{n}_1 + \beta\mathbf{n}_2$ is also an eigenvector with the same eigenvalue λ. In other words, if there are two distinct eigenvectors with the same eigenvalue, then there are infinitely many eigenvectors (which form a plane) with the same eigenvalue. This situation arises when the characteristic equation has a repeated root. Suppose the characteristic equation has roots $\lambda_1, \lambda_2 = \lambda_3 = \lambda\ (\lambda_1 \neq \lambda)$. Let $\mathbf{n}_1$ be the eigenvector corresponding to λ_1, then $\mathbf{n}_1$ is perpendicular to any eigenvector of λ. Now, corresponding to λ, the equations

$$(T_{11} - \lambda)\alpha_1 + T_{12}\alpha_2 + T_{13}\alpha_3 = 0,$$

$$T_{21}\alpha_1 + (T_{22} - \lambda)\alpha_2 + T_{23}\alpha_3 = 0,$$

$$T_{31}\alpha_1 + T_{32}\alpha_2 + (T_{33} - \lambda)\alpha_3 = 0$$

degenerate to one independent equation (*see* Example 2.16) so that there are infinitely many eigenvectors lying on the plane whose normal is $\mathbf{n}_1$. Therefore, though not unique, there again exist three mutually perpendicular principal directions.

In the case of a triple root the above three equations will be automatically satisfied for whatever values of $(\alpha_1, \alpha_2, \alpha_3)$ so that any vector is an eigenvector.†

Thus for every real symmetric tensor there exists at least one triad of principal directions which are mutually perpendicular.

2B14. Matrix of a Tensor with Respect to Principal Directions

We have shown that for a real symmetric tensor, there always exist three principal directions which are mutually perpendicular. Let $\mathbf{n}_1$, $\mathbf{n}_2$, and $\mathbf{n}_3$ be unit vectors in these directions. Then, using $\mathbf{n}_1$, $\mathbf{n}_2$, $\mathbf{n}_3$ as base vectors, the components of the tensor are

$$T_{11} = \mathbf{n}_1 \cdot \mathbf{T}\mathbf{n}_1 = \mathbf{n}_1 \cdot (\lambda_1 \mathbf{n}_1) = \lambda_1,$$

$$T_{22} = \mathbf{n}_2 \cdot \mathbf{T}\mathbf{n}_2 = \mathbf{n}_2 \cdot (\lambda_2 \mathbf{n}_2) = \lambda_2,$$

$$T_{33} = \mathbf{n}_3 \cdot \mathbf{T}\mathbf{n}_3 = \mathbf{n}_3 \cdot (\lambda_3 \mathbf{n}_3) = \lambda_3,$$

$$T_{12} = \mathbf{n}_1 \cdot \mathbf{T}\mathbf{n}_2 = \mathbf{n}_1 \cdot (\lambda_2 \mathbf{n}_2) = 0 = T_{21},$$

$$T_{13} = \mathbf{n}_1 \cdot \mathbf{T}\mathbf{n}_3 = \mathbf{n}_1 \cdot (\lambda_3 \mathbf{n}_3) = 0 = T_{31},$$

and

$$T_{23} = \mathbf{n}_2 \cdot \mathbf{T}\mathbf{n}_3 = \mathbf{n}_2 \cdot (\lambda_3 \mathbf{n}_3) = 0 = T_{32}.$$

That is

$$[\mathbf{T}]_{\mathbf{n}_1, \mathbf{n}_2, \mathbf{n}_3} = \begin{bmatrix} \lambda_1 & 0 & 0 \\ 0 & \lambda_2 & 0 \\ 0 & 0 & \lambda_3 \end{bmatrix}. \tag{B39}$$

The matrix is diagonal and the diagonal elements are the eigenvalues of **T**.

We now show that the principal values of a tensor **T** include the maximum and minimum values that the diagonal elements of any matrix of **T** can have.

First, for any unit vector $\mathbf{e}_1' = \alpha \mathbf{n}_1 + \beta \mathbf{n}_2 + \gamma \mathbf{n}_3$

$$T_{11}' = \mathbf{e}_1' \cdot \mathbf{T}\mathbf{e}_1' = [\alpha, \beta, \gamma] \begin{bmatrix} \lambda_1 & 0 & 0 \\ 0 & \lambda_2 & 0 \\ 0 & 0 & \lambda_3 \end{bmatrix} \begin{bmatrix} \alpha \\ \beta \\ \gamma \end{bmatrix},$$

i.e.,

$$T_{11}' = \lambda_1 \alpha^2 + \lambda_2 \beta^2 + \lambda_3 \gamma^2.$$

†In fact, it is not hard to show that the only tensors that have every vector as its eigenvector is of the form $\alpha\mathbf{I}$ where α is a scalar and $\mathbf{I}$, an identity tensor. The characteristic equation of $\alpha\mathbf{I}$ obviously has α as its triple root.

Without loss of generality, let

$$\lambda_1 \geqslant \lambda_2 \geqslant \lambda_3,$$

then noting that $\alpha^2+\beta^2+\gamma^2 = 1$, we have

$$\lambda_1 = \lambda_1(\alpha^2+\beta^2+\gamma^2) \geqslant \lambda_1\alpha^2+\lambda_2\beta^2+\lambda_3\gamma^2,$$

i.e.,

$$\lambda_1 \geqslant T'_{11}.$$

Also

$$\lambda_1\alpha^2+\lambda_2\beta^2+\lambda_3\gamma^2 \geqslant \lambda_3(\alpha^2+\beta^2+\gamma^2) = \lambda_3,$$

i.e.,

$$T'_{11} \geqslant \lambda_3.$$

Thus the $\left\{\begin{matrix}\text{maximum}\\\text{minimum}\end{matrix}\right\}$ value of the principal values of **T** is the $\left\{\begin{matrix}\text{maximum}\\\text{minimum}\end{matrix}\right\}$ value of the diagonal elements of all [**T**] of **T**.

2B15. Scalar Invariants of a Tensor

The characteristic equation of a tensor **T**, $|T_{ij}-\lambda\delta_{ij}| = 0$ is a cubic equation in λ. It can be written as

$$\lambda^3 - I_1\lambda^2 + I_2\lambda - I_3 = 0, \tag{B40}$$

where

$$I_1 = T_{11}+T_{22}+T_{33} = T_{ii},$$

$$I_2 = \begin{vmatrix} T_{11} & T_{12} \\ T_{21} & T_{22} \end{vmatrix} + \begin{vmatrix} T_{22} & T_{23} \\ T_{32} & T_{33} \end{vmatrix} + \begin{vmatrix} T_{11} & T_{13} \\ T_{31} & T_{33} \end{vmatrix} = \tfrac{1}{2}(T_{ii}T_{jj} - T_{ij}T_{ji}), \tag{B41}$$

$$I_3 = \begin{vmatrix} T_{11} & T_{12} & T_{13} \\ T_{21} & T_{22} & T_{23} \\ T_{31} & T_{32} & T_{33} \end{vmatrix}.$$

Since by definition the eigenvalues of **T** does *not* depend on the choices of the base vectors $\{\mathbf{e}_1, \mathbf{e}_2, \mathbf{e}_3\}$, therefore the coefficients of the cubic equation should be the same for all $\{\mathbf{e}_1, \mathbf{e}_2, \mathbf{e}_3\}$.† They are called **scalar invariants of T**.

We note that, in terms of the eigenvalues of **T** which are the roots of

†Direct proof of this fact can also be made. *See* Example 2.7 for I_1 and Problems B20 and B21.

Eq. (B40), the I_i's take the simpler form

$$I_1 = \lambda_1 + \lambda_2 + \lambda_3,$$

$$I_2 = \lambda_1\lambda_2 + \lambda_2\lambda_3 + \lambda_3\lambda_1, \qquad \text{(B42)}\dagger$$

and

$$I_3 = \lambda_1\lambda_2\lambda_3.$$

EXAMPLE 2.17

For the tensor of Example 2.14, first find the scalar invariants and then evaluate the eigenvalues using Eq. (B40).

Solution. The matrix of T is

$$[\mathbf{T}] = \begin{bmatrix} 2 & 0 & 0 \\ 0 & 3 & 4 \\ 0 & 4 & -3 \end{bmatrix}$$

and therefore the scalar invariants are

$$I_1 = 2 + 3 - 3 = 2,$$

$$I_2 = \begin{vmatrix} 2 & 0 \\ 0 & 3 \end{vmatrix} + \begin{vmatrix} 3 & 4 \\ 4 & -3 \end{vmatrix} + \begin{vmatrix} 2 & 0 \\ 0 & -3 \end{vmatrix} = -25,$$

$$I_3 = |\mathbf{T}| = 2 \begin{vmatrix} 3 & 4 \\ 4 & -3 \end{vmatrix} = -50.$$

These values give the characteristic equation

$$\lambda^3 - 2\lambda^2 - 25\lambda + 50 = 0$$

or

$$(\lambda - 2)(\lambda - 5)(\lambda + 5) = 0.$$

Thus, the eigenvalues are $\lambda = 2, 5, -5$.

2B16. Tensor-Valued Functions of a Scalar

Let $\mathbf{T} = \mathbf{T}(t)$ be a tensor-valued function of a scalar t (such as time). The derivative of $\mathbf{T}$ with respect to t is defined to be a second-order tensor given by

$$\frac{d\mathbf{T}}{dt} = \lim_{\Delta t \to 0} \frac{\mathbf{T}(t + \Delta t) - \mathbf{T}(t)}{\Delta t}. \qquad \text{(B43)}$$

The following identities can be easily established [only Eq. (B44d) will be proved here]:

$$\frac{d}{dt}(\mathbf{T} + \mathbf{S}) = \frac{d\mathbf{T}}{dt} + \frac{d\mathbf{S}}{dt}, \qquad \text{(B44a)}$$

†These are invariants for nonsymmetric tensors as well.

$$\frac{d}{dt}\{\alpha(t)\mathbf{T}\} = \frac{d\alpha}{dt}\mathbf{T} + \alpha\frac{d\mathbf{T}}{dt}, \tag{B44b}$$

$$\frac{d}{dt}(\mathbf{TS}) = \frac{d\mathbf{T}}{dt}\mathbf{S} + \mathbf{T}\frac{d\mathbf{S}}{dt}, \tag{B44c}$$

$$\frac{d}{dt}(\mathbf{Ta}) = \frac{d\mathbf{T}}{dt}\mathbf{a} + \mathbf{T}\frac{d\mathbf{a}}{dt}, \tag{B44d}$$

$$\frac{d}{dt}(\mathbf{T}^T) = \left(\frac{d\mathbf{T}}{dt}\right)^T. \tag{B44e}$$

To prove Eq. (B44d), we use the definition (B43)

$$\begin{aligned}\frac{d}{dt}(\mathbf{Ta}) &= \lim_{\Delta t\to 0}\frac{\mathbf{T}(t+\Delta t)\mathbf{a}(t+\Delta t) - \mathbf{T}(t)\mathbf{a}(t)}{\Delta t}\\ &= \lim_{\Delta t\to 0}\frac{\mathbf{T}(t+\Delta t)\mathbf{a}(t+\Delta t) - \mathbf{T}(t)\mathbf{a}(t) + \mathbf{T}(t)\mathbf{a}(t+\Delta t) - \mathbf{T}(t)\mathbf{a}(t+\Delta t)}{\Delta t}\\ &= \lim_{\Delta t\to 0}\frac{\{\mathbf{T}(t+\Delta t) - \mathbf{T}(t)\}\mathbf{a}(t+\Delta t)}{\Delta t} + \lim_{\Delta t\to 0}\frac{\mathbf{T}(t)\{\mathbf{a}(t+\Delta t) - \mathbf{a}(t)\}}{\Delta t}.\end{aligned}$$

Thus

$$\frac{d}{dt}(\mathbf{Ta}) = \frac{d\mathbf{T}}{dt}\mathbf{a} + \mathbf{T}\frac{d\mathbf{a}}{dt}.$$

EXAMPLE 2.18

Show that in Cartesian coordinates the components of $d\mathbf{T}/dt$, i.e., $(d\mathbf{T}/dt)_{ij}$ are given by the derivatives of the components, dT_{ij}/dt.

Solution.

$$T_{ij} = \mathbf{e}_i \cdot \mathbf{T}\mathbf{e}_j.$$

Since

$$\frac{d\mathbf{e}_1}{dt} = \frac{d\mathbf{e}_2}{dt} = \frac{d\mathbf{e}_3}{dt} = \mathbf{0}.$$

Therefore,

$$\frac{dT_{ij}}{dt} = \mathbf{e}_i \cdot \frac{d}{dt}(\mathbf{T}\mathbf{e}_j) = \mathbf{e}_i \cdot \left(\frac{d\mathbf{T}}{dt}\right)\mathbf{e}_j = \left(\frac{d\mathbf{T}}{dt}\right)_{ij}.$$

EXAMPLE 2.19

Show that for an orthogonal tensor $\mathbf{Q}(t)$, $(d\mathbf{Q}/dt)\mathbf{Q}^T$ is an antisymmetric tensor.

Solution. Since $\mathbf{QQ}^T = \mathbf{I}$, we have

$$\frac{d\mathbf{Q}}{dt}\mathbf{Q}^T + \mathbf{Q}\frac{d\mathbf{Q}^T}{dt} = \mathbf{0},$$

that is,

$$\mathbf{Q}\frac{d\mathbf{Q}^T}{dt} = -\frac{d\mathbf{Q}}{dt}\mathbf{Q}^T.$$

Since

$$\frac{d\mathbf{Q}^T}{dt} = \left(\frac{d\mathbf{Q}}{dt}\right)^T, \qquad \text{[\textit{see} Eq. (B44e)]}$$

therefore,

$$\mathbf{Q}\left(\frac{d\mathbf{Q}}{dt}\right)^T = -\frac{d\mathbf{Q}}{dt}\mathbf{Q}^T.$$

But

$$\mathbf{Q}\left(\frac{d\mathbf{Q}}{dt}\right)^T = \left(\frac{d\mathbf{Q}}{dt}\mathbf{Q}^T\right)^T, \qquad \text{[\textit{see} Eq. (B24)]}$$

therefore,

$$\left\{\frac{d\mathbf{Q}}{dt}\mathbf{Q}^T\right\}^T = -\frac{d\mathbf{Q}}{dt}\mathbf{Q}^T.$$

EXAMPLE 2.20

A time-dependent rigid body rotation about a fixed point can be represented by a rotation tensor $\mathbf{R}(t)$, so that a position vector $\mathbf{r}_0$ is transformed through rotation into $\mathbf{r}(t) = \mathbf{R}(t)\mathbf{r}_0$. Derive the equation

$$\frac{d\mathbf{r}}{dt} = \boldsymbol{\omega} \times \mathbf{r}.$$

Solution. From $\mathbf{r} = \mathbf{R}(t)\mathbf{r}_0$,

$$\frac{d\mathbf{r}}{dt} = \frac{d\mathbf{R}}{dt}\mathbf{r}_0 = \left(\frac{d\mathbf{R}}{dt}\mathbf{R}^T\right)\mathbf{r}.$$

But $(d\mathbf{R}/dt)\mathbf{R}^T$ is an antisymmetric tensor, (*see* Example 2.19) so that

$$\frac{d\mathbf{r}}{dt} = \left(\frac{d\mathbf{R}}{dt}\mathbf{R}^T\right)\mathbf{r} = \boldsymbol{\omega} \times \mathbf{r},$$

where $\boldsymbol{\omega}$ is the dual vector of

$$\left(\frac{d\mathbf{R}}{dt}\mathbf{R}^T\right).$$

From the well-known equation in rigid body kinematics, we can identify $\boldsymbol{\omega}$ as the angular velocity vector of the body.

2B17. Scalar Field, Gradient of a Scalar Function

Let $\phi(\mathbf{r})$ be a scalar-valued function of the position vector $\mathbf{r}$. That is, for each position $\mathbf{r}$, ϕ gives the value of a scalar, such as density, temperature or electric potential at the point. In other words, $\phi(\mathbf{r})$ describes a scalar field. Associated with a scalar field, there is a vector field, called the **gradient** of ϕ, which is of considerable importance. The gradient of ϕ at a point $\mathbf{r}$ is defined to be a vector, denoted by $\nabla\phi$, such that its dot product with $d\mathbf{r}$ gives the difference of the values of the scalar at $\mathbf{r}+d\mathbf{r}$ and $\mathbf{r}$, i.e.,

$$d\phi = \phi(\mathbf{r}+d\mathbf{r}) - \phi(\mathbf{r}) \equiv \nabla\phi \cdot d\mathbf{r}. \tag{B45}$$

Let dr denote the magnitude of $d\mathbf{r}$, and $\mathbf{e}$ the unit vector in the direction of $d\mathbf{r}$ (note $\mathbf{e} = d\mathbf{r}/dr$). Then the above equation gives

$$\left(\frac{d\phi}{dr}\right)_{\text{in } \mathbf{e}\text{-direction}} = \nabla\phi \cdot \mathbf{e}. \tag{B46}$$

That is, the component of $\nabla\phi$ in the direction of $\mathbf{e}$ gives the rate of change of ϕ in that direction (directional derivative).

Since

$$\left(\frac{d\phi}{dr}\right)_{\text{in } \mathbf{e}_1\text{-direction}} \equiv \frac{\partial\phi}{\partial x_1} = \nabla\phi \cdot \mathbf{e}_1 = (\nabla\phi)_1,$$

$$\left(\frac{d\phi}{dr}\right)_{\text{in } \mathbf{e}_2\text{-direction}} \equiv \frac{\partial\phi}{\partial x_2} = \nabla\phi \cdot \mathbf{e}_2 = (\nabla\phi)_2,$$

$$\left(\frac{d\phi}{dr}\right)_{\text{in } \mathbf{e}_3\text{-direction}} \equiv \frac{\partial\phi}{\partial x_3} = \nabla\phi \cdot \mathbf{e}_3 = (\nabla\phi)_3,$$

therefore, the Cartesian components of $\nabla\phi$ are $\partial\phi/\partial x_i$, that is,

$$\nabla\phi = \frac{\partial\phi}{\partial x_1}\mathbf{e}_1 + \frac{\partial\phi}{\partial x_2}\mathbf{e}_2 + \frac{\partial\phi}{\partial x_3}\mathbf{e}_3 = \frac{\partial\phi}{\partial x_i}\mathbf{e}_i. \tag{B47}$$

The gradient vector has a simple geometrical interpretation. For any surface on which the value of ϕ is constant, $d\phi = 0$ for any $d\mathbf{r}$ tangent to the surface. Thus $\nabla\phi \cdot d\mathbf{r} = 0$, so that $\nabla\phi$ is normal to the surface of constant ϕ. Then $d\phi$ is greatest if $d\mathbf{r}$ is normal to the surface of constant ϕ.

EXAMPLE 2.21

If $\phi = xy + z$, find a unit vector $\mathbf{n}$ normal to the surface of a constant ϕ passing through (2, 1, 0).

Solution.

$$\nabla\phi = \frac{\partial\phi}{\partial x}\mathbf{e}_1 + \frac{\partial\phi}{\partial y}\mathbf{e}_2 + \frac{\partial\phi}{\partial z}\mathbf{e}_3 = y\mathbf{e}_1 + x\mathbf{e}_2 + \mathbf{e}_3.$$

At the point (2, 1, 0), $\nabla\phi = \mathbf{e}_1 + 2\mathbf{e}_2 + \mathbf{e}_3$. Thus

$$\mathbf{n} = \frac{1}{\sqrt{6}}(\mathbf{e}_1 + 2\mathbf{e}_2 + \mathbf{e}_3).$$

EXAMPLE 2.22

If $\mathbf{q}$ denotes the heat flux vector (rate of heat transfer/area), the Fourier heat conduction law states that

$$\mathbf{q} = -k\nabla\theta,$$

where θ is the temperature field and k is thermal conductivity. If $\theta = 2(x^2 + y^2)$, find $\mathbf{q}$ at $A(1, 0)$ and $B(1, 1)$. Sketch curves of constant θ (isotherms) and indicate the vectors $\mathbf{q}$ at the two points.

Solution.

$$\nabla\theta = \frac{\partial\theta}{\partial x}\mathbf{e}_1 + \frac{\partial\theta}{\partial y}\mathbf{e}_2 = 4x\mathbf{e}_1 + 4y\mathbf{e}_2.$$

Thus

$$\mathbf{q} = -4k(x\mathbf{e}_1 + y\mathbf{e}_2).$$

At A

$$\mathbf{q}_A = -4k\mathbf{e}_1$$

and B

$$\mathbf{q}_B = -4k(\mathbf{e}_1 + \mathbf{e}_2).$$

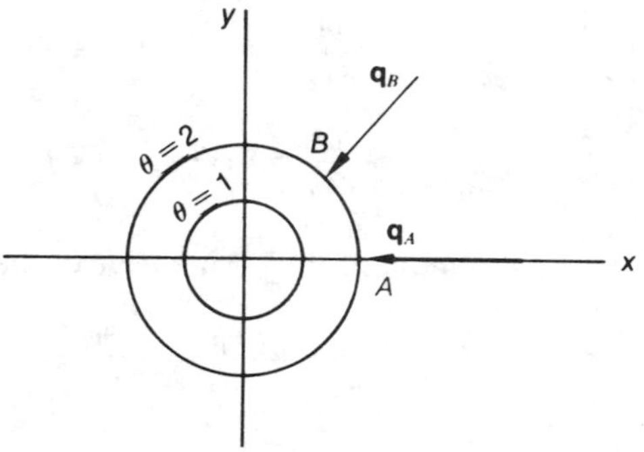

Fig. 2.5

EXAMPLE 2.23

A more general heat conduction law can be given in the following form:

$$\mathbf{q} = -\mathbf{K}\nabla\theta,$$

where $\mathbf{K}$ is a tensor known as thermal conductivity tensor.

(a) What tensor $\mathbf{K}$ corresponds to the Fourier heat conduction law mentioned in the previous example?

(b) If it is known that $\mathbf{K}$ is symmetric, show that there are at least three directions in which heat flow is normal to the surface of constant temperature.

(c) If $\theta = 2x + 3y$ and

$$[\mathbf{K}] = \begin{bmatrix} 2 & -1 & 0 \\ -1 & 2 & 0 \\ 0 & 0 & 3 \end{bmatrix},$$

find $\mathbf{q}$ and sketch its direction.

Solution. (a) $\mathbf{K} = +k\mathbf{I}$, so that $\mathbf{q} = -k\mathbf{I}\nabla\theta = -k\nabla\theta$.
(b) For symmetric $\mathbf{K}$, we know from Section 2B13 that there exist at least three principal directions $(\nabla\theta)^{(1)}$ $(\nabla\theta)^{(2)}$ and $(\nabla\theta)^{(3)}$ such that

$$\mathbf{q}^{(1)} = -\mathbf{K}(\nabla\theta)^{(1)} = -k_1(\nabla\theta)^{(1)},$$

$$\mathbf{q}^{(2)} = -\mathbf{K}(\nabla\theta)^{(2)} = -k_2(\nabla\theta)^{(2)},$$

$$\mathbf{q}^{(3)} = -\mathbf{K}(\nabla\theta)^{(3)} = -k_3(\nabla\theta)^{(3)},$$

where k_1, k_2, and k_3 are eigenvalues of $\mathbf{K}$. If $k_1 \neq k_2 \neq k_3$, the equations indicate different thermal conductivities in the three principle directions and the material is said to be anisotropic with respect to heat conduction.

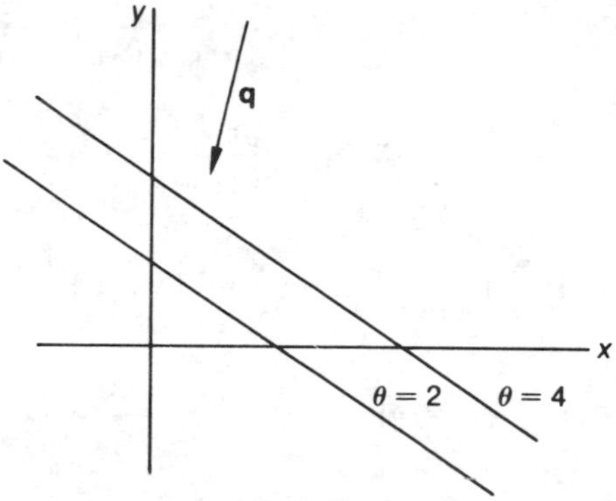

Fig. 2.6

(c) $\nabla\theta = 2\mathbf{e}_1 + 3\mathbf{e}_2$ and

$$[\mathbf{q}] = \begin{bmatrix} -2 & 1 & 0 \\ 1 & -2 & 0 \\ 0 & 0 & -3 \end{bmatrix} \begin{bmatrix} 2 \\ 3 \\ 0 \end{bmatrix} = \begin{bmatrix} -1 \\ -4 \\ 0 \end{bmatrix},$$

i.e.,

$$\mathbf{q} = -\mathbf{e}_1 - 4\mathbf{e}_2.$$

2B18. Vector Field, Gradient of a Vector Field

Let $\mathbf{v}(\mathbf{r})$ be a vector-valued function of position, describing, for example, displacement or velocity fields. Associated with $\mathbf{v}(\mathbf{r})$, there is a tensor field, called the gradient of $\mathbf{v}$, which is of considerable importance. The **gradient** of $\mathbf{v}$ (denoted by $\nabla\mathbf{v}$) is defined to be the second-order tensor which, when operating on $d\mathbf{r}$ gives the difference of $\mathbf{v}$ at $\mathbf{r} + d\mathbf{r}$ and $\mathbf{r}$. That is,

$$d\mathbf{v} = \mathbf{v}(\mathbf{r} + d\mathbf{r}) - \mathbf{v}(\mathbf{r}) \equiv (\nabla\mathbf{v})d\mathbf{r}. \qquad \text{(B48)}$$

Again, let dr denote $|d\mathbf{r}|$ and $\mathbf{e}$ denote $d\mathbf{r}/dr$, we have

$$\left(\frac{d\mathbf{v}}{dr}\right)_{\text{in e-direction}} = (\nabla\mathbf{v})\mathbf{e}. \tag{B49}$$

Thus, the second-order tensor† $(\nabla\mathbf{v})$ transforms a unit vector $\mathbf{e}$ into the vector describing the rate of change $\mathbf{v}$ in that direction. Since

$$\left(\frac{d\mathbf{v}}{dr}\right)_{\text{in } \mathbf{e}_1\text{-direction}} \equiv \frac{\partial \mathbf{v}}{\partial x_1} = (\nabla\mathbf{v})\mathbf{e}_1,$$

thus, in Cartesian coordinates,

$$(\nabla\mathbf{v})_{11} = \mathbf{e}_1 \cdot (\nabla\mathbf{v})\mathbf{e}_1 = \mathbf{e}_1 \cdot \frac{\partial \mathbf{v}}{\partial x_1} = \frac{\partial}{\partial x_1}(\mathbf{e}_1 \cdot \mathbf{v}).$$

That is,

$$(\nabla\mathbf{v})_{11} = \frac{\partial v_1}{\partial x_1}.$$

Or, in general

$$\left(\frac{d\mathbf{v}}{dr}\right)_{\text{in } \mathbf{e}_j\text{-direction}} \equiv \frac{\partial \mathbf{v}}{\partial x_j} = (\nabla\mathbf{v})\mathbf{e}_j, \tag{B50}$$

thus

$$(\nabla\mathbf{v})_{ij} = \mathbf{e}_i \cdot (\nabla\mathbf{v})\mathbf{e}_j = \mathbf{e}_i \cdot \frac{\partial \mathbf{v}}{\partial x_j} = \frac{\partial}{\partial x_j}(\mathbf{e}_i \cdot \mathbf{v}), \tag{B51}$$

so that the Cartesian components of $(\nabla\mathbf{v})$ are‡

$$(\nabla\mathbf{v})_{ij} = \frac{\partial v_i}{\partial x_j}. \tag{B52a}$$

That is,

$$[\nabla\mathbf{v}] = \begin{bmatrix} \dfrac{\partial v_1}{\partial x_1} & \dfrac{\partial v_1}{\partial x_2} & \dfrac{\partial v_1}{\partial x_3} \\ \dfrac{\partial v_2}{\partial x_1} & \dfrac{\partial v_2}{\partial x_2} & \dfrac{\partial v_2}{\partial x_3} \\ \dfrac{\partial v_3}{\partial x_1} & \dfrac{\partial v_3}{\partial x_2} & \dfrac{\partial v_3}{\partial x_3} \end{bmatrix}. \tag{B52b}$$

Geometrical interpretation of $\nabla\mathbf{v}$ will be given later in connection with the kinematics of deformation.

†It can be shown that $(\nabla\mathbf{v})$, thus defined, is a linear transformation. We omit the proof.

‡The notation $\nabla\mathbf{v}$ used here follows that used in the Encyclopedia of Physics 3/III (1965). It is simply a short-hand notation for (**grad v**), not to be interpreted as the dyadic product of the operation $\nabla_d \equiv \mathbf{e}_i\, \partial/\partial x_i$ and $\mathbf{v}$. In fact, $\nabla\mathbf{v} = (\nabla_d \mathbf{v})^T \equiv \mathbf{v}\nabla_d$.

2B19. Trace of a Second-Order Tensor

The **trace** of any dyad **ab** is defined to be a scalar given by $\mathbf{a} \cdot \mathbf{b}$. That is,

$$\text{tr}\, \mathbf{ab} = \mathbf{a} \cdot \mathbf{b}. \tag{B53}$$

Furthermore, the trace of the sum of dyads is defined to be the sum of the traces. That is,

$$\text{tr}\, (\alpha\, \mathbf{ab} + \beta \mathbf{cd}) = \alpha\, \text{tr}\, \mathbf{ab} + \beta\, \text{tr}\, \mathbf{cd}. \tag{B54}$$

Letting $\mathbf{a} = a_i \mathbf{e}_i$ and $\mathbf{b} = b_i \mathbf{e}_i$, where $\mathbf{e}_i$ are rectangular Cartesian base vectors, we have

$$\text{tr}\, \mathbf{ab} = a_1 b_1 + a_2 b_2 + a_3 b_3 = a_i b_i = \text{sum of the diagonal elements of } [\mathbf{ab}]_{\mathbf{e}_i}. \tag{B55}$$

Since any second-order tensor **T** can be written (*see* Eq. (B12))

$$\mathbf{T} = T_{ij} \mathbf{e}_i \mathbf{e}_j,$$

thus, the trace of **T** can be obtained as

$$\text{tr}\, \mathbf{T} = \text{tr}\, (T_{ij} \mathbf{e}_i \mathbf{e}_j) = T_{ij}\, \text{tr}\, (\mathbf{e}_i \mathbf{e}_j) = T_{ij} \delta_{ij},$$

that is,

$$\text{tr}\, \mathbf{T} = T_{ii} = T_{11} + T_{22} + T_{33}. \tag{B56}$$

It is obvious that $\text{tr}\, \mathbf{T}^T = \text{tr}\, \mathbf{T}$.

EXAMPLE 2.24

Show that for any second-order tensor **A** and **B**, $\text{tr}\, (\mathbf{AB}) = \text{tr}\, (\mathbf{BA})$.
Solution. Let $\mathbf{C} = \mathbf{AB}$, then $C_{ij} = A_{im} B_{mj}$. Thus

$$\text{tr}\, \mathbf{AB} = \text{tr}\, \mathbf{C} = C_{ii} = A_{im} B_{mi}.$$

Let $\mathbf{D} = \mathbf{BA}$, then $D_{ij} = B_{im} A_{mj}$, and

$$\text{tr}\, \mathbf{BA} = \text{tr}\, \mathbf{D} = D_{ii} = B_{im} A_{mi}.$$

But $B_{im} A_{mi} = B_{mi} A_{im}$ (change of dummy indices), that is, $\text{tr}\, \mathbf{BA} = \text{tr}\, \mathbf{AB}$.

EXAMPLE 2.25

Show that if **T** is symmetric and **W** is antisymmetric, then $\text{tr}\, (\mathbf{TW}) = 0$.
Solution. We have

$$\text{tr}\, (\mathbf{TW}) = \text{tr}\, (\mathbf{TW})^T = \text{tr}\, (\mathbf{W}^T \mathbf{T}^T).$$

Since **T** is symmetric and **W** is antisymmetric; therefore $\mathbf{T} = \mathbf{T}^T$, $\mathbf{W} = -\mathbf{W}^T$. Thus

$$\text{tr}\, (\mathbf{TW}) = -\text{tr}\, (\mathbf{WT}) = -\text{tr}\, (\mathbf{TW}) \qquad (\textit{see}\ \text{previous example}).$$

Consequently,

$$2\operatorname{tr}(\mathbf{TW}) = 0.$$

That is

$$\operatorname{tr}(\mathbf{TW}) = 0.$$

2B20. Divergence of a Vector Field and Divergence of a Tensor Field

Let $\mathbf{v}(\mathbf{r})$ be a vector field. The **divergence** of $\mathbf{v}(\mathbf{r})$ is defined to be a scalar field given by the trace of the gradient of $\mathbf{v}$. That is,

$$\operatorname{div}\mathbf{v} \equiv \operatorname{tr}(\nabla\mathbf{v}). \tag{B57}$$

With reference to rectangular Cartesian basis the diagonal elements of $\nabla\mathbf{v}$ are $\partial v_1/\partial x_1$, $\partial v_2/\partial x_2$ and $\partial v_3/\partial x_3$. Thus

$$\operatorname{div}\mathbf{v} = \frac{\partial v_1}{\partial x_1} + \frac{\partial v_2}{\partial x_2} + \frac{\partial v_3}{\partial x_3} = \frac{\partial v_i}{\partial x_i}. \tag{B58}$$

Let $\mathbf{T}(\mathbf{r})$ be a second-order tensor field. The **divergence** of $\mathbf{T}$ is defined to be a vector field, denoted by div $\mathbf{T}$, such that for any vector $\mathbf{a}$

$$(\mathbf{div}\,\mathbf{T})\cdot\mathbf{a} \equiv \operatorname{div}(\mathbf{T}^T\mathbf{a}) - \operatorname{tr}(\mathbf{T}^T(\nabla\mathbf{a})). \tag{B59}\dagger$$

To find the Cartesian components of the vector **div T**, let $\mathbf{b} \equiv \mathbf{div}\,\mathbf{T}$, then (note, $\nabla\mathbf{e}_i = 0$ for Cartesian coordinates),

$$b_i = \mathbf{b}\cdot\mathbf{e}_i = \operatorname{div}(\mathbf{T}^T\mathbf{e}_i) - \operatorname{tr}(\mathbf{T}^T\nabla\mathbf{e}_i) = \operatorname{div}(T_{im}\mathbf{e}_m) - 0 = \frac{\partial T_{im}}{\partial x_m} - 0.$$

In other words,

$$\mathbf{div}\,\mathbf{T} = \mathbf{b} = \frac{\partial T_{im}}{\partial x_m}\mathbf{e}_i. \tag{B60}$$

EXAMPLE 2.26

If $\alpha = \alpha(\mathbf{r})$ and $\mathbf{a} = \mathbf{a}(\mathbf{r})$, show that $\operatorname{div}(\alpha\mathbf{a}) = \alpha\operatorname{div}\mathbf{a} + (\nabla\alpha)\cdot\mathbf{a}$.

Solution. Let $\mathbf{b} = \alpha\mathbf{a}$. Then $b_i = \alpha a_i$ and

$$\operatorname{div}\mathbf{b} = \frac{\partial b_i}{\partial x_i} = \frac{\partial}{\partial x_i}(\alpha a_i) = \alpha\frac{\partial a_i}{\partial x_i} + \left(\frac{\partial\alpha}{\partial x_i}\right)a_i$$
$$= \alpha\operatorname{div}\mathbf{a} + (\nabla\alpha)\cdot\mathbf{a}.$$

†Some authors define **div T** as follows: for any constant $\mathbf{a}$, $(\mathbf{div}\,\mathbf{T})\cdot\mathbf{a} \equiv \operatorname{div}(\mathbf{T}^T\mathbf{a})$. Also, in dyadic notation, $\mathbf{div}\,\mathbf{T} = \mathbf{T}\cdot\nabla_d$, where $\nabla_d \equiv \mathbf{e}_i(\partial/\partial x_i)$.

EXAMPLE 2.27

Given $\alpha(\mathbf{r})$ and $\mathbf{T}(\mathbf{r})$ show that

$$\mathbf{div}\,(\alpha\mathbf{T}) = \mathbf{T}(\nabla\alpha) + \alpha\,\mathbf{div}\,\mathbf{T}.$$

Solution. We have, from Eq. (B60)

$$\mathbf{div}\,(\alpha\mathbf{T}) = \frac{\partial}{\partial x_j}(\alpha T_{ij})\mathbf{e}_i = \left(\frac{\partial\alpha}{\partial x_j}\right)T_{ij}\mathbf{e}_i + \alpha\frac{\partial T_{ij}}{\partial x_j}\mathbf{e}_i.$$

But

$$T_{ij}\frac{\partial\alpha}{\partial x_j}\mathbf{e}_i = \mathbf{T}(\nabla\alpha)$$

and

$$\alpha\frac{\partial T_{ij}}{\partial x_j}\mathbf{e}_i = \alpha\,\mathbf{div}\,\mathbf{T},$$

thus the desired result follows.

2B21. Curl of a Vector Field

Let $\mathbf{v}(\mathbf{r})$ be a vector field. The **curl** of $\mathbf{v}$ is defined to be the vector field given by twice the dual vector of the antisymmetric part of $\nabla\mathbf{v}$. That is,

$$\mathbf{curl}\,\mathbf{v} \equiv 2\mathbf{t}^A, \tag{B61}$$

where $\mathbf{t}^A$ is the dual vector of $(\nabla\mathbf{v})^A$.

In rectangular Cartesian basis

$$[\nabla\mathbf{v}]^A = \begin{bmatrix} 0 & \frac{1}{2}\left(\frac{\partial v_1}{\partial x_2}-\frac{\partial v_2}{\partial x_1}\right) & \frac{1}{2}\left(\frac{\partial v_1}{\partial x_3}-\frac{\partial v_3}{\partial x_1}\right) \\ -\frac{1}{2}\left(\frac{\partial v_1}{\partial x_2}-\frac{\partial v_2}{\partial x_1}\right) & 0 & \frac{1}{2}\left(\frac{\partial v_2}{\partial x_3}-\frac{\partial v_3}{\partial x_2}\right) \\ -\frac{1}{2}\left(\frac{\partial v_1}{\partial x_3}-\frac{\partial v_3}{\partial x_1}\right) & -\frac{1}{2}\left(\frac{\partial v_2}{\partial x_3}-\frac{\partial v_3}{\partial x_2}\right) & 0 \end{bmatrix}.$$

Thus the curl of $\mathbf{v}$ is given by (*see* Eq. (B36a)):

$$\mathbf{curl}\,\mathbf{v} = 2\mathbf{t}^A = \left(\frac{\partial v_3}{\partial x_2}-\frac{\partial v_2}{\partial x_3}\right)\mathbf{e}_1 + \left(\frac{\partial v_1}{\partial x_3}-\frac{\partial v_3}{\partial x_1}\right)\mathbf{e}_2 + \left(\frac{\partial v_2}{\partial x_1}-\frac{\partial v_1}{\partial x_2}\right)\mathbf{e}_3. \tag{B62}$$

2B22. Polar Coordinates

In this section the invariant definitions of ∇f, $\nabla\mathbf{v}$, div $\mathbf{v}$, curl $\mathbf{v}$, and div $\mathbf{T}$ will be utilized in order to determine their components in plane polar coordinates. Similar derivations can easily be made to obtain their components in other orthogonal curvilinear coordinate systems (i.e., cylindrical, spherical, etc.).

Let r, ϕ denote, *see* Fig. 2.7, plane polar coordinates such that

$$r = (x_1^2 + x_2^2)^{1/2},$$
$$\phi = \tan^{-1} x_2/x_1.$$

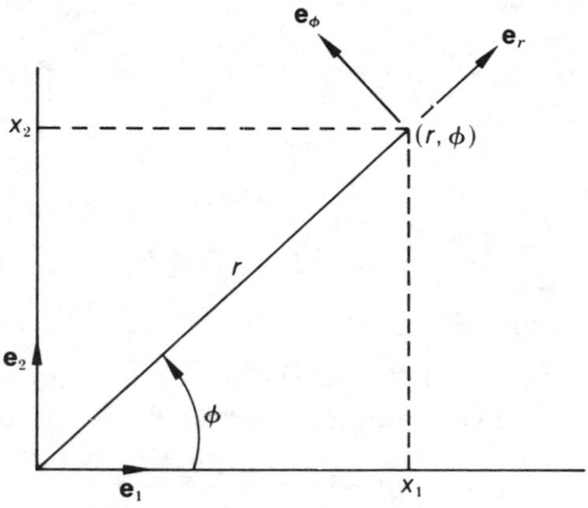

Fig. 2.7

The unit base vectors $\mathbf{e}_r$ and $\mathbf{e}_\phi$ can be expressed in terms of the Cartesian base vectors $\mathbf{e}_1$ and $\mathbf{e}_2$ as:

$$\mathbf{e}_r = \cos\phi \mathbf{e}_1 + \sin\phi \mathbf{e}_2, \tag{B63a}$$

$$\mathbf{e}_\phi = -\sin\phi \mathbf{e}_1 + \cos\phi \mathbf{e}_2. \tag{B63b}$$

These unit base vectors vary in direction as ϕ changes. In fact, from Eqs. (B63a) and (B63b), it is easily derived that

$$d\mathbf{e}_r = d\phi \mathbf{e}_\phi, \tag{B64a}$$

$$d\mathbf{e}_\phi = -d\phi \mathbf{e}_r. \tag{B64b}$$

From the position vector $\mathbf{r} = r\mathbf{e}_r$, we have

$$d\mathbf{r} = dr\mathbf{e}_r + rd\mathbf{e}_r.$$

Using Eq. (B64), we get

$$d\mathbf{r} = dr\mathbf{e}_r + rd\phi \mathbf{e}_\phi. \tag{B65}$$

The component of ∇f, $\nabla \mathbf{v}$, etc., will now be obtained.

(i) *Components of* ∇f

Let $f(r,\phi)$ be a scalar field. By definition of the gradient of f, we have

$$df=(\nabla f)\cdot d\mathbf{r}=[(\nabla f)_r\mathbf{e}_r+(\nabla f)_\phi\mathbf{e}_\phi]\cdot[dr\mathbf{e}_r+rd\phi\mathbf{e}_\phi].$$

That is,

$$df=(\nabla f)_r dr+(\nabla f)_\phi r d\phi. \tag{B66}$$

But, from calculus

$$df=\frac{\partial f}{\partial r}dr+\frac{\partial f}{\partial \phi}d\phi. \tag{B67}$$

Since Eqs. (B66) and (B67) must yield the same result for all increments dr, $d\phi$, we have

$$\nabla f=\frac{\partial f}{\partial r}\mathbf{e}_r+\frac{1}{r}\frac{\partial f}{\partial \phi}\mathbf{e}_\phi. \tag{B68}$$

(ii) *Components of* $(\nabla\mathbf{v})$

Let

$$\mathbf{v}(r,\phi)=v_r(r,\phi)\mathbf{e}_r+v_\phi(r,\phi)\mathbf{e}_\phi. \tag{B69}$$

By definition of $\nabla\mathbf{v}$ we have

$$d\mathbf{v}\equiv(\nabla\mathbf{v})\,d\mathbf{r}=(\nabla\mathbf{v})\,(dr\mathbf{e}_r+rd\phi\mathbf{e}_\phi)=dr(\nabla\mathbf{v})\mathbf{e}_r+rd\phi(\nabla\mathbf{v})\mathbf{e}_\phi.$$

Now,†

$$(\nabla\mathbf{v})\mathbf{e}_r=(\nabla\mathbf{v})_{rr}\mathbf{e}_r+(\nabla\mathbf{v})_{\phi r}\mathbf{e}_\phi \quad\text{and}\quad (\nabla\mathbf{v})\mathbf{e}_\phi=(\nabla\mathbf{v})_{r\phi}\mathbf{e}_r+(\nabla\mathbf{v})_{\phi\phi}\mathbf{e}_\phi.$$

Therefore,

$$d\mathbf{v}=[(\nabla\mathbf{v})_{rr}dr+(\nabla\mathbf{v})_{r\phi}rd\phi]\mathbf{e}_r+[(\nabla\mathbf{v})_{\phi r}dr+(\nabla\mathbf{v})_{\phi\phi}rd\phi]\mathbf{e}_\phi. \tag{B70}$$

But from Eq. (B69)

$$d\mathbf{v}=dv_r\mathbf{e}_r+v_r d\mathbf{e}_r+dv_\phi\mathbf{e}_\phi+v_\phi d\mathbf{e}_\phi,$$

and from calculus,

$$dv_r=\frac{\partial v_r}{\partial r}dr+\frac{\partial v_r}{\partial \phi}d\phi \quad\text{and}\quad dv_\phi=\frac{\partial v_\phi}{\partial r}dr+\frac{\partial v_\phi}{\partial \phi}d\phi,$$

we have

$$d\mathbf{v}=\left[\frac{\partial v_r}{\partial r}dr+\left(\frac{\partial v_r}{\partial \phi}-v_\phi\right)d\phi\right]\mathbf{e}_r+\left[\frac{\partial v_\phi}{\partial r}dr+\left(\frac{\partial v_\phi}{\partial \phi}+v_r\right)d\phi\right]\mathbf{e}_\phi. \tag{B71}$$

†$\mathbf{T}\mathbf{e}_i=T_{ji}\mathbf{e}_j$.

In order that Eqs. (B70) and (B71) agree for all increments dr, $d\phi$, we have

$$(\nabla \mathbf{v})_{rr} = \frac{\partial v_r}{\partial r}, \quad (\nabla \mathbf{v})_{\phi r} = \frac{1}{r}\left(\frac{\partial v_r}{\partial \phi} - v_\phi\right), \text{ etc.}$$

In matrix form

$$[\nabla \mathbf{v}]_{(\mathbf{e}_r, \mathbf{e}_\phi)} = \begin{bmatrix} \dfrac{\partial v_r}{\partial r} & \dfrac{1}{r}\left(\dfrac{\partial v_r}{\partial \phi} - v_\phi\right) \\ \dfrac{\partial v_\phi}{\partial r} & \dfrac{1}{r}\left(\dfrac{\partial v_\phi}{\partial \phi} + v_r\right) \end{bmatrix}. \tag{B72}$$

(iii) *Div* **v**

Using the components of $\nabla \mathbf{v}$ obtained in (ii), we have

$$\operatorname{div} \mathbf{v} \equiv \operatorname{tr}(\nabla \mathbf{v}) = (\nabla \mathbf{v})_{rr} + (\nabla \mathbf{v})_{\phi\phi} = \frac{\partial v_r}{\partial r} + \frac{1}{r}\left(\frac{\partial v_\phi}{\partial \phi} + v_r\right). \tag{B73}$$

(iv) **Curl v**

The antisymmetric part of $\nabla \mathbf{v}$ is

$$[(\nabla \mathbf{v})^A]_{(\mathbf{e}_r, \mathbf{e}_\phi)} = \begin{bmatrix} 0 & \dfrac{1}{2}\left(\dfrac{1}{r}\dfrac{\partial v_r}{\partial \phi} - \dfrac{v_\phi}{r} - \dfrac{\partial v_\phi}{\partial r}\right) \\ -\dfrac{1}{2}\left(\dfrac{1}{r}\dfrac{\partial v_r}{\partial \phi} - \dfrac{v_\phi}{r} - \dfrac{\partial v_\phi}{\partial r}\right) & 0 \end{bmatrix}.$$

Therefore, from the definition that curl $\mathbf{v} \equiv$ twice the dual vector of $(\nabla \mathbf{v})^A$, we have

$$\operatorname{curl} \mathbf{v} = \left(\frac{\partial v_\phi}{\partial r} + \frac{v_\phi}{r} - \frac{1}{r}\frac{\partial v_r}{\partial \phi}\right) \mathbf{e}_3. \tag{B74}$$

(v) *Components of* **div T**

The invariant definition of the divergence of a second-order tensor is

$$(\mathbf{div\,T}) \cdot \mathbf{a} = \operatorname{div}(\mathbf{T}^T\mathbf{a}) - \operatorname{tr}\{(\nabla \mathbf{a})\mathbf{T}^T\}. \tag{B75}$$

Since, $(\mathbf{div\,T})_r \equiv (\mathbf{div\,T}) \cdot \mathbf{e}_r$, therefore from Eq. (B75)

$$(\mathbf{div\,T})_r = \operatorname{div}(\mathbf{T}^T\mathbf{e}_r) - \operatorname{tr}\{(\nabla \mathbf{e}_r)\mathbf{T}^T\}. \tag{B76}$$

Using Eq. (B73) we evaluate the first term in the right-hand side of Eq. (B76) as

$$\operatorname{div}(\mathbf{T}^T\mathbf{e}_r) = \operatorname{div}(T_{rr}\mathbf{e}_r + T_{r\phi}\mathbf{e}_\phi) = \frac{\partial T_{rr}}{\partial r} + \frac{1}{r}\frac{\partial T_{r\phi}}{\partial \phi} + \frac{T_{rr}}{r}.$$

To evaluate the second term we first use Eq. (B72) to obtain $(\nabla \mathbf{e}_r)$. In fact†,

$$[\nabla \mathbf{e}_r]_{(\mathbf{e}_r, \mathbf{e}_\phi)} = \begin{bmatrix} 0 & 0 \\ 0 & \frac{1}{r} \end{bmatrix}.$$

Thus

$$\operatorname{tr}\{[\nabla \mathbf{e}_r]\,[\mathbf{T}^T]\} = \operatorname{tr}\begin{bmatrix} 0 & 0 \\ \frac{T_{r\phi}}{r} & \frac{T_{\phi\phi}}{r} \end{bmatrix} = \frac{T_{\phi\phi}}{r}.$$

From Eq. (B76), we have

$$(\mathbf{div}\ \mathbf{T})_r = \frac{\partial T_{rr}}{\partial r} + \frac{1}{r}\frac{\partial T_{r\phi}}{\partial \phi} + \frac{T_{rr} - T_{\phi\phi}}{r}. \tag{B77}$$

In a similar manner, we find

$$(\mathbf{div}\ \mathbf{T})_\phi = \frac{\partial T_{\phi r}}{\partial r} + \frac{1}{r}\frac{\partial T_{\phi\phi}}{\partial \phi} + \frac{T_{r\phi} + T_{\phi r}}{r}. \tag{B78}$$

PROBLEMS

B1. A tensor **T** transforms every vector into its mirror image with respect to the plane whose normal is

$$\mathbf{n} = \frac{\sqrt{2}}{2}(\mathbf{e}_1 + \mathbf{e}_2).$$

(a) Find the matrix of **T**.

(b) Use this linear transformation to find the mirror image of a vector $\mathbf{a} = \mathbf{e}_1 + 2\mathbf{e}_2$.

B2. A tensor **T** transforms every vector **a** into a vector $\mathbf{b} = \alpha\mathbf{a}$. Find the matrix of **T**. How will the components of **T** change when referred to a different basis?

B3. Find the matrix of the tensor **T** which transforms any vector **a** into a vector $\mathbf{b} = \mathbf{m}(\mathbf{a} \cdot \mathbf{n})$ where

$$\mathbf{m} = \frac{\sqrt{2}}{2}(\mathbf{e}_1 + \mathbf{e}_2)$$

and

$$\mathbf{n} = \frac{\sqrt{2}}{2}(-\mathbf{e}_1 + \mathbf{e}_3).$$

Write this tensor as a dyadic product.

B4. Find the matrix of the tensor **T** that transforms every vector **a** into a vector $\mathbf{b} = \mathbf{m} \times \mathbf{a}$, where $\mathbf{m} = \mathbf{e}_1 + \mathbf{e}_2$. Prove that **T** is a linear transformation.

B5. A transformation **T** operates on a vector **a** to give $\mathbf{Ta} = \mathbf{a}/|\mathbf{a}|$, where $|\mathbf{a}|$ is the magnitude of **a**. Is **T** a linear transformation?

†Note, $\mathbf{e}_r = (1)\mathbf{e}_r + (0)\mathbf{e}_\phi$.

B6. (a) Find the matrix of the tensor $\mathbf{T}_1$ that transforms every vector into its mirror image in a plane whose normal is $\mathbf{e}_2$. Find the matrix of a second tensor $\mathbf{T}_2$ which transforms every vector by a 45° right-hand rotation about the $\mathbf{e}_1$-axis.
(b) Find the matrix of the tensor that transforms every vector by the combination of first the reflection and then the rotation of part (a).
(c) Do part (b) for the rotation followed by the reflection.

B7. Given $\mathbf{T}$ a linear transformation, show that $\mathbf{T}^T$ is also a linear transformation.

B8. For arbitrary tensors $\mathbf{T}$ and $\mathbf{S}$, without relying on the component form, prove that

(a) $\mathbf{T}^T+\mathbf{S}^T=(\mathbf{T}+\mathbf{S})^T$,
(b) $(\mathbf{TS})^T=\mathbf{S}^T\mathbf{T}^T$.

B9. Indicate the validity of the inequality $\mathbf{T}(\mathbf{a}\times\mathbf{b})\neq\mathbf{a}\times\mathbf{Tb}+\mathbf{Ta}\times\mathbf{b}$ by choosing specific vectors $\mathbf{a}$, $\mathbf{b}$, and a linear transformation $\mathbf{T}$.

B10. Suppose that a linear transformation $\mathbf{Q}$ satisfies the identity $\mathbf{Q}^T\mathbf{Q}=\mathbf{I}$.
(a) Consider $\mathbf{Qa}=\mathbf{b}$, where $\mathbf{b}$ is an arbitrary vector and show that $\mathbf{QQ}^T=\mathbf{I}$.
(b) Show that both $\mathbf{Q}$ and $\mathbf{Q}^T$ are orthogonal transformations.

B11. Rigid body rotations that are small can be described by an orthogonal transformation $\mathbf{R}=\mathbf{I}+\epsilon\mathbf{R}$,* where $\epsilon\to 0$ as the rotation angle approaches zero. Considering two successive rotations $\mathbf{R}_1$ and $\mathbf{R}_2$, show that for small rotations (so that terms of ϵ^2 can be neglected as compared with terms of ϵ) the final result does not depend on the order of the rotations.

B12. Let $\mathbf{Q}$ define an orthogonal transformation of coordinates, so that $\mathbf{e}_i'=Q_{mi}\mathbf{e}_m$. Consider $\mathbf{e}_i'\cdot\mathbf{e}_j'$ and verify that $Q_{mi}Q_{mj}=\delta_{ij}$.

B13. The basis $\mathbf{e}_i'$ is obtained by a 30° counterclockwise rotation of the $\mathbf{e}_i$ basis about $\mathbf{e}_3$.
(a) Find the orthogonal transformation $\mathbf{Q}$ that defines this change of basis, i.e., $\mathbf{e}_i'=Q_{mi}\mathbf{e}_m$.
(b) By using the vector transformation law, find the components of $\mathbf{a}=\sqrt{3}\,\mathbf{e}_1+\mathbf{e}_2$ in the primed basis (i.e., find a_i').
(c) Do part (b) geometrically.

B14. Do Problem B13 for $\mathbf{e}_i'$ obtained by a 30° clockwise rotation of the $\mathbf{e}_i$-basis about $\mathbf{e}_3$.

B15. Show that if the components of a tensor are equal to zero in respect to one basis, then they vanish in respect to any basis.

B16. The matrix of a tensor $\mathbf{T}$ in respect to the basis $(\mathbf{e}_1, \mathbf{e}_2, \mathbf{e}_3)$ is

$$[\mathbf{T}]=\begin{bmatrix}1 & 5 & -5\\ 5 & 0 & 0\\ -5 & 0 & 1\end{bmatrix}.$$

Find T_{11}', T_{12}', and T_{31}' in respect to a right-hand basis $\mathbf{e}_i'$ where $\mathbf{e}_1'$ is in the direction of $-\mathbf{e}_2+2\mathbf{e}_3$ and $\mathbf{e}_2'=\mathbf{e}_1$.

B17. (a) For the tensor of the previous problem, find $[T'_{ij}]$ if $\mathbf{e}'_i$ is obtained by a 90° right-hand rotation about the $\mathbf{e}_3$-axis.
(b) Compare both the sum of the diagonal elements and the determinant of $[\mathbf{T}]$ and $[\mathbf{T}]'$.

B18. Do Problem B17 for a 180° rotation about the $\mathbf{e}_3$-axis.

B19. The dot product of two vectors $\mathbf{a} = a_i\mathbf{e}_i$ and $\mathbf{b} = b_i\mathbf{e}_i$ is equal to a_ib_i. Show that the dot product is a scalar invariant with respect to an orthogonal transformation of coordinates.

B20. (a) If T_{ij} are the components of a tensor, show that $T_{ij}T_{ij}$ is a scalar invariant with respect to an orthogonal transformation of coordinates.
(b) Evaluate $T_{ij}T_{ij}$ if in respect to the basis $\mathbf{e}_i$

$$[\mathbf{T}] = \begin{bmatrix} 1 & 0 & 0 \\ 1 & 2 & 5 \\ 1 & 2 & 3 \end{bmatrix}_{\mathbf{e}_i}.$$

(c) Find $[\mathbf{T}]'$ if $\mathbf{e}'_i = \mathbf{Q}\mathbf{e}_i$ and

$$[\mathbf{Q}] = \begin{bmatrix} 0 & 0 & 1 \\ 1 & 0 & 0 \\ 0 & 1 & 0 \end{bmatrix}_{\mathbf{e}_i}.$$

(d) Show for this specific $[\mathbf{T}]$ and $[\mathbf{T}]'$ that

$$T'_{mn}T'_{mn} = T_{ij}T_{ij}.$$

B21. Let $[\mathbf{T}]$ and $[\mathbf{T}]'$ be two matrices of the same tensor $\mathbf{T}$. Show that

$$\det[\mathbf{T}] = \det[\mathbf{T}]'.$$

B22. For any vector $\mathbf{a}$ and arbitrary tensor $\mathbf{T}$, show that
(a) $\mathbf{a} \cdot \mathbf{T}^A\mathbf{a} = 0$,
(b) $\mathbf{a} \cdot \mathbf{T}\mathbf{a} = \mathbf{a} \cdot \mathbf{T}^S\mathbf{a}$.

B23. (a) The components of a third-order tensor are R_{ijk}. Show that R_{iik} are components of a vector.
(b) Generalize the result of part (a) by considering the components of a tensor of nth order $R_{ijk\ldots}$. Show that $R_{iik\ldots}$ are components of an $(n-2)$th order tensor.

B24. The components of an arbitrary vector $\mathbf{a}$ and an arbitrary second-order tensor $\mathbf{T}$ are related by a triply subscripted quantity R_{ijk} in the manner $a_i = R_{ijk}T_{jk}$ for any basis $\{\mathbf{e}_1, \mathbf{e}_2, \mathbf{e}_3\}$. Prove that R_{jik} are the components of a third-order tensor.

B25. Any tensor may be decomposed into a symmetric and antisymmetric part. Prove that the decomposition is unique. (Hint: Assume that it is not unique.)

B26. Given that a tensor **T** has a matrix

$$[\mathbf{T}] = \begin{bmatrix} 1 & 2 & 3 \\ 4 & 5 & 6 \\ 7 & 8 & 9 \end{bmatrix},$$

(a) find the symmetric and antisymmetric part of **T**, and (b) find the dual vector of the antisymmetric part of **T**.

B27. Prove that the only possible eigenvalues of an orthogonal tensor are $\lambda = +1$ or $\lambda = -1$.

B28. Tensors **T**, **R**, and **S** are related by

$$\mathbf{T} = \mathbf{RS}.$$

Tensors **R** and **S** have the same eigenvector **n** and corresponding eigenvalues r_1 and s_1. Find an eigenvalue and eigenvector of **T**.

B29. If **n** is a real eigenvector of an antisymmetric tensor **T**, then show that the corresponding eigenvalue vanishes.

B30. Let **F** be an arbitrary tensor.

(a) Show that $\mathbf{FF}^T$ and $\mathbf{F}^T\mathbf{F}$ are symmetric tensors.

It can be shown (Polar Decomposition Theorem) that any invertible tensor **F** can be expressed as $\mathbf{F} = \mathbf{VQ} = \mathbf{QU}$, where **Q** is an orthogonal tensor and **U** and **V** are symmetric tensors.

(b) Show that $\mathbf{VV} = \mathbf{FF}^T$ and $\mathbf{UU} = \mathbf{F}^T\mathbf{F}$.

(c) If λ_i and $\mathbf{n}_i$ are the eigenvalues and eigenvectors of **U**, find the eigenvalues and eigenvectors of **V**.

B31. (a) What is an obvious eigenvector of the dyadic product **ab**?

(b) What does the first scalar invariant of **ab** correspond to?

(c) Show that the third scalar invariant of **ab** vanishes.

B32. A rotation tensor **R** is defined by the relations

$$\mathbf{Re}_1 = \mathbf{e}_2, \quad \mathbf{Re}_2 = \mathbf{e}_3, \quad \mathbf{Re}_3 = \mathbf{e}_1.$$

(a) Find the matrix of **R** and verify that $\mathbf{RR}^T = \mathbf{I}$ and $\det|\mathbf{R}| = 1$.

(b) Find the single axis of rotation that could have been used to effect this rotation.

B33. For any rotation transformation a basis $\mathbf{e}_i'$ may be chosen so that $\mathbf{e}_3'$ is along the axis of rotation.

(a) Verify that for an angle of rotation θ, the rotation matrix in respect to the primed basis is

$$[\mathbf{R}]' = \begin{bmatrix} \cos\theta & -\sin\theta & 0 \\ \sin\theta & \cos\theta & 0 \\ 0 & 0 & 1 \end{bmatrix}_{\mathbf{e}_i'}.$$

(b) Find the symmetric and antisymmetric parts of $[\mathbf{R}]'$.

(c) Find the eigenvalues and eigenvectors of $\mathbf{R}^S$.
(d) Find the first scalar invariant of $\mathbf{R}$ (that is, R'_{ii}).
(e) Find the dual vector of $\mathbf{R}^A$.
(f) Use the results of (d) and (e) to find the angle of rotation and the axis of rotation for Problem B32.

B34. (a) If $\mathbf{Q}$ is an improper orthogonal transformation (corresponding to a reflection), what are the eigenvalues and corresponding eigenvectors of $\mathbf{Q}$?
(b) If the matrix of $\mathbf{Q}$ is

$$[\mathbf{Q}] = \begin{bmatrix} \frac{1}{3} & -\frac{2}{3} & -\frac{2}{3} \\ -\frac{2}{3} & \frac{1}{3} & -\frac{2}{3} \\ -\frac{2}{3} & -\frac{2}{3} & \frac{1}{3} \end{bmatrix},$$

find the normal to the plane of reflection.

B35. Show that the second scalar invariant of $\mathbf{T}$ is

$$I_2 = \frac{T_{ii}T_{jj}}{2} - \frac{T_{ij}T_{ji}}{2}$$

by expanding this equation.

B36. Using the matrix transformation law for second-order tensors, show that the third scalar invariant is independent of the particular basis.

B37. A tensor $\mathbf{T}$ has a matrix

$$[\mathbf{T}] = \begin{bmatrix} 5 & 4 & 0 \\ 4 & -1 & 0 \\ 0 & 0 & 3 \end{bmatrix}.$$

(a) Find the scalar invariants, the principal values and corresponding principal directions of the tensor $\mathbf{T}$.
(b) If $\mathbf{n}_1, \mathbf{n}_2, \mathbf{n}_3$ are the principal directions, write $[\mathbf{T}]_{\mathbf{n}_1,\mathbf{n}_2,\mathbf{n}_3}$.
(c) Could the following matrix represent the tensor $\mathbf{T}$ in respect to some basis?

$$\begin{bmatrix} 7 & 2 & 0 \\ 2 & 1 & 0 \\ 0 & 0 & -1 \end{bmatrix}.$$

B38. Do Problem B37 for the matrix

$$[\mathbf{T}] = \begin{bmatrix} 3 & 0 & 0 \\ 0 & 0 & 4 \\ 0 & 4 & 0 \end{bmatrix}.$$

B39. A tensor $\mathbf{T}$ has a matrix

$$[\mathbf{T}] = \begin{bmatrix} 1 & 1 & 0 \\ 1 & 1 & 0 \\ 0 & 0 & 2 \end{bmatrix}.$$

Find the principal values and three mutually orthogonal principal directions.

B40. The inertia tensor $\bar{\mathbf{I}}_0$ of a rigid body with respect to a point 0, is defined by

$$\bar{\mathbf{I}}_0 = \int (r^2\mathbf{I} - \mathbf{rr})\rho\, dV,$$

where $\mathbf{r}$ is the position vector, $r = |\mathbf{r}|$, ρ = mass density, $\mathbf{I}$ is identity tensor, and dV is differential volume. The moment of inertia, with respect to an axis passing through 0, is given by $\bar{\mathbf{I}}_{nn} = \mathbf{n} \cdot \bar{\mathbf{I}}_0\mathbf{n}$, (no sum on n), where $\mathbf{n}$ is a unit vector in the direction of the axis of interest.

(a) Show that $\bar{\mathbf{I}}_0$ is symmetric.

(b) Letting $\mathbf{r} = x\mathbf{e}_1 + y\mathbf{e}_2 + z\mathbf{e}_3$, write out all components of $\bar{\mathbf{I}}_0$.

(c) The diagonal terms of the inertia matrix are termed moments of inertia and the off-diagonal terms are products of inertia. For what axes will the products of inertia be zero? For which axis will the moments of inertia be greatest (or least)?

Let a coordinate frame $\mathbf{e}_1$, $\mathbf{e}_2$, $\mathbf{e}_3$ be attached to a rigid body which is spinning with an angular velocity $\boldsymbol{\omega}$. Then, the angular momentum vector $\mathbf{H}_c$, in respect to the mass center, is given by

$$\mathbf{H}_c = \bar{\mathbf{I}}_c\boldsymbol{\omega}$$

and

$$\frac{d\mathbf{e}_i}{dt} = \boldsymbol{\omega} \times \mathbf{e}_i.$$

(d) Let $\boldsymbol{\omega} = \omega_i\mathbf{e}_i$ and demonstrate that

$$\dot{\boldsymbol{\omega}} = \frac{d\boldsymbol{\omega}}{dt} = \frac{d\omega_i}{dt}\mathbf{e}_i$$

and that

$$\dot{\mathbf{H}}_c = \frac{d}{dt}\mathbf{H}_c = \bar{\mathbf{I}}_c\dot{\boldsymbol{\omega}} + \boldsymbol{\omega} \times (\bar{\mathbf{I}}_c\boldsymbol{\omega}).$$

(e) Euler's Law for a rigid body in motion requires that the moment of the external forces about the mass center is equal to the time rate of change of the angular momentum, or

$$\mathbf{M}_c = \frac{d}{dt}\mathbf{H}_c.$$

Show that

$$\mathbf{M}_c \cdot \boldsymbol{\omega} = \frac{d}{dt}\left(\frac{\boldsymbol{\omega} \cdot \mathbf{H}_c}{2}\right).$$

This equation can be interpreted as an equality between the rate of work of the external moment and the rate of change of the rotational kinetic energy.

B41. Prove the identities (B44a–e) of Section 2B16.

B42. Consider the scalar field defined by $\phi = x^2 + 3yx + 2z$.

(a) Find a unit normal to the surface of constant ϕ at the origin (0, 0, 0).

(b) What is the maximum value of the directional derivative of ϕ at the origin?

(c) Evaluate $d\phi/dr$ at the origin if $d\mathbf{r} = ds(\mathbf{e}_1 + \mathbf{e}_3)$.

B43. Consider the ellipsoid defined by the equation $x^2/a^2+y^2/b^2+z^2/c^2=1$. Find the unit normal vector at a given position (x, y, z).

B44. The isotherms in a plane heat conduction problem are given by $\theta = 3xy$.
(a) Find the heat flux at the point A (1, 1, 1) if $\mathbf{q} = -k\nabla\theta$.
(b) Find the heat flux at the same point as part (a) if $\mathbf{q} = -\mathbf{K}(\nabla\theta)$, where

$$[\mathbf{K}] = \begin{bmatrix} k & 0 & 0 \\ 0 & 2k & 0 \\ 0 & 0 & 3k \end{bmatrix}.$$

B45. Consider an electrostatic potential $\psi = \alpha[x\cos\theta + y\sin\theta]$, where α and θ are constants.
(a) Find the electric field $\mathbf{E}$, $\mathbf{E} = -\nabla\phi$.
(b) Find the electric displacement $\mathbf{D} = \epsilon\mathbf{E}$, where the matrix of ϵ is

$$[\epsilon] = \begin{bmatrix} \epsilon_1 & 0 & 0 \\ 0 & \epsilon_2 & 0 \\ 0 & 0 & \epsilon_3 \end{bmatrix}.$$

(c) Find the angle θ for which $|\mathbf{D}|$ is a maximum.

B46. Let $\phi(x, y, z)$ and $\psi(x, y, z)$ be scalar functions of positions, and let $\mathbf{v}(x, y, z)$ and $\mathbf{w}(x, y, z)$ be vector functions of position. By writing the subscripted component form, verify the following identities:
(a) $\nabla(\phi+\psi) = \nabla\phi + \nabla\psi$.
Sample solution:

$$[\nabla(\phi+\psi)]_i = \frac{\partial}{\partial x_i}(\phi+\psi) = \frac{\partial\phi}{\partial x_i} + \frac{\partial\psi}{\partial x_i} = (\nabla\phi)_i + (\nabla\psi)_i.$$

(b) div $(\mathbf{v}+\mathbf{w})$ = div $\mathbf{v}$ + div $\mathbf{w}$,
(c) div $(\phi\mathbf{v}) = (\nabla\phi)\cdot\mathbf{v} + \phi(\text{div }\mathbf{v})$,
(d) **curl** $(\nabla\phi) = 0$,
(e) div (**curl** $\mathbf{v}$) = 0.

B47. In cylindrical coordinates, with $\mathbf{e}_r$, $\mathbf{e}_\phi$, $\mathbf{e}_z$ denoting unit vectors in the coordinate direction of r, ϕ, z, we have $\mathbf{r} = r\mathbf{e}_r + z\mathbf{e}_z$ and $d\mathbf{r} = dr\mathbf{e}_r + (r d\phi)\mathbf{e}_\phi + dz\mathbf{e}_z$. From $df \equiv (\nabla f)\cdot d\mathbf{r}$ obtain the components of (∇f) in cylindrical coordinates as

$$(\nabla f)_r = \frac{\partial f}{\partial r}, \quad (\nabla f)_\phi = \frac{1}{r}\frac{\partial f}{\partial \phi}, \quad (\nabla f)_z = \frac{\partial f}{\partial z}.$$

B48. Consider the vector field $\mathbf{v} = x^2\mathbf{e}_1 + z^2\mathbf{e}_2 + y^2\mathbf{e}_3$. For the point (1, 1, 0):
(a) Find the matrix of $\nabla\mathbf{v}$.
(b) Find the vector $(\nabla\mathbf{v})\mathbf{v}$.
(c) Find div $\mathbf{v}$ and **curl** $\mathbf{v}$.
(d) If $d\mathbf{r} = ds(\mathbf{e}_1+\mathbf{e}_2+\mathbf{e}_3)$, find the differential $d\mathbf{v}$.

B49. (a) Consider a general vector field **v** and the relation $d\mathbf{v} = (\nabla\mathbf{v})d\mathbf{r}$. If $d\mathbf{r} = (dr)\mathbf{n}$, show that

$$\left|\frac{d\mathbf{v}}{dr}\right|^2 = \mathbf{n}\cdot[(\nabla\mathbf{v})^T(\nabla\mathbf{v})]\mathbf{n}.$$

How would you find the direction of **n** that gives a maximum value for dv/dr.
(b) Find $|dv/dr|_{\max}$ if $\mathbf{v} = x^3\mathbf{e}_1 + y^2\mathbf{e}_2 + z^2\mathbf{e}_3$ at $A(2, 2, 2)$.

B50. In cylindrical coordinates, with $\mathbf{e}_r$, $\mathbf{e}_\phi$, $\mathbf{e}_z$ denoting unit vectors in the direction of r, ϕ, z, a vector **v** can be written $\mathbf{v} = v_r\mathbf{e}_r + v_\phi\mathbf{e}_\phi + v_z\mathbf{e}_z$. Using the relation $d\mathbf{e}_r = (d\phi)\mathbf{e}_\phi$, $d\mathbf{e}_\phi = -(d\phi)\mathbf{e}_r$, $d\mathbf{r} = dr\mathbf{e}_r + rd\phi\mathbf{e}_\phi + dz\mathbf{e}_z$ and $d\mathbf{v} = (\nabla\mathbf{v})d\mathbf{r}$, obtain the components of $(\nabla\mathbf{v})$ in cylindrical coordinates as

$$[\nabla\mathbf{v}]_{\mathbf{e}_r,\mathbf{e}_\phi,\mathbf{e}_z} = \begin{bmatrix} \dfrac{\partial v_r}{\partial r} & \dfrac{1}{r}\dfrac{\partial v_r}{\partial\phi} - \dfrac{v_\phi}{r} & \dfrac{\partial v_r}{\partial z} \\ \dfrac{\partial v_\phi}{\partial r} & \dfrac{1}{r}\dfrac{\partial v_\phi}{\partial\phi} + \dfrac{v_r}{r} & \dfrac{\partial v_\phi}{\partial z} \\ \dfrac{\partial v_z}{\partial r} & \dfrac{1}{r}\dfrac{\partial v_z}{\partial\phi} & \dfrac{\partial v_z}{\partial z} \end{bmatrix}.$$

B51. Using the result of the preceding problem, write in cylindrical coordinates expressions for
(a) $(\nabla\mathbf{v})^S$,
(b) div **v**,
(c) **curl v**.

B52. Using the result of Problem B50 write in cylindrical coordinates expressions for
(a) **div T**,
(b) **div** $(\nabla\mathbf{v})$.

3

Kinematics of a Continuum

The branch of mechanics in which materials are treated as continuous is known as "**continuum mechanics**."† Thus, in this theory, one speaks of an infinitesimal volume of material, the totality of which forms a "**body**." One also speaks of a "**particle**" in a continuum, meaning, in fact, an infinitesimal volume of material. This chapter is concerned with the kinematics of such particles.

3.1 DESCRIPTION OF MOTIONS OF A CONTINUUM

Let us suppose that a body, at certain time $t = t_0$ occupies a certain region of the physical space. The position of a particle at this time can be described by its position vector $\mathbf{X}$ measured from some fixed point 0 (*see* Fig. 3.1). Let the position vector of the particle be $\mathbf{x}$ at time t. Then an equation of the form

$$\mathbf{x} = \mathbf{x}(\mathbf{X}, t) \qquad \text{with } \mathbf{x}(\mathbf{X}, t_0) = \mathbf{X} \tag{3.1}$$

describes the path of every particle, which, at $t = t_0$ is located at $\mathbf{X}$ (different $\mathbf{X}$ for different particles).

Let $\mathbf{X} = X_1\mathbf{e}_1 + X_2\mathbf{e}_2 + X_3\mathbf{e}_3$ and $\mathbf{x} = x_1\mathbf{e}_1 + x_2\mathbf{e}_2 + x_3\mathbf{e}_3$. Then Eq. (3.1) takes the component form

$$\begin{aligned} x_1 &= x_1(X_1, X_2, X_3, t), \\ x_2 &= x_2(X_1, X_2, X_3, t), \\ x_3 &= x_3(X_1, X_2, X_3, t), \end{aligned} \tag{3.2}$$

†*See* Chapter 1 for a discussion of the model of "Continuum Mechanics."

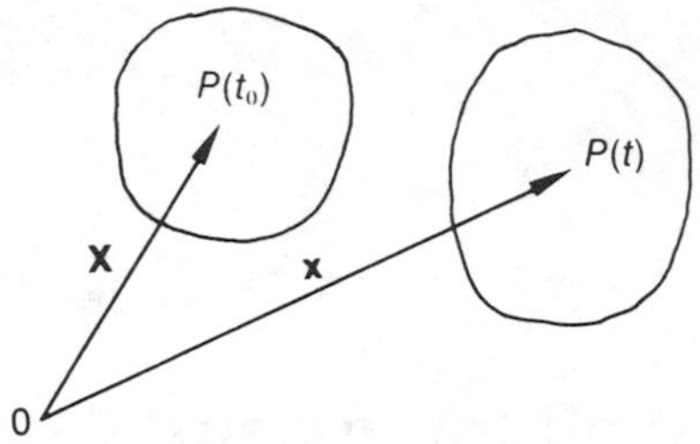

Fig. 3.1

or $x_i = x_i(X_1, X_2, X_3, t)$ with $X_i = x_i(X_1, X_2, X_3, t_0)$. In Eq. (3.2) the triple (X_1, X_2, X_3) serves to identify the different particles of the body and is known as **material coordinates**. Equation (3.1) or Eq. (3.2) is said to define a **motion** for a continuum. We note that for a specific particle the equation defines the **path line** (trajectory) of the particle.

EXAMPLE 3.1

Consider the motion $\mathbf{x} = \mathbf{X} + ktX_2\mathbf{e}_1$, or

$$x_1 = X_1 + ktX_2,$$

$$x_2 = X_2,$$

$$x_3 = X_3,$$

where the material coordinates (X_1, X_2, X_3) give the position of a particle at $t = 0$. Sketch the configurations at time t for a body, which, at $t = 0$, has the shape of a cube of unit sides in Fig. 3.2.

Solution. At $t = 0$, the particle O is located at $(X_1, X_2, X_3)_O = (0, 0, 0)$. Thus, for this particle $(x_1, x_2, x_3)_O = (0, 0, 0)$ for all t, i.e., the particle remains at $(0, 0, 0)$ at all times. Similarly, $(X_1, X_2, X_3)_A = (1, 0, 0)$ and $(x_1, x_2, x_3)_A = (1, 0, 0)$ so that the particle A also

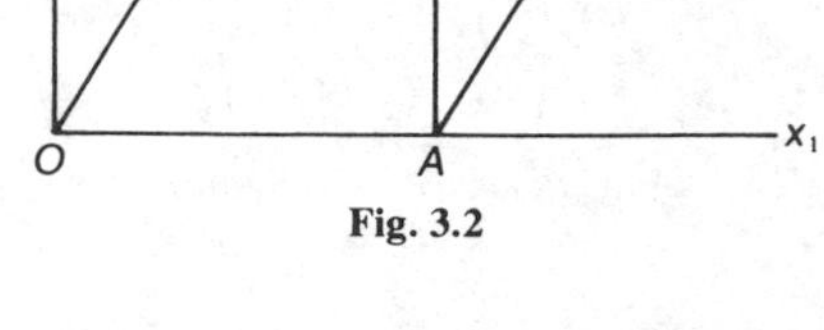

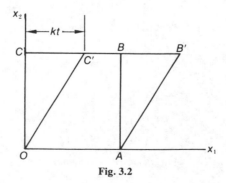

Fig. 3.2

does not move. In fact, for any particle on OA, $(X_1, X_2, X_3)_{OA} = (X_1, 0, 0)$ and $(x_1, x_2, x_3)_{OA} = (X_1, 0, 0)$. That is, the material line OA is a fixed line.

For the material line CB, $(X_1, X_2, X_3)_{CB} = (X_1, 1, 0)$ and $(x_1, x_2, x_3)_{CB} = (X_1 + kt, 1, 0)$, so that every particle on the line is displaced horizontally to the right through a distance kt.

For the material line OC, $(X_1, X_2, X_3)_{OC} = (0, X_2, 0)$ and $(x_1, x_2, x_3)_{OC} = (ktX_2, X_2, 0)$, so that every particle on the line moves horizontally to the right through a distance linearly proportional to its height, i.e., it remains a straight line. A similar situation prevails for the material line BA.

Thus at time t, the side view of the cube changes from that of a square to a parallelogram as shown. The motion given in this example is known as simple shearing motion.

3.2 MATERIAL DESCRIPTION AND SPATIAL DESCRIPTION

When a continuum is in motion, tensor quantities that are associated with specific particles (temperature θ, velocity $\mathbf{v}$, etc.) change with time. We can describe these changes by:

I. Following the particles, i.e., we express θ, v_i, as functions of the particles (identified by the material coordinates X_1, X_2, X_3) and time. In other words, we express

$$\theta = \theta(X_1, X_2, X_3, t),$$
$$\mathbf{v} = \mathbf{v}(X_1, X_2, X_3, t).$$

Such a description is known as **material description**. Other names for it are: **Lagrangian description** and **reference description**.

II. Observing the changes at fixed locations, i.e., we express θ, v_i, etc., as functions of fixed position and time. Thus

$$\theta = \theta(x_1, x_2, x_3, t),$$
$$\mathbf{v} = \mathbf{v}(x_1, x_2, x_3, t).$$

Such a description is known as **spatial** or, **Eulerian description**. The triple (x_1, x_2, x_3) locates the fixed positions of points in the physical space and is known as a **spatial coordinate**. The spatial coordinates x_i of a particle at any time t is related to the material coordinates X_i of the particle by Eq. (3.2). We note that in this description, what is described (or measured) is the change of quantities at a fixed location as a function of time. Spatial positions are occupied by different particles at different times. Therefore, the spatial description does not provide direct information regarding changes in particle properties as they move about. The material and spatial descriptions are, of course, related by the motion. That is, if the motion is known then, one description can be obtained from the other as illustrated from the following example.

EXAMPLE 3.2

Given the motion of a body to be

$$x_1 = X_1 + ktX_2,$$
$$x_2 = X_2, \tag{i}$$
$$x_3 = X_3.$$

If the temperature field is given by the spatial description

$$\theta = x_1 + x_2. \tag{ii}$$

(a) Find the material description of temperature.
(b) Obtain the velocity and rate of change of temperature for particular material particles. Express the answer in both a material and spatial description.

Solution. (a) Substituting (i) into (ii), we obtain

$$\theta = x_1 + x_2 = X_1 + ktX_2 + X_2,$$

i.e.,

$$\theta = X_1 + (kt + 1)X_2.$$

(b) Since a particular material particle is designated by a specific X_i, its velocity will be given by

$$v_i = \left.\frac{\partial x_i}{\partial t}\right|_{X_i\text{-fixed}},$$

so that

$$v_1 = kX_2, \quad v_2 = v_3 = 0.$$

In view of the second equation of Eqs. (i), the spatial description is

$$v_1 = kx_2, \quad v_2 = v_3 = 0.$$

The rate of change of temperature for particular material particles

$$\left.\frac{\partial \theta}{\partial t}\right|_{X_i\text{-fixed}} = kX_2 = kx_2.$$

Note that even though temperature is independent of time in the spatial description, each particle experiences changes in temperature, since it flows from one spatial position to another.

3.3 MATERIAL DERIVATIVE

The time rate of change of a quantity (such as temperature or velocity, etc.) of a material particle, is known as a **material derivative**. We shall denote the material derivative by D/Dt.

(i) When a material description of the quantity is used,

$$\theta = \theta(X_1, X_2, X_3, t),$$

then

$$\frac{D\theta}{Dt} = \left.\frac{\partial \theta}{\partial t}\right|_{X_i\text{-fixed}}. \tag{3.3}$$

(ii) When a spatial description of the quantity is used,

$$\theta = \theta(x_1, x_2, x_3, t),$$

where x_i, the present positions of material particles, are related to the material coordinates by the known motion $x_i = x_i(X_1, X_2, X_3, t)$. Then

$$\frac{D\theta}{Dt} = \left(\frac{\partial\theta}{\partial t}\right)_{X_i\text{-fixed}} = \frac{\partial\theta}{\partial x_1}\frac{\partial x_1}{\partial t} + \frac{\partial\theta}{\partial x_2}\frac{\partial x_2}{\partial t} + \frac{\partial\theta}{\partial x_3}\frac{\partial x_3}{\partial t} + \left(\frac{\partial\theta}{\partial t}\right)_{x_i\text{-fixed}},$$

where $\partial x_1/\partial t$, $\partial x_2/\partial t$, and $\partial x_3/\partial t$ are to be obtained with fixed values of the X_i's. When rectangular Cartesian coordinates are used, $(\partial x_i/\partial t)_{X_i\text{-fixed}}$ gives the component v_1 of the velocity of a particle etc. and therefore the material derivative in rectangular coordinates is

$$\frac{D\theta}{Dt} = \frac{\partial\theta}{\partial t} + v_1\frac{\partial\theta}{\partial x_1} + v_2\frac{\partial\theta}{\partial x_2} + v_3\frac{\partial\theta}{\partial x_3} \tag{3.4a}$$

or

$$\frac{D\theta}{Dt} = \frac{\partial\theta}{\partial t} + \mathbf{v}\cdot\nabla\theta, \tag{3.4b}$$

where it should be emphasized that these equations are for θ in a spatial description, i.e., $\theta = \theta(x_1, x_2, x_3, t)$. Note that if the temperature field is independent of time and if the velocity of a particle is perpendicular to $\nabla\theta$ (i.e., the particle is moving along the path of constant θ) then, as expected, $D\theta/dt = 0$.

Note that Eq. (3.4a) is valid only for rectangular Cartesian coordinates, whereas Eq. (3.4b) has the advantage that it is valid for all coordinate systems. For a specific coordinate system, all that is needed is the appropriate expression for the gradient.†

EXAMPLE 3.3

Use Eq. (3.4), find $D\theta/Dt$ for the motion and temperature of the previous example.

Solution. From Example 3.2 we have

$$\mathbf{v} = (kx_2)\mathbf{e}_1,$$

and

$$\theta = x_1 + x_2.$$

The gradient of θ is simply $\nabla\theta = \mathbf{e}_1 + \mathbf{e}_2$ and therefore

$$\frac{D\theta}{Dt} = 0 + (kx_2\mathbf{e}_1)\cdot(\mathbf{e}_1 + \mathbf{e}_2) = kx_2.$$

†For example, the components of $\nabla\theta$ in cylindrical coordinates are given in Problem B47 of Chapter 2.

3.4 FINDING ACCELERATION OF A PARTICLE FROM A GIVEN VELOCITY FIELD

The acceleration of a particle is the rate of change of velocity of the particle. It is therefore the material derivative of velocity. If the motion of a continuum is given by Eq. (3.1), i.e.,

$$\mathbf{x} = \mathbf{x}(\mathbf{X}, t) \quad \text{with} \quad \mathbf{x}(\mathbf{X}, t_0) = \mathbf{X},$$

then the velocity $\mathbf{v}$, at time t, of a particle $\mathbf{X}$ is given by

$$\mathbf{v} \equiv \left(\frac{\partial \mathbf{x}}{\partial t}\right)_{\mathbf{X}\text{-fixed}} \equiv \frac{D\mathbf{x}}{Dt}, \tag{i}$$

and the acceleration $\mathbf{a}$, at time t, of a particle $\mathbf{X}$ is given by

$$\mathbf{a} \equiv \left(\frac{\partial \mathbf{v}}{\partial t}\right)_{\mathbf{X}\text{-fixed}} \equiv \frac{D\mathbf{v}}{Dt}. \tag{ii}$$

Thus, if the material description of velocity $\mathbf{v}(\mathbf{X}, t)$ is known [or is obtained from Eq. (i)], then the acceleration is very easily computed, simply taking the partial derivative with respect to time of the function $\mathbf{v}(\mathbf{X}, t)$. On the other hand, if only the spatial description of velocity [i.e., $\mathbf{v} = \mathbf{v}(\mathbf{x}, t)$] is known, the computation of acceleration is not as simple. We derive the formula for its computation in the following. Let $\mathbf{v} = v_1(x_1, x_2, x_3, t)\mathbf{e}_1 + v_2(x_1, x_2, x_3, t)\mathbf{e}_2 + v_3(x_1, x_2, x_3, t)\mathbf{e}_3$ be the spatial description of velocity with respect to *rectangular Cartesian coordinates* (where $\mathbf{e}_i$ are fixed), then the acceleration is given by

$$\mathbf{a} = \frac{D\mathbf{v}}{Dt} = \frac{Dv_1}{Dt}\mathbf{e}_1 + \frac{Dv_2}{Dt}\mathbf{e}_2 + \frac{Dv_3}{Dt}\mathbf{e}_3,$$

where

$$\frac{Dv_i}{Dt} = \frac{\partial v_i}{\partial t} + v_1\frac{\partial v_i}{\partial x_1} + v_2\frac{\partial v_i}{\partial x_2} + v_3\frac{\partial v_i}{\partial x_3}. \tag{3.5a}$$

In other words, in *Cartesian rectangular coordinates*, the acceleration components are given by

$$a_i = \frac{\partial v_i}{\partial t} + v_j\frac{\partial v_i}{\partial x_j} \tag{3.5b}$$

or in a form valid for all coordinate systems

$$\mathbf{a} = \frac{\partial \mathbf{v}}{\partial t} + (\nabla \mathbf{v})\mathbf{v}. \tag{3.5c}†$$

†This equation can also be written as

$$\mathbf{a} = \frac{\partial \mathbf{v}}{\partial t} + (\mathbf{v}\cdot\nabla_d)\mathbf{v},$$

where $\nabla_d \equiv \mathbf{e}_m(\partial/\partial x_m)$. *See* footnote on p. 44.

For example, the components of **a** in cylindrical coordinates are given in Problem 3.46. Note that the second term in this equation is just the tensor $\nabla\mathbf{v}$ operating on the velocity $\mathbf{v}$.

EXAMPLE 3.4

(a) Find the velocity field associated with the motion of a rigid body rotating with angular velocity $\boldsymbol{\omega} = \omega\mathbf{e}_3$.
(b) Using the velocity field of part (a), evaluate the acceleration field.

Solution. (a) The velocity $\mathbf{v} = \boldsymbol{\omega}\times\mathbf{x} = \omega\mathbf{e}_3 \times (x_i\mathbf{e}_i)$. Carrying out the cross-product, we obtain

$$v_1 = -x_2\omega, \quad v_2 = x_1\omega, \quad v_3 = 0.$$

(b) Here,

$$\frac{\partial v_i}{\partial t} = 0$$

and

$$[\nabla\mathbf{v}]\,[\mathbf{v}] = \begin{bmatrix} 0 & -\omega & 0 \\ \omega & 0 & 0 \\ 0 & 0 & 0 \end{bmatrix}\begin{bmatrix} -x_2\omega \\ x_1\omega \\ 0 \end{bmatrix} = \begin{bmatrix} -x_1\omega^2 \\ -x_2\omega^2 \\ 0 \end{bmatrix}.$$

Therefore,

$$\mathbf{a} = -\,\omega^2(x_1\mathbf{e}_1 + x_2\mathbf{e}_2)$$

as expected.

EXAMPLE 3.5

Given the velocity field

$$v_1 = \frac{x_1}{1+t}, \quad v_2 = \frac{x_2}{1+t}, \quad v_3 = \frac{x_3}{1+t},$$

(a) find the acceleration field, (b) find the path line $\mathbf{x} = \mathbf{x}(\mathbf{X}, t)$.

Solution. (a) Since

$$v_i = \frac{x_i}{1+t}, \quad \frac{\partial v_i}{\partial t} = \frac{-x_i}{(1+t)^2}$$

and

$$[(\nabla\mathbf{v})\mathbf{v}] = [\nabla\mathbf{v}]\,[\mathbf{v}] = \begin{bmatrix} \frac{1}{1+t} & 0 & 0 \\ 0 & \frac{1}{1+t} & 0 \\ 0 & 0 & \frac{1}{1+t} \end{bmatrix}\begin{bmatrix} v_1 \\ v_2 \\ v_3 \end{bmatrix} = \begin{bmatrix} \frac{x_1}{(1+t)^2} \\ \frac{x_2}{(1+t)^2} \\ \frac{x_3}{(1+t)^2} \end{bmatrix},$$

the components of the acceleration are given by

$$a_1 = \frac{Dv_1}{Dt} = \frac{\partial v_1}{\partial t} + ((\nabla\mathbf{v})\mathbf{v})_1 = -\frac{x_1}{(1+t)^2} + \frac{x_1}{(1+t)^2} = 0$$

and similarly,

$$\frac{Dv_2}{Dt} = \frac{Dv_3}{Dt} = 0.$$

Even though at a fixed position the velocity is observed to be changing, the actual velocity of a particular particle is constant.

(b) Since

$$v_1 = \left(\frac{\partial x_1}{\partial t}\right)_{X_i\text{-fixed}} = \frac{x_1}{1+t},$$

therefore,

$$\int_{X_1}^{x_1} \frac{dx_1}{x_1} = \int_0^t \frac{dt}{1+t},$$

i.e.,

$$\ln x_1 \Big]_{X_1}^{x} = \ln(1+t)\Big]_0^t.$$

Thus

$$x_1 = X_1(1+t).$$

Similarly, $x_2 = X_2(1+t)$ and $x_3 = X_3(1+t)$.

3.5 DEFORMATION

Suppose a body having a particular configuration at some reference time t_0 changes to another configuration at time t. Referring to Fig. 3.3, a typical material point P undergoes a displacement $\mathbf{u}$, so that it arrives at the position

$$\mathbf{x} = \mathbf{X} + \mathbf{u}(\mathbf{X}, t). \tag{3.6}$$

A neighboring point $\mathbf{Q}$ at $\mathbf{X} + d\mathbf{X}$ arrives at $\mathbf{x} + d\mathbf{x} = \mathbf{X} + d\mathbf{X} + \mathbf{u}(\mathbf{X} + d\mathbf{X}, t)$. We have

$$d\mathbf{x} = d\mathbf{X} + \mathbf{u}(\mathbf{X} + d\mathbf{X}, t) - \mathbf{u}(\mathbf{X}, t).$$

We may write this equation as [*see* Eq. (B48)]

$$d\mathbf{x} = d\mathbf{X} + (\nabla\mathbf{u})d\mathbf{X}, \tag{3.7}$$

where $\nabla\mathbf{u}$ is a second-order tensor known as the **displacement gradient** (with respect to $\mathbf{X}$). The matrix of $\nabla\mathbf{u}$ with respect to rectangular Carte-

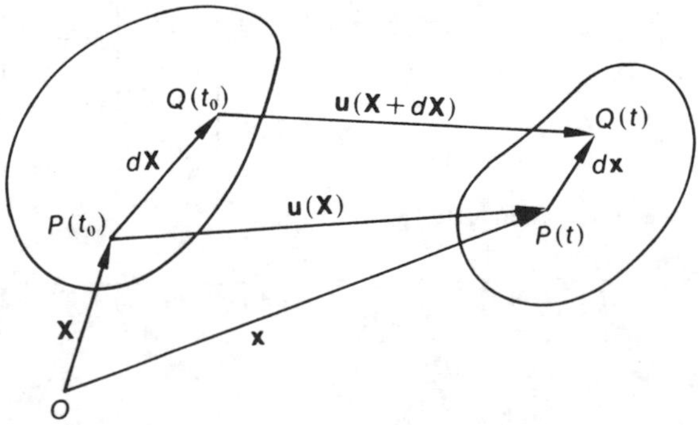

Fig. 3.3

sian coordinates (with $\mathbf{X} = X_i\mathbf{e}_i$ and $\mathbf{u} = u_i\mathbf{e}_i$) is

$$[\nabla \mathbf{u}] = \begin{bmatrix} \frac{\partial u_1}{\partial X_1} & \frac{\partial u_1}{\partial X_2} & \frac{\partial u_1}{\partial X_3} \\ \frac{\partial u_2}{\partial X_1} & \frac{\partial u_2}{\partial X_2} & \frac{\partial u_2}{\partial X_3} \\ \frac{\partial u_3}{\partial X_1} & \frac{\partial u_3}{\partial X_2} & \frac{\partial u_3}{\partial X_3} \end{bmatrix}. \tag{3.8}$$

EXAMPLE 3.6

Given the following displacement components

$$u_1 = kX_2^2, \quad u_2 = u_3 = 0,$$

(a) sketch the deformed shape of the unit square $OABC$ in Fig. 3.4;

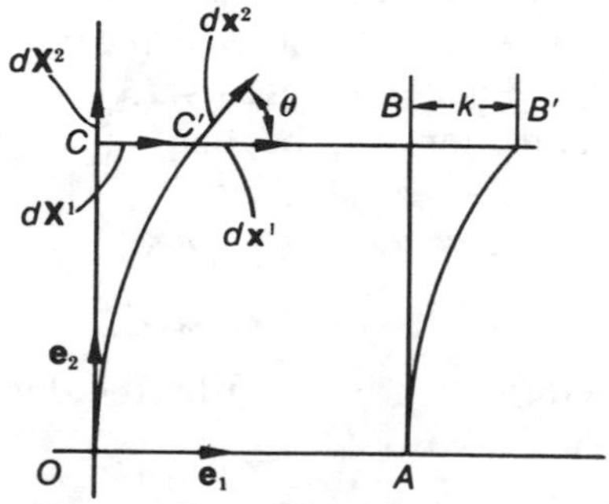

Fig. 3.4

(b) find the deformed vector (i.e., $d\mathbf{x}^1$ and $d\mathbf{x}^2$) of the material elements $d\mathbf{X}^1 = dX_1\mathbf{e}_1$ and $d\mathbf{X}^2 = dX_2\mathbf{e}_2$ which were at point C:
(c) determine the ratio of the deformed to undeformed lengths of the differential elements (known as stretch) of part (b) and the change in angle between these elements.

Solution. (a) For the material line OA, $X_2 = 0$, therefore $u_1 = u_2 = u_3 = 0$. That is, the line is not displaced. For the material line CB, $X_2 = 1$, $u_1 = k$, the line is displaced by k units to the right. For the material lines OC and AB, $u_1 = kX_2^2$, the lines become parabolic in shape. Thus, the deformed shape is given by $OAB'C'$ in Fig. 3.4.
(b) For the material point C, the matrix of the displacement gradient is

$$[\nabla \mathbf{u}]_{\text{at}\,C} = \begin{bmatrix} 0 & 2kX_2 & 0 \\ 0 & 0 & 0 \\ 0 & 0 & 0 \end{bmatrix}_{X_2=1} = \begin{bmatrix} 0 & 2k & 0 \\ 0 & 0 & 0 \\ 0 & 0 & 0 \end{bmatrix}.$$

Therefore,

$$d\mathbf{x}^1 = d\mathbf{X}^1 + (\nabla \mathbf{u})\, d\mathbf{X}^1 = dX_1\mathbf{e}_1 + 0 = dX_1\mathbf{e}_1,$$
$$d\mathbf{x}^2 = d\mathbf{X}^2 + (\nabla \mathbf{u})\, d\mathbf{X}^2 = dX_2\mathbf{e}_2 + 2k\,dX_2\mathbf{e}_1.$$

(c) From the results of part (b), we have

$$\frac{|d\mathbf{x}^1|}{|d\mathbf{X}^1|} = 1 \quad \text{and} \quad \frac{|d\mathbf{x}^2|}{|d\mathbf{X}^2|} = (1+4k^2)^{1/2}$$

and

$$\cos\theta = \frac{d\mathbf{x}^1 \cdot d\mathbf{x}^2}{|d\mathbf{x}^1|\,|d\mathbf{x}^2|} = \frac{2k}{(1+4k^2)^{1/2}}.$$

If γ denotes the decrease in angle, then

$$\cos\theta = \cos\left(\frac{\pi}{2}-\gamma\right) = \sin\gamma = \frac{2k}{(1+4k^2)^{1/2}},$$

that is,

$$\gamma = \sin^{-1}\frac{2k}{(1+4k^2)^{1/2}}.$$

From Eq. (3.7), we see that if $\nabla\mathbf{u} = 0$, then $d\mathbf{x} = d\mathbf{X}$ and the motion of the neighborhood of particle P in Fig. 3.3 is that of rigid body translation. Whatever deformation there is, is embodied in the transformation $\nabla\mathbf{u}$.

Let us now consider two material vectors $d\mathbf{X}^1$ and $d\mathbf{X}^2$ issuing from P in Fig. 3.3. Through the motion, $d\mathbf{X}^1$ becomes $d\mathbf{x}^1$ and $d\mathbf{X}^2$ becomes $d\mathbf{x}^2$ in such a way that

$$d\mathbf{x}^1 = d\mathbf{X}^1 + (\nabla\mathbf{u})d\mathbf{X}^1,$$

$$d\mathbf{x}^2 = d\mathbf{X}^2 + (\nabla\mathbf{u})d\mathbf{X}^2.$$

A measure of the deformation is given by the dot product of $d\mathbf{x}^1$ and $d\mathbf{x}^2$. Now

$$d\mathbf{x}^1 \cdot d\mathbf{x}^2 = d\mathbf{X}^1 \cdot d\mathbf{X}^2 + d\mathbf{X}^1 \cdot (\nabla\mathbf{u})d\mathbf{X}^2 + d\mathbf{X}^2 \cdot (\nabla\mathbf{u})d\mathbf{X}^1 + \{(\nabla\mathbf{u})d\mathbf{X}^1\} \cdot \{(\nabla\mathbf{u})d\mathbf{X}^2\}.$$

Since, by definition of transpose,

$$d\mathbf{X}^2 \cdot (\nabla\mathbf{u})d\mathbf{X}^1 = d\mathbf{X}^1 \cdot (\nabla\mathbf{u})^T d\mathbf{X}^2$$

and

$$\{(\nabla\mathbf{u})\,d\mathbf{X}^1\} \cdot \{(\nabla\mathbf{u})\,d\mathbf{X}^2\} = d\mathbf{X}^1 \cdot (\nabla\mathbf{u})^T(\nabla\mathbf{u})\,d\mathbf{X}^2,$$

thus

$$d\mathbf{x}^1 \cdot d\mathbf{x}^2 = d\mathbf{X}^1 \cdot d\mathbf{X}^2 + d\mathbf{X}^1 \cdot \{(\nabla\mathbf{u}) + (\nabla\mathbf{u})^T + (\nabla\mathbf{u})^T(\nabla\mathbf{u})\}d\mathbf{X}^2.$$

Let

$$\mathbf{E}^* = \tfrac{1}{2}\{(\nabla\mathbf{u}) + (\nabla\mathbf{u})^T + (\nabla\mathbf{u})^T(\nabla\mathbf{u})\}, \tag{3.9}$$

then

$$d\mathbf{x}^1 \cdot d\mathbf{x}^2 = d\mathbf{X}^1 \cdot d\mathbf{X}^2 + 2d\mathbf{X}^1 \cdot \mathbf{E}^* d\mathbf{X}^2. \tag{3.10}$$

We see from Eq. (3.10) that if $\mathbf{E}^* = 0$, then $d\mathbf{x}^1 \cdot d\mathbf{x}^2 = d\mathbf{X}^1 \cdot d\mathbf{X}^2$ so that lengths of, and angles between material line elements remain unchanged. In other words, the second-order tensor $\mathbf{E}^*$ characterizes the deformations of the neighborhood of the particle P, and is known as the (finite) **Lagrangian strain tensor**. The components of $\mathbf{E}^*$ with respect to rectangular Cartesian coordinates are given by

$$\mathbf{E}^*_{ij} = \frac{1}{2}\left(\frac{\partial u_i}{\partial X_j} + \frac{\partial u_j}{\partial X_i} + \frac{\partial u_k}{\partial X_i}\frac{\partial u_k}{\partial X_j}\right). \tag{3.11}$$

In many practical problems, the deformations of a body are such that the magnitudes of the components of the displacement gradient $\partial u_i/\partial X_j$ are all very much smaller than unity (often of the order of 10^{-4}), so that product terms in $\mathbf{E}^*$ [i.e., $(\nabla\mathbf{u})^T(\nabla\mathbf{u})$] can be neglected. When this is the case, from Eq. (3.9), we see

$$\mathbf{E}^* \approx \tfrac{1}{2}\{(\nabla\mathbf{u}) + (\nabla\mathbf{u})^T\} = (\nabla\mathbf{u})^S.$$

Then, for small deformation, we have

$$d\mathbf{x}^1 \cdot d\mathbf{x}^2 = d\mathbf{X}^1 \cdot d\mathbf{X}^2 + 2d\mathbf{X}^1 \cdot \mathbf{E}d\mathbf{X}^2. \tag{3.12}$$

where

$$\mathbf{E} = \tfrac{1}{2}\{(\nabla\mathbf{u}) + (\nabla\mathbf{u})^T\}. \tag{3.13a}$$

If $(\nabla\mathbf{u})$ is antisymmetric, then $\mathbf{E} = \mathbf{0}$ and $d\mathbf{x}^1 \cdot d\mathbf{x}^2 = d\mathbf{X}^1 \cdot d\mathbf{X}^2$, thus an antisymmetric infinitesimal displacement gradient tensor characterizes an infinitesimal rigid body rotation of the neighborhood of the particle P. The tensor $\Omega = \tfrac{1}{2}\{(\nabla\mathbf{u}) - (\nabla\mathbf{u})^T\}$ is called the (infinitesimal) rotation tensor. The dual vector of this antisymmetric tensor actually defines the axis and angle of this rotation (*see* Example 2.12). The tensor $\mathbf{E}$ is called the **(infinitesimal) strain tensor**. The components of $\mathbf{E}$ with respect to rectangular Cartesian coordinates are given by

$$E_{ij} = \frac{1}{2}\left(\frac{\partial u_i}{\partial X_j} + \frac{\partial u_j}{\partial X_i}\right) \tag{3.13b}$$

or

$$[\mathbf{E}] = \begin{bmatrix} \dfrac{\partial u_1}{\partial X_1} & \dfrac{1}{2}\left(\dfrac{\partial u_1}{\partial X_2} + \dfrac{\partial u_2}{\partial X_1}\right) & \dfrac{1}{2}\left(\dfrac{\partial u_1}{\partial X_3} + \dfrac{\partial u_3}{\partial X_1}\right) \\ \dfrac{1}{2}\left(\dfrac{\partial u_1}{\partial X_2} + \dfrac{\partial u_2}{\partial X_1}\right) & \dfrac{\partial u_2}{\partial X_2} & \dfrac{1}{2}\left(\dfrac{\partial u_2}{\partial X_3} + \dfrac{\partial u_3}{\partial X_2}\right) \\ \dfrac{1}{2}\left(\dfrac{\partial u_1}{\partial X_3} + \dfrac{\partial u_3}{\partial X_1}\right) & \dfrac{1}{2}\left(\dfrac{\partial u_2}{\partial X_3} + \dfrac{\partial u_3}{\partial X_2}\right) & \dfrac{\partial u_3}{\partial X_3} \end{bmatrix}. \tag{3.13c}$$

The infinitesimal strain components can be given simple geometrical interpretation as follows.

To interpret the diagonal elements, consider the single material element $d\mathbf{X}^1 = d\mathbf{X}^2 = (dS)\mathbf{n}$, where $\mathbf{n}$ is an arbitrary unit vector and dS the length of $d\mathbf{X}^1$. Let ds denote the deformed length of $d\mathbf{X}^1$, i.e., $ds = |d\mathbf{x}^1|$. Then Eq. (3.12) gives

$$(ds)^2 - (dS)^2 = 2(dS)^2\mathbf{n}\cdot\mathbf{E}\mathbf{n}.$$

Now for small deformation $(ds)^2 - (dS)^2 = (ds + dS)\,(ds - dS) \approx 2dS\,(ds - dS)$.† Thus,

$$\frac{ds - dS}{dS} = \mathbf{n}\cdot\mathbf{E}\mathbf{n} = E_{(n)(n)}. \quad \text{(no sum on } n\text{)} \tag{3.14}$$

In other words, the change of length per unit length, known as **unit elongation**, or **normal strain**, of every material element emanating from point P is easily determined from the strain tensor $\mathbf{E}$. For example, if we let $\mathbf{n} = \mathbf{e}_1$, then Eq. (3.14) gives

$$E_{11} = \mathbf{e}_1 \cdot \mathbf{E}\mathbf{e}_1$$

as the unit elongation for an element originally in the x_1-direction.

Similarly, E_{22} and E_{33} give the unit elongations in the x_2- and x_3-direction respectively.

To interpret the off-diagonal elements of the strain matrix, let $d\mathbf{X}^1 = (dS_1)\mathbf{m}$ and $d\mathbf{X}^2 = (dS_2)\mathbf{n}$, where the unit vectors $\mathbf{m}$ and $\mathbf{n}$ are perpendicular. Then Eq. (3.12) gives

$$(ds_1)\,(ds_2)\cos\theta = 2(dS_1)\,(dS_2)\mathbf{m}\cdot\mathbf{E}\mathbf{n},$$

where θ is the angle between $d\mathbf{x}^1$ and $d\mathbf{x}^2$. If we let $\theta = (\pi/2) - \gamma$, then γ will measure the small decrease in angle between $d\mathbf{X}^1$ and $d\mathbf{X}^2$ (known as **shear strain**). Since

$$\cos\left(\frac{\pi}{2} - \gamma\right) = \sin\gamma$$

and for small strain $\sin\gamma \approx \gamma$, $ds_1/dS_1 \approx 1$, $ds_2/dS_2 \approx 1$, then

$$\gamma = 2\mathbf{m}\cdot\mathbf{E}\mathbf{n}. \tag{3.15}$$

For the particular choice of $\mathbf{m} = \mathbf{e}_1$ and $\mathbf{n} = \mathbf{e}_2$, $2E_{12}$ gives the decrease in angle between two elements initially in the x_1- and x_2-direction. Similarly $2E_{13}$ gives the decrease in angle between elements which were in the x_1- and x_3-direction, etc.

†For example, $(1.0001 + 1)(1.0001 - 1) \approx (2)(0.0001)$.

EXAMPLE 3.7

Given the displacement components

$$u_1 = kX_2^2, \quad u_2 = u_3 = 0, \quad k = 10^{-4},$$

(a) obtain the infinitesimal strain tensor **E**;
(b) using the strain tensor **E**, find the unit elongation for the material elements $d\mathbf{X}^1 = dX_1\mathbf{e}_1$ and $d\mathbf{X}^2 = dX_2\mathbf{e}_2$ which were at point $C(0, 1, 0)$ (*see* Fig. 3.4). Also find the decrease in angle between these two elements;
(c) compare the results with those of Example 3.6.

Solution. (a) We have

$$[\nabla \mathbf{u}] = \begin{bmatrix} 0 & 2kX_2 & 0 \\ 0 & 0 & 0 \\ 0 & 0 & 0 \end{bmatrix},$$

therefore,

$$[\mathbf{E}] = [\nabla \mathbf{u}]^S = \begin{bmatrix} 0 & kX_2 & 0 \\ kX_2 & 0 & 0 \\ 0 & 0 & 0 \end{bmatrix}.$$

(b) At the point C, $X_2 = 1$, therefore,

$$[\mathbf{E}] = \begin{bmatrix} 0 & k & 0 \\ k & 0 & 0 \\ 0 & 0 & 0 \end{bmatrix}.$$

For the element $d\mathbf{X}^1 = dX_1\mathbf{e}_1$, the unit elongation is E_{11}, which is zero. For the element $d\mathbf{X}^2 = dX_2\mathbf{e}_2$, the unit elongation is E_{22}, which is zero. The decrease in angle is given by $2E_{12}$, that is, $2k = 2 \times 10^{-4}$.

(c) From the results of Example 3.6, we have

$$\frac{|d\mathbf{x}^1| - |d\mathbf{X}^1|}{|d\mathbf{X}^1|} = 0, \quad \frac{|d\mathbf{x}^2| - |d\mathbf{X}^2|}{|d\mathbf{X}^2|} = (1 + 4k^2)^{1/2} - 1$$

and

$$\sin\gamma = \frac{2k}{(1 + 4k^2)^{1/2}}.$$

Now with $k = 10^{-4}$, we have

$$(1 + 4k^2)^{1/2} - 1 \approx 1 + 2k^2 - 1 = 2k^2 = 2 \times 10^{-8}$$

and $\sin\gamma \approx 2k = 2 \times 10^{-4}$ so that $\gamma \approx 2 \times 10^{-4}$. Since 10^{-8} is negligible compared to 10^{-4}, we see that the results of Example 3.6 reduce to those of this example for small values of k.

EXAMPLE 3.8

Given the displacement field

$$u_1 = k(2X_1 + X_2^2), \quad u_2 = k(X_1^2 - X_2^2), \quad u_3 = 0,$$

where $k = 10^{-4}$,

(a) find the unit elongations and the change of angle for two material elements $d\mathbf{X}^1 = dX_1\mathbf{e}_1$ and $d\mathbf{X}^2 = dX_2\mathbf{e}_2$ that emanate from a particle designated by $\mathbf{X} = \mathbf{e}_1 - \mathbf{e}_2$;
(b) find the deformed position of these two elements $d\mathbf{X}^1$ and $d\mathbf{X}^2$.

Solution. (a) We evaluate $[\nabla \mathbf{u}]$ at $X_1 = 1, X_2 = -1, X_3 = 0$ as

$$[\nabla \mathbf{u}] = k\begin{bmatrix} 2 & -2 & 0 \\ 2 & 2 & 0 \\ 0 & 0 & 0 \end{bmatrix},$$

and therefore the strain matrix is

$$[\mathbf{E}] = k\begin{bmatrix} 2 & 0 & 0 \\ 0 & 2 & 0 \\ 0 & 0 & 0 \end{bmatrix}.$$

Since $E_{11} = E_{22} = 2k$, both elements have a unit elongation of 2×10^{-4}. Further, since $E_{12} = 0$, these lines remain perpendicular to each other.

(b) From

$$d\mathbf{x} = d\mathbf{X} + (\nabla \mathbf{u})\, d\mathbf{X},$$

we have

$$[d\mathbf{x}^1] = [d\mathbf{X}^1] + [\nabla \mathbf{u}]\,[d\mathbf{X}^1] = \begin{bmatrix} dX_1 \\ 0 \\ 0 \end{bmatrix} + k\begin{bmatrix} 2 & -2 & 0 \\ 2 & 2 & 0 \\ 0 & 0 & 0 \end{bmatrix}\begin{bmatrix} dX_1 \\ 0 \\ 0 \end{bmatrix} = dX_1\begin{bmatrix} 1+2k \\ 2k \\ 0 \end{bmatrix}$$

and similarly,

$$[d\mathbf{x}^2] = dX_2\begin{bmatrix} -2k \\ 1+2k \\ 0 \end{bmatrix}.$$

The deformed position of these elements is sketched in Fig. 3.5. Note from the diagram that

$$\alpha \approx \tan \alpha \approx \frac{2k\,dX_1}{dX_1} = 2k$$

and

$$\beta \approx \tan \beta \approx \frac{2k\,dX_2}{dX_2} = 2k.$$

Thus, as previously obtained, there is no change of angle between $d\mathbf{X}^1$ and $d\mathbf{X}^2$.

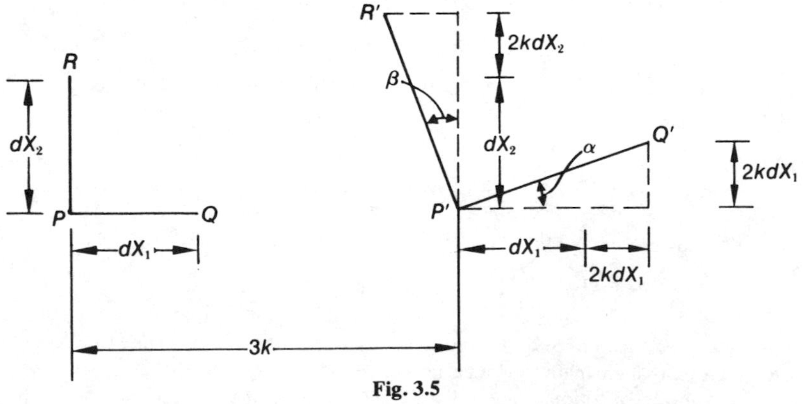

Fig. 3.5

EXAMPLE 3.9

A unit cube, with edges parallel to the coordinate axes, is given a displacement field

$$u_1 = kX_1, \quad u_2 = u_3 = 0, \quad k = 10^{-4}.$$

Find the increase in length of the diagonal AB (*see* Fig. 3.6), (a) by using the strain tensor and (b) by geometry.

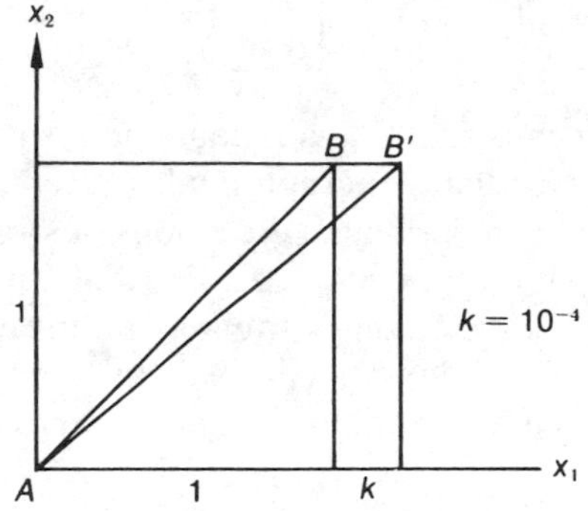

Fig. 3.6

Solution. (a) The strain tensor is easily obtained to be

$$[\mathbf{E}] = \begin{bmatrix} k & 0 & 0 \\ 0 & 0 & 0 \\ 0 & 0 & 0 \end{bmatrix}.$$

Since the diagonal AB was originally in the direction $\mathbf{n} = \sqrt{2}/2\ (\mathbf{e}_1 + \mathbf{e}_2)$, its unit elongation is given by

$$E_{(n)(n)} = \mathbf{n} \cdot \mathbf{En} = [\sqrt{2}/2, \sqrt{2}/2, 0] \begin{bmatrix} k & 0 & 0 \\ 0 & 0 & 0 \\ 0 & 0 & 0 \end{bmatrix} \begin{bmatrix} \sqrt{2}/2 \\ \sqrt{2}/2 \\ 0 \end{bmatrix} = \frac{k}{2}. \quad \text{(no sum on } n\text{)}$$

Since $AB = \sqrt{2}$,

$$\Delta AB = \left(\frac{k}{2}\right)\sqrt{2}.$$

(b) Geometrically,

$$AB' - AB = \{1 + (1+k)^2\}^{1/2} - \sqrt{2}$$

or

$$\Delta AB = \sqrt{2}\,[(1 + k + k^2/2)^{1/2} - 1].$$

To take advantage of the smallness of k we expand the first term as

$$\left(1 + k + \frac{k^2}{2}\right)^{1/2} = 1 + \frac{1}{2}\left(k + \frac{k^2}{2}\right) + \cdots \approx 1 + \frac{k}{2}.$$

Therefore, in agreement with part (a)

$$\Delta AB = \sqrt{2}\left(\frac{k}{2}\right).$$

3.6 PRINCIPAL STRAIN

Since the strain tensor **E** is symmetric, therefore (*see* Section 2B13) there exists at least three mutually perpendicular directions $\mathbf{n}_1$, $\mathbf{n}_2$, $\mathbf{n}_3$, with respect to which the matrix of **E** is diagonal. That is,

$$[\mathbf{E}]_{\mathbf{n}_1,\mathbf{n}_2,\mathbf{n}_3} = \begin{bmatrix} E_1 & 0 & 0 \\ 0 & E_2 & 0 \\ 0 & 0 & E_3 \end{bmatrix}.$$

Geometrically, this means that infinitesimal line elements in the direction of $\mathbf{n}_1$, $\mathbf{n}_2$, $\mathbf{n}_3$ remain mutually perpendicular after deformation. These directions are known as the principal directions of strain. The unit elongation along the principal direction (i.e., E_1, E_2, and E_3) are the eigenvalues or principal strains of **E**, they include the maximum and the minimum normal strains among all directions emanating from the particle. From a given **E**, the principal strains are to be found from the characteristic equation of **E**, i.e.,

$$\lambda^3 - I_1\lambda^2 + I_2\lambda - I_3 = 0, \tag{3.16}$$

where

$$I_1 = E_{11} + E_{22} + E_{33},$$

$$I_2 = \begin{vmatrix} E_{11} & E_{12} \\ E_{21} & E_{22} \end{vmatrix} + \begin{vmatrix} E_{11} & E_{13} \\ E_{31} & E_{33} \end{vmatrix} + \begin{vmatrix} E_{22} & E_{23} \\ E_{32} & E_{33} \end{vmatrix}, \tag{3.17}$$

and

$$I_3 = |E_{ij}|,$$

where I_1, I_2, I_3 are called the **scalar invariants** of the **strain tensor**.

3.7 DILATATION

The first scalar invariant of the strain tensor has a simple geometric meaning. For a specific deformation, consider the three material lines that emanate from a single point P and are in the principal directions. These lines define a rectangular parallelepiped whose sides have been elongated from the initial dimension dS_1, dS_2, dS_3 to $dS_1(1+E_1)$, dS_2-$(1+E_2)$, $dS_3(1+E_3)$, where E_1, E_2, E_3 are the principal strains. Hence, the change $\Delta(dV)$ in this material volume dV is

$$\begin{aligned}\Delta(dV) &= (dS_1)(dS_2)(dS_3)(1+E_1)(1+E_2)(1+E_3) - (dS_1)(dS_2)(dS_3)\\ &= (dV)(E_1+E_2+E_3)\end{aligned}$$

plus higher-order terms in the E_i's. Thus, for small deformation

$$e \equiv \frac{\Delta(dV)}{dV} = E_1 + E_2 + E_3 = E_{11} + E_{22} + E_{33} = E_{ii}. \tag{3.18}$$

This unit volume change is known as **dilatation**. Note that

$$E_{ii} = \frac{\partial u_i}{\partial X_i} = \operatorname{div} \mathbf{u}.$$

3.8 RATE OF DEFORMATION

Let us consider a material element $d\mathbf{x}$ emanating from a material point P located at $\mathbf{x}$ at time t. We wish to compute $(D/Dt)d\mathbf{x}$, the rate of change of length and direction of the material element $d\mathbf{x}$. From $\mathbf{x} = \mathbf{x}(\mathbf{X}, t)$ we have,

$$d\mathbf{x} = \mathbf{x}(\mathbf{X} + d\mathbf{X}, t) - \mathbf{x}(\mathbf{X}, t).$$

Thus,

$$\frac{D}{Dt}(d\mathbf{x}) = \mathbf{v}(\mathbf{X} + d\mathbf{X}, t) - \mathbf{v}(\mathbf{X}, t) = (\nabla_{\mathbf{X}} \mathbf{v})\, d\mathbf{X}, \tag{3.19}$$

where $\nabla_{\mathbf{X}} \mathbf{v}$ is the gradient of $\mathbf{v}$ with respect to the material coordinate $\mathbf{X}$.

Equation (3.19) expressed $(D/Dt)d\mathbf{x}$ in a material description. To obtain $(D/Dt)d\mathbf{x}$ in a spatial description, we note that since $\mathbf{v}(\mathbf{X}, t)$ is the velocity of the material point P presently at the position $\mathbf{x}$, therefore if a spatial description of $\mathbf{v}$ is employed, this velocity is given by $\mathbf{v} = \mathbf{v}(\mathbf{x}, t)$. [Note $\mathbf{v}(\mathbf{x}, t)$ and $\mathbf{v}(\mathbf{X}, t)$ are different functions.] Thus,

$$\frac{D}{Dt}(d\mathbf{x}) = \mathbf{v}(\mathbf{x} + d\mathbf{x}, t) - \mathbf{v}(\mathbf{x}, t) = (\nabla_{\mathbf{x}} \mathbf{v})\, d\mathbf{x}, \tag{3.20†}$$

where $\nabla_{\mathbf{x}} \mathbf{v}$ is the spatial velocity gradient. In the following, the spatial description will be used exclusively, the notation $(\nabla \mathbf{v})$ will be understood to mean $(\nabla_{\mathbf{x}} \mathbf{v})$. With respect to rectangular Cartesian coordinates, the components of $(\nabla \mathbf{v})$ are given by

$$[\nabla \mathbf{v}] = \begin{bmatrix} \frac{\partial v_1}{\partial x_1} & \frac{\partial v_1}{\partial x_2} & \frac{\partial v_1}{\partial x_3} \\ \frac{\partial v_2}{\partial x_1} & \frac{\partial v_2}{\partial x_2} & \frac{\partial v_2}{\partial x_3} \\ \frac{\partial v_3}{\partial x_1} & \frac{\partial v_3}{\partial x_2} & \frac{\partial v_3}{\partial x_3} \end{bmatrix}. \tag{3.21}$$

†Or, $(\nabla_{\mathbf{X}} \mathbf{v})\, d\mathbf{X} = \frac{\partial v_i}{\partial X_j} dX_j \mathbf{e}_i = \frac{\partial v_i}{\partial x_m} \frac{\partial x_m}{\partial X_j} dX_j \mathbf{e}_i = \frac{\partial v_i}{\partial x_m} dx_m \mathbf{e}_i = (\nabla_{\mathbf{x}} \mathbf{v}) d\mathbf{x}$

If $\nabla \mathbf{v}$ happens to be antisymmetric, we know from Section 2B11, that there exists a vector $\boldsymbol{\omega}$ such that

$$\mathbf{v}(\mathbf{x}+d\mathbf{x}, t)-\mathbf{v}(\mathbf{x}, t) = (\nabla \mathbf{v})\, d\mathbf{x} = \boldsymbol{\omega} \times d\mathbf{x}.$$

Thus, an antisymmetric velocity gradient represents the angular velocity $\boldsymbol{\omega}$ of a local rigid body rotation.

For a general $\nabla \mathbf{v}$, we decompose it into the sum of a symmetric and antisymmetric part. Then,

$$\nabla \mathbf{v} = \mathbf{D}+\mathbf{W}, \tag{3.22}$$

where

$$\mathbf{D} = \frac{(\nabla \mathbf{v})+(\nabla \mathbf{v})^T}{2} \quad \text{(the symmetric part)} \tag{3.23}$$

is known as the **rate of deformation tensor** and

$$\mathbf{W} = \frac{(\nabla \mathbf{v})-(\nabla \mathbf{v})^T}{2} \quad \text{(the antisymmetric part)} \tag{3.24}$$

is known as the **spin tensor.** Its dual vector $\boldsymbol{\omega}$ is the angular velocity of that part of motion, representing rigid body rotation.† An alternative interpretation of $\boldsymbol{\omega}$ is given in Section 6.13.

With respect to rectangular Cartesian coordinates, the components of $\mathbf{D}$ and $\mathbf{W}$ are given by

$$[\mathbf{D}] = \begin{bmatrix} \dfrac{\partial v_1}{\partial x_1} & \dfrac{1}{2}\left(\dfrac{\partial v_1}{\partial x_2}+\dfrac{\partial v_2}{\partial x_1}\right) & \dfrac{1}{2}\left(\dfrac{\partial v_1}{\partial x_3}+\dfrac{\partial v_3}{\partial x_1}\right) \\ \dfrac{1}{2}\left(\dfrac{\partial v_1}{\partial x_2}+\dfrac{\partial v_2}{\partial x_1}\right) & \dfrac{\partial v_2}{\partial x_2} & \dfrac{1}{2}\left(\dfrac{\partial v_2}{\partial x_3}+\dfrac{\partial v_3}{\partial x_2}\right) \\ \dfrac{1}{2}\left(\dfrac{\partial v_1}{\partial x_3}+\dfrac{\partial v_3}{\partial x_1}\right) & \dfrac{1}{2}\left(\dfrac{\partial v_2}{\partial x_3}+\dfrac{\partial v_3}{\partial x_2}\right) & \dfrac{\partial v_3}{\partial x_3} \end{bmatrix}, \tag{3.25}$$

$$[\mathbf{W}] = \begin{bmatrix} 0 & \dfrac{1}{2}\left(\dfrac{\partial v_1}{\partial x_2}-\dfrac{\partial v_2}{\partial x_1}\right) & \dfrac{1}{2}\left(\dfrac{\partial v_1}{\partial x_3}-\dfrac{\partial v_3}{\partial x_1}\right) \\ -\dfrac{1}{2}\left(\dfrac{\partial v_1}{\partial x_2}-\dfrac{\partial v_2}{\partial x_1}\right) & 0 & \dfrac{1}{2}\left(\dfrac{\partial v_2}{\partial x_3}-\dfrac{\partial v_3}{\partial x_2}\right) \\ -\dfrac{1}{2}\left(\dfrac{\partial v_1}{\partial x_3}-\dfrac{\partial v_3}{\partial x_1}\right) & -\dfrac{1}{2}\left(\dfrac{\partial v_2}{\partial x_3}-\dfrac{\partial v_3}{\partial x_2}\right) & 0 \end{bmatrix}. \tag{3.26}$$

In the following, we give a geometrical interpretation of the diagonal

†The tensor $2\mathbf{W}$ and the vector $2\boldsymbol{\omega} = \mathbf{curl\ v}$ are called the **vorticity tensor** and the **vorticity vector** respectively by many Hydrodynamicists.

elements of $\mathbf{D}$. Let $d\mathbf{x} = (ds)\mathbf{n}$, where $\mathbf{n}$ is a unit vector in the direction of $d\mathbf{x}$. Then

$$d\mathbf{x} \cdot d\mathbf{x} = (ds)^2.$$

Thus

$$\frac{D}{Dt}(d\mathbf{x} \cdot d\mathbf{x}) = \frac{D}{Dt}(ds)^2,$$

i.e.,

$$2d\mathbf{x} \cdot \frac{D(d\mathbf{x})}{Dt} = 2ds \frac{D(ds)}{Dt}.$$

But

$$\frac{D(d\mathbf{x})}{Dt} = (\nabla \mathbf{v})\, d\mathbf{x}. \quad (\textit{See } \text{Eq. } (3.20).)$$

Therefore

$$d\mathbf{x} \cdot (\nabla \mathbf{v})\, d\mathbf{x} = ds \frac{D(ds)}{Dt}$$

or

$$(ds)^2 \mathbf{n} \cdot (\nabla \mathbf{v})\mathbf{n} = (ds) \frac{D(ds)}{Dt}.$$

i.e.,

$$\frac{1}{ds}\frac{D(ds)}{Dt} = \mathbf{n} \cdot (\nabla \mathbf{v})\mathbf{n} = \mathbf{n} \cdot \mathbf{D}\mathbf{n} + \mathbf{n} \cdot \mathbf{W}\mathbf{n}.$$

Since

$$\mathbf{n} \cdot \mathbf{W}\mathbf{n} = \mathbf{n} \cdot \mathbf{W}^T\mathbf{n} \quad \text{(definition of transpose)}$$

$$= -\mathbf{n} \cdot \mathbf{W}\mathbf{n} \quad (\mathbf{W} \text{ is antisymmetric}).$$

Thus $\mathbf{n} \cdot \mathbf{W}\mathbf{n} = 0$ and

$$\frac{1}{ds}\frac{D(ds)}{Dt} = \mathbf{n} \cdot \mathbf{D}\mathbf{n} = D_{(n)(n)}. \quad (\text{no sum on } n) \tag{3.27}$$

In other words, $D_{(n)(n)}$ gives the rate of change of length per unit length known as **stretching** or **rate of extension** for a material element in the direction of $\mathbf{n}$. In particular, D_{11} gives the stretching for a material element in the x_1-direction and similar interpretations for D_{22} and D_{33}. Actually if we note that $\mathbf{v}dt$ gives the infinitesimal displacement suffered by a particle during the time interval dt, the interpretations just given can be inferred from those for the strain components. Thus, we obviously will have the following results: $2D_{12}$ gives the rate of decrease of angle (from $\pi/2$) of two elements in the direction of $\mathbf{e}_1$ and $\mathbf{e}_2$, known as **shearing** or **rate of shear**, etc.† Also the first scalar invariant of the rate of deformation

†*See also* Problem 3.33.

tensor **D** gives the rate of change of volume per unit volume, i.e., denoting the infinitesimal volume of a particle by dV. We have

$$\Delta \equiv \frac{1}{dV}\frac{D(dV)}{Dt} = D_{11} + D_{22} + D_{33} = \frac{\partial v_1}{\partial x_1} + \frac{\partial v_2}{\partial x_2} + \frac{\partial v_3}{\partial x_3} = \frac{\partial v_i}{\partial x_i}. \quad (3.28)$$

Since **D** is symmetric, we also have the result that there always exists three mutually perpendicular directions (eigenvectors of **D**) along which the stretching (eigenvalues of **D**) include a maximum and a minimum value among all differential elements extending from the particle P.

EXAMPLE 3.10

Given the velocity field (*see also* Example 3.2)

$$v_1 = kx_2, \quad v_2 = v_3 = 0,$$

(a) find the rate of deformation and spin tensor;
(b) determine the rate of extension of the material elements

$$d\mathbf{x}^1 = (ds_1)\mathbf{e}_1, \quad d\mathbf{x}^2 = (ds_2)\mathbf{e}_2, \text{ and } d\mathbf{x} = dl(\mathbf{e}_1 + 2\mathbf{e}_2);$$

(c) find the maximum and minimum rates of extension.

Solution. (a) The matrix of the vector gradient is

$$[\nabla \mathbf{v}] = \begin{bmatrix} 0 & k & 0 \\ 0 & 0 & 0 \\ 0 & 0 & 0 \end{bmatrix},$$

so that

$$[\mathbf{D}] = [\nabla \mathbf{v}]^S = \begin{bmatrix} 0 & \frac{k}{2} & 0 \\ \frac{k}{2} & 0 & 0 \\ 0 & 0 & 0 \end{bmatrix}$$

and

$$[\mathbf{W}] = [\nabla \mathbf{v}]^A = \begin{bmatrix} 0 & \frac{k}{2} & 0 \\ -\frac{k}{2} & 0 & 0 \\ 0 & 0 & 0 \end{bmatrix}$$

(b) The material element $d\mathbf{x}^1$ is currently in the $\mathbf{e}_1$-direction and therefore its rate of extension is equal to $D_{11} = 0$. Similarly, the rate of extension of $d\mathbf{x}^2$ is equal to $D_{22} = 0$.

For the element $d\mathbf{x} = (ds)\mathbf{n}$, where $\mathbf{n} = (1/\sqrt{5})\ (\mathbf{e}_1 + 2\mathbf{e}_2)$ and $ds = \sqrt{5}\, dl$, we have

$$\frac{\frac{D}{Dt}(ds)}{ds} = \mathbf{n}\cdot\mathbf{D}\mathbf{n} = \frac{1}{5}[1, 2, 0]\begin{bmatrix} 0 & \frac{k}{2} & 0 \\ \frac{k}{2} & 0 & 0 \\ 0 & 0 & 0 \end{bmatrix}\begin{bmatrix} 1 \\ 2 \\ 0 \end{bmatrix} = \frac{2}{5}k.$$

(c) From the characteristic equation

$$|\mathbf{D}-\lambda \mathbf{I}| = -\lambda(\lambda^2 - k^2/4) = 0,$$

we determine the eigenvalues of the tensor **D** as $\lambda = 0, \pm k/2$. Therefore, $k/2$ is the maximum and $-k/2$ the minimum rate of extension. The eigenvector $\mathbf{n}_1 = (\sqrt{2}/2)(\mathbf{e}_1 + \mathbf{e}_2)$ and $\mathbf{n}_2 = (\sqrt{2}/2)(\mathbf{e}_1 - \mathbf{e}_2)$ give the directions of the elements having the maximum and the minimum stretching respectively.

3.9 EQUATION OF CONSERVATION OF MASS

If we follow a particle† through its motion, its volume may change, but its total mass will remain unchanged. Let ρ and dV denote the density and volume of a particle, then we have

$$\frac{D}{Dt}(\rho dV) = 0,$$

i.e.,

$$\rho \frac{D(dV)}{Dt} + dV \frac{D\rho}{Dt} = 0.$$

Using Eq. (3.28), we obtain

$$\rho \frac{\partial v_i}{\partial x_i} + \frac{D\rho}{Dt} = 0. \tag{3.29a}$$

Or, in invariant form,

$$\rho \operatorname{div} \mathbf{v} + \frac{D\rho}{Dt} = 0, \tag{3.29b}$$

where in spatial description

$$\frac{D\rho}{Dt} = \frac{\partial \rho}{\partial t} + v_i \frac{\partial \rho}{\partial x_i} = \frac{\partial \rho}{\partial t} + \mathbf{v} \cdot \boldsymbol{\nabla} \rho.$$

Equation (3.29) is the **equation** of **conservation of mass**, also known as **equation of continuity**.

An important special case of Eq. (3.29) is that of an incompressible material. By definition, the material derivative of the density is zero, and the mass conservation equation reduces to $\partial v_i/\partial x_i = 0$.

EXAMPLE 3.11

For the velocity field of Example 3.5, $v_i = x_i/(1+t)$, find the density of a material particle as a function of time.

†By a particle is meant a small volume of material, as mentioned earlier.

Solution. From the mass conservation equation,

$$\frac{D\rho}{Dt} = -\rho\frac{\partial v_i}{\partial x_i} = -\rho\left[\frac{1}{1+t}+\frac{1}{1+t}+\frac{1}{1+t}\right] = -\frac{3\rho}{1+t}.$$

Integrating, we have

$$\ln\rho = \int\frac{D\rho}{\rho} = -3\int\frac{Dt}{1+t} = -3\ln(1+t)+A,$$

where A is a constant of integration.

If $\rho=\rho_0$ at $t=0$, then we have $A=\ln\rho_0$ and therefore

$$\rho = \frac{\rho_0}{(1+t)^3}.$$

3.10 COMPATIBILITY CONDITIONS FOR INFINITESIMAL STRAIN COMPONENTS

When any three displacement functions u_1, u_2, and u_3 are given, one can always determine the six strain components in any region where the partial derivatives $\partial u_i/\partial X_j$ exist. On the other hand, when the six strain components E_{ij} are arbitrarily prescribed in some region, in general, there does not exist any displacement field u_i, satisfying the equations

$$\frac{1}{2}\left(\frac{\partial u_i}{\partial X_j}+\frac{\partial u_j}{\partial X_i}\right) = E_{ij}. \tag{3.30}$$

For example if we let

$$E_{11} = X_2^2, \quad E_{22}=E_{33}=E_{12}=E_{13}=E_{23}=0. \tag{3.31}$$

Then from $E_{11}=\partial u_1/\partial X_1 = X_2^2$ and $E_{22}=\partial u_2/\partial X_2=0$, we get by integration,

$$u_1 = X_1X_2^2+f(X_2,X_3), \tag{3.32}$$

$$u_2 = g(X_1,X_3), \tag{3.33}$$

where f and g are arbitrary integration functions.

Now, since $E_{12}=0$, we must have

$$\frac{\partial u_1}{\partial X_2}+\frac{\partial u_2}{\partial X_1} = 0. \tag{3.34}$$

Using Eqs. (3.32) and (3.33), we get

$$2X_1X_2+\frac{\partial f(X_2,X_3)}{\partial X_2}+\frac{\partial g(X_1,X_3)}{\partial X_1} = 0. \tag{3.35}$$

Since the second or third term cannot have terms of the form X_1X_2, the above equation can never be satisfied. In other words, there is no displacement field corresponding to the given E_{ij}. In the terminology of the theory of elasticity, we say that the given E_{ij}'s are not compatible.

We now state the following theorem: If $E_{ij}(X_1, X_2, X_3)$ are continuous functions having continuous second partial derivatives in a simply-connected region,† then the necessary and sufficient conditions for the existence of single-valued continuous solutions $u_i(X_1, X_2, X_3)$ of Eq. (3.30) are

$$\frac{\partial^2 E_{11}}{\partial X_2{}^2} + \frac{\partial^2 E_{22}}{\partial X_1{}^2} = 2\frac{\partial^2 E_{12}}{\partial X_1 \partial X_2}, \tag{3.36a}$$

$$\frac{\partial^2 E_{22}}{\partial X_3{}^2} + \frac{\partial^2 E_{33}}{\partial X_2{}^2} = 2\frac{\partial^2 E_{23}}{\partial X_2 \partial X_3}, \tag{3.36b}$$

$$\frac{\partial^2 E_{33}}{\partial X_1{}^2} + \frac{\partial^2 E_{11}}{\partial X_3{}^2} = 2\frac{\partial^2 E_{31}}{\partial X_3 \partial X_1}, \tag{3.36c}$$

$$\frac{\partial^2 E_{11}}{\partial X_2 \partial X_3} = \frac{\partial}{\partial X_1}\left(-\frac{\partial E_{23}}{\partial X_1} + \frac{\partial E_{31}}{\partial X_2} + \frac{\partial E_{12}}{\partial X_3}\right), \tag{3.36d}$$

$$\frac{\partial^2 E_{22}}{\partial X_3 \partial X_1} = \frac{\partial}{\partial X_2}\left(-\frac{\partial E_{31}}{\partial X_2} + \frac{\partial E_{12}}{\partial X_3} + \frac{\partial E_{23}}{\partial X_1}\right), \tag{3.36e}$$

$$\frac{\partial^2 E_{33}}{\partial X_1 \partial X_2} = \frac{\partial}{\partial X_3}\left(-\frac{\partial E_{12}}{\partial X_3} + \frac{\partial E_{23}}{\partial X_1} + \frac{\partial E_{31}}{\partial X_2}\right). \tag{3.36f}$$

These six equations are known as the **equations of compatibility** (or **integrability conditions**).

That these conditions are necessary can be easily proved as follows:

From
$$\frac{\partial u_1}{\partial X_1} = E_{11} \quad \text{and} \quad \frac{\partial u_2}{\partial X_2} = E_{22},$$
we get
$$\frac{\partial^2 E_{11}}{\partial X_2{}^2} = \frac{\partial^3 u_1}{\partial X_2{}^2 \partial X_1} \quad \text{and} \quad \frac{\partial^2 E_{22}}{\partial X_1{}^2} = \frac{\partial^3 u_2}{\partial X_1{}^2 \partial X_2}.$$

Now since the left-hand sides of the above equations are, by postulate, continuous, therefore the right-hand sides are continuous, and so the order of the differentiations is immaterial, so that

$$\frac{\partial^2 E_{11}}{\partial X_2{}^2} = \frac{\partial^2}{\partial X_1 \partial X_2}\left(\frac{\partial u_1}{\partial X_2}\right) \quad \text{and} \quad \frac{\partial^2 E_{22}}{\partial X_1{}^2} = \frac{\partial^2}{\partial X_1 \partial X_2}\left(\frac{\partial u_2}{\partial X_1}\right).$$

†A region of space is said to be simply-connected if every closed curve drawn in the region can be shrunk to a point, by continuous deformation, without passing out of the boundaries of the region. For example, the solid prismatical bar represented in Fig. 3.7a is simply-connected whereas, the prismatical tube represented in Fig. 3.7b is not simply-connected.

Thus

$$\frac{\partial^2 E_{11}}{\partial X_2^2}+\frac{\partial^2 E_{22}}{\partial X_1^2}=\frac{\partial^2}{\partial X_1 \partial X_2}\left(\frac{\partial u_1}{\partial X_2}+\frac{\partial u_2}{\partial X_1}\right)=2\frac{\partial^2}{\partial X_1 \partial X_2}E_{12}.$$

The other five conditions can be similarly established. We omit the proof that the conditions are also sufficient (under the conditions stated in the theorem). In Example 3.14 we shall give an instance where the conditions are not sufficient for a region which is not simply-connected.

EXAMPLE 3.12

Does the following strain field

$$[\mathbf{E}]=\begin{bmatrix} 2X_1 & X_1+2X_2 & 0 \\ X_1+2X_2 & 2X_1 & 0 \\ 0 & 0 & 2X_3 \end{bmatrix}$$

represent a compatible strain field?

Solution. Since each term of the compatibility equations involves second derivatives of the strain components with respect to the coordinates, the above strain tensor, with each component a linear function of X_1, X_2, and X_3 will obviously satisfy them. The given strain components are obviously continuous functions having continuous second derivatives (in fact continuous derivatives of all orders) in any bounded region. Thus the existence of single-valued continuous displacement field in any bounded simply-connected region is ensured by the theorem stated above. In fact it can be easily verified that

$$u_1 = X_1^2+X_2^2, \quad u_2 = 2X_1X_2+X_1^2, \quad u_3 = X_3^2,$$

(to which, of course, can be added any rigid body displacements) which is a single-valued continuous displacement field in any bounded region, including multiply-connected regions.

EXAMPLE 3.13

Will the strain components obtained from the displacements

$$u_1 = X_1^5, \quad u_2 = e^{X_1}, \quad u_3 = \sin X_2$$

be compatible?

Solution. Yes. There is no need to check, because the displacement $\mathbf{u}$ is given (and therefore exists!).

EXAMPLE 3.14

For the following strain field

$$E_{11} = -\frac{X_2}{X_1^2+X_2^2}, \quad E_{12} = \frac{\frac{1}{2}X_1}{X_1^2+X_2^2}, \quad E_{22}=E_{33}=E_{23}=E_{13}=0, \tag{i}$$

does there exist single-valued continuous displacement fields for the cylindrical body with the normal cross-section shown in Fig. 3.7a? for the body with the normal cross-section shown in Fig. 3.7b?

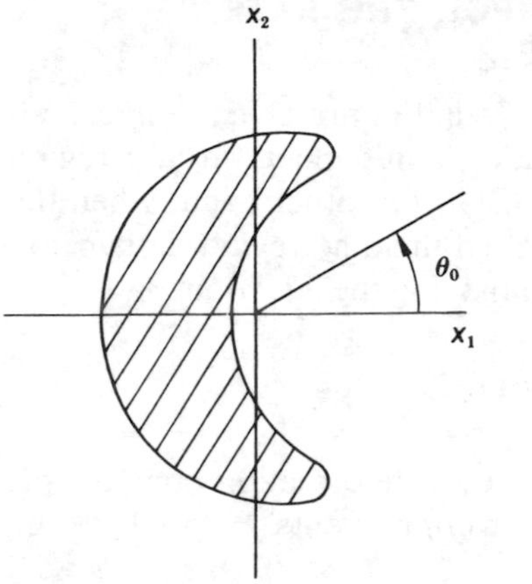

Fig. 3.7a

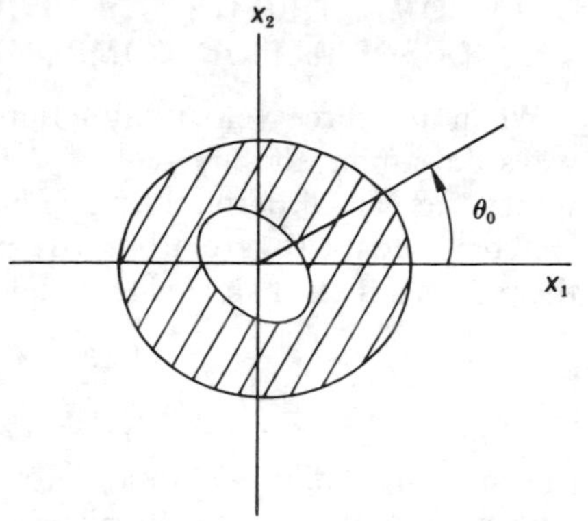

Fig. 3.7b

Solution. Out of the six compatible conditions only the first one needs to be checked, the others are automatically satisfied. Now

$$\frac{\partial E_{11}}{\partial X_2} = -\frac{(X_1^2+X_2^2)-X_2(2X_2)}{(X_1^2+X_2^2)^2} = \frac{X_2^2-X_1^2}{(X_1^2+X_2^2)^2},$$

$$2\frac{\partial E_{12}}{\partial X_1} = \frac{(X_1^2+X_2^2)-X_1(2X_1)}{(X_1^2+X_2^2)^2} = \frac{X_2^2-X_1^2}{(X_1^2+X_2^2)^2},$$

and

$$\frac{\partial^2 E_{22}}{\partial X_1^2} = 0.$$

Thus the equation

$$\frac{\partial^2 E_{11}}{\partial X_2^2} + \frac{\partial^2 E_{22}}{\partial X_1^2} = 2\frac{\partial^2 E_{12}}{\partial X_2 \partial X_1}$$

is satisfied, and the existence of solutions is assured. In fact it can be easily verified that for the given E_{ij},

$$u_1 = \arctan\frac{X_2}{X_1}, \quad u_2 = 0, \quad u_3 = 0, \tag{ii}$$

(to which, of course, any rigid body displacement field can be added). Now arctan X_2/X_1 is a multiple-valued function, having infinitely many values corresponding to a point (X_1, X_2, X_3). For example, for the point $(X_1, X_2, X_3) = (1, 0, 0)$, arctan $X_2/X_1 = 0, 2\pi, 4\pi$, etc. It can be made a single-valued function by the restriction $\theta_0 \leq \arctan X_2/X_1 < \theta_0 + 2\pi$ for any θ_0. For a simply-connected region as that shown in Fig. 3.7a, a θ_0 can be chosen so that such a restriction makes Eq. (ii) a single-valued *continuous* displacement for the region. But, for the body shown in Fig. 3.7b, the function $u_1 = \arctan X_1/X_2$, under the same restriction, is discontinuous along the line $\theta = \theta_0$ in the body (in fact u_1 jumps by the value of 2π in crossing the line). Thus for this so-called double-connected region, there does not exist single-valued continuous u_1 corresponding to the given E_{ij}, even though the compatibility equations are satisfied.

3.11 COMPATIBILITY CONDITIONS FOR THE RATE OF DEFORMATION COMPONENTS

When any three velocity functions v_1, v_2, and v_3 are given, one can always determine the six rate of deformation components in any region where the partial derivatives $\partial v_i/\partial v_j$ exist. On the other hand, when the six components D_{ij} are arbitrarily prescribed in some region, in general, there does not exist any velocity field v_i, satisfying the equations

$$\frac{1}{2}\left(\frac{\partial v_i}{\partial x_j}+\frac{\partial v_j}{\partial x_i}\right)=D_{ij}.$$

The compatibility conditions for the rate of deformation components are similar to those of the infinitesimal strain components [Eq. (3.36)], i.e.,

$$\frac{\partial^2 D_{11}}{\partial {x_2}^2}+\frac{\partial^2 D_{22}}{\partial {x_1}^2}=2\frac{\partial^2 D_{12}}{\partial x_1 \partial x_2}, \text{ etc.}$$

It should be emphasized that if one deals directly with differentiable velocity functions $v_i(x_1, x_2, x_3, t)$, (as is often the case in fluid mechanics), the question of compatibility does not arise.

PROBLEMS

3.1. Consider the motion

$$x_1 = kt + X_1,$$
$$x_2 = X_2,$$
$$x_3 = X_3,$$

where the material coordinates X_i designate the position of a particle at $t=0$.
(a) Determine the velocity and acceleration of a particle in both a material and spatial description.
(b) If in a spatial description there is a temperature field $\theta = Ax_1$, then find the material derivative $D\theta/Dt$.
(c) Do part (b) if instead the temperature is given by $\theta = Bx_2$.

3.2. Consider the motion

$$x_1 = X_1,$$
$$x_2 = k{X_1}^2 t^2 + X_2,$$
$$x_3 = X_3,$$

where X_i are the material coordinates.
(a) At $t=0$ the corners of a unit square are at $A(0,0,0)$, $B(0,1,0)$, $C(1,1,0)$,

and $D(1, 0, 0)$. Determine the positions of A, B, C, D at $t = 1$, and sketch the new shape of the square.
(b) Find the velocity $\mathbf{v}$ and the acceleration $D\mathbf{v}/Dt$ in a material description.
(c) Show that the spatial velocity field is given by

$$v_1 = v_3 = 0, \quad v_2 = 2kx_1{}^2 t.$$

3.3. Consider the motion

$$x_1 = kX_2{}^2 t^2 + X_1,$$
$$x_2 = kX_2 t + X_2,$$
$$x_3 = X_3.$$

(a) At $t = 0$ the corners of a unit square are at $A(0, 0, 0)$, $B(0, 1, 0)$, $C(1, 1, 0)$, and $D(1, 0, 0)$. Sketch the shape of the square at $t = 2$.
(b) Obtain the spatial description of the velocity field.
(c) Obtain the spatial description of the acceleration field.

3.4. Consider the motion

$$x_1 = (k + X_1)t + X_1,$$
$$x_2 = X_2,$$
$$x_3 = X_3.$$

(a) For this motion repeat part (a) of the previous problem.
(b) Find the velocity and acceleration as a function of time of a particle that is initially at the origin.
(c) Find the velocity and acceleration as a function of time of particles that are passing by the origin.

3.5. Consider the motion

$$x_1 = \frac{1+t}{1+t_0} X_1, \quad x_2 = X_2, \quad x_3 = X_3.$$

(a) Show that reference time is $t = t_0$.
(b) Find the velocity field in spatial coordinates.
(c) Show that the velocity field is identical to that of the motion

$$x_1 = (1+t)X_1, \quad x_2 = X_2, \quad x_3 = X_3.$$

3.6. (a) Show that the velocity field

$$v_i = \frac{x_i}{1+t}$$

corresponds to the motion

$$x_i = X_i(1+t).$$

(b) Find the acceleration of this motion in the material description.

3.7. In a spatial description the equation to evaluate the acceleration

$$\frac{D\mathbf{v}}{Dt} = \frac{\partial \mathbf{v}}{\partial t} + (\nabla \mathbf{v})\,\mathbf{v}$$

is nonlinear. That is, if we consider two velocity fields $\mathbf{v}^A$ and $\mathbf{v}^B$, then

$$\mathbf{a}^A + \mathbf{a}^B \neq \mathbf{a}^{A+B},$$

where $\mathbf{a}^A$ and $\mathbf{a}^B$ denote respectively the acceleration fields corresponding to the velocity fields $\mathbf{v}^A$ and $\mathbf{v}^B$ each existing alone, $\mathbf{a}^{A+B}$ denotes the acceleration field corresponding to the combined velocity field $\mathbf{v}^A + \mathbf{v}^B$. Verify this inequality for the velocity fields

$$\mathbf{v}^A = -2x_2\mathbf{e}_1 + 2x_1\mathbf{e}_2,$$

$$\mathbf{v}^B = 2x_2\mathbf{e}_1 - 2x_1\mathbf{e}_2.$$

3.8. Consider the motion

$$x_1 = X_1,$$

$$x_2 = X_2 + (\sin \pi t)(\sin \pi X_1),$$

$$x_3 = X_3.$$

(a) At $t = 0$ a material filament coincides with the straight line that extends from (0, 0, 0) to (1, 0, 0). Sketch the deformed shape of this filament at $t = \frac{1}{2}$, $t = 1$, and $t = \frac{3}{2}$.
(b) Find the velocity and acceleration in a material and spatial description.

3.9. Consider the following velocity and temperature fields:

$$\mathbf{v} = \frac{x_1\mathbf{e}_1 + x_2\mathbf{e}_2}{x_1^2 + x_2^2}, \quad \theta = k(x_1^2 + x_2^2).$$

(a) Determine the velocity at several positions and indicate the general nature of this velocity field. What do the isotherms look like?
(b) At the point $A(1, 1)$, determine the acceleration and the material derivative of the temperature field.

3.10. Do Problem 3.9 for the temperature and velocity fields

$$\mathbf{v} = \frac{-x_2\mathbf{e}_1 + x_1\mathbf{e}_2}{x_1^2 + x_2^2}, \quad \theta = k(x_1^2 + x_2^2).$$

3.11. Consider the motion $\mathbf{x} = \mathbf{X} + X_1 k\mathbf{e}_1$ and let $d\mathbf{X}^1 = (dS_1/\sqrt{2})(\mathbf{e}_1 + \mathbf{e}_2)$ and $d\mathbf{X}^2 = (dS_2/\sqrt{2})(-\mathbf{e}_1 + \mathbf{e}_2)$ be differential material elements in the undeformed configuration.
(a) Find the deformed elements $d\mathbf{x}^1$ and $d\mathbf{x}^2$.
(b) Evaluate both the stretch of these elements, ds_1/dS_1 and ds_2/dS_2, and the change in the angle between them.
(c) Do part (b) if $k = 1$ and $k = 10^{-2}$.
(d) Compare the results of part (c) to that predicted by the small strain tensor $\mathbf{E}$.

3.12. The motion of a continuum from initial position $\mathbf{X}$ to current position $\mathbf{x}$ is given by

$$\mathbf{x} = (\mathbf{I} + \mathbf{B})\mathbf{X},$$

where **I** is the identity tensor and **B** is a tensor whose components B_{ij} are constants and small compared to unity. If the components of **x** are x_i and those of **X** are X_j, find (a) the components of the displacement vector **u**, (b) the small strain tensor **E**.

3.13. At some time t, the position of a particle initially at (X_1, X_2, X_3) is defined by

$$x_1 = X_1 + kX_3,$$
$$x_2 = X_2 + kX_2,$$
$$x_3 = X_3,$$

where $k = 10^{-5}$.

(a) Find the components of the strain tensor.

(b) Find the unit elongation of an element initially in the direction of $\mathbf{e}_1 + \mathbf{e}_2$.

3.14. Consider the displacement field

$$u_1 = k(2X_1^2 + X_1X_2), \quad u_2 = kX_2^2, \quad u_3 = 0,$$

where $k = 10^{-4}$.

(a) Find the unit elongations and the change of angle for two material elements $d\mathbf{X}^1 = dX_1\mathbf{e}_1$ and $d\mathbf{X}^2 = dX_2\mathbf{e}_2$ that emanate from a particle designated by $\mathbf{X} = \mathbf{e}_1 + \mathbf{e}_2$.

(b) Find the deformed shape of these two elements.

3.15. For the displacement field of Example 3.9, consider instead, the unit cube diagonal in the direction $\mathbf{e}_1 + \mathbf{e}_2 + \mathbf{e}_3$ and determine the increase in length of this diagonal (a) by using the strain tensor and (b) by geometry.

3.16. With reference to an x_1-, x_2-, x_3-coordinate system, the state of strain at a point is given by the matrix

$$[\mathbf{E}] = \begin{bmatrix} 5 & 3 & 0 \\ 3 & 4 & -1 \\ 0 & -1 & 2 \end{bmatrix} \times 10^{-4}.$$

(a) What is the unit elongation in the direction $2\mathbf{e}_1 + 2\mathbf{e}_2 + \mathbf{e}_3$?

(b) What is the change of angle between two perpendicular lines (in the undeformed state) emanating from the point and in the direction of $2\mathbf{e}_1 + 2\mathbf{e}_2 + \mathbf{e}_3$ and $3\mathbf{e}_1 - 6\mathbf{e}_3$?

3.17. Do the previous problem for (a) the unit elongation in the direction $3\mathbf{e}_1 - 4\mathbf{e}_2$, (b) the change in angle between two elements in the direction $3\mathbf{e}_1 - 4\mathbf{e}_3$ and $4\mathbf{e}_1 + 3\mathbf{e}_3$.

3.18. (a) For Problem 3.16, determine the scalar invariants of the state of strain.

(b) Show that the following matrix

$$\begin{bmatrix} 3 & 0 & 0 \\ 0 & 6 & 0 \\ 0 & 0 & 2 \end{bmatrix} \times 10^{-4}$$

cannot represent the same state of strain as in Problem 3.16.

3.19. For the displacement field

$$u_1 = kX_1^2, \quad u_2 = kX_2X_3, \quad u_3 = k(2X_1X_3 + X_1^2), \quad k = 10^{-6},$$

find the maximum unit elongation for an element that is initially at (1, 0, 0).

3.20. Given the matrix of a strain field

$$[\mathbf{E}] = \begin{bmatrix} k_1X_2 & 0 & 0 \\ 0 & -k_2X_2 & 0 \\ 0 & 0 & -k_2X_2 \end{bmatrix},$$

(a) find the location of the particle that does not suffer any volume change.
(b) What should the relation between k_1 and k_2 be, such that no element changes volume?

3.21. For any motion the mass of a particle remains a constant. Consider the mass to be a product of its volume times its mass density and show that (a) for infinitesimal deformation $\rho(1+E_{kk}) = \rho_0$, where ρ_0 refers to the initial density and ρ to the current density.
(b) Use the smallness of E_{kk} to show that the current density is given by

$$\rho = \rho_0(1 - E_{kk}).$$

3.22. The unit elongations at a certain point on the surface of a body are measured experimentally by means of strain gages that are arranged 45° apart (called the 45° strain rosette) in the directions $\mathbf{e}_1$, $(\sqrt{2}/2)(\mathbf{e}_1 + \mathbf{e}_2)$, and $\mathbf{e}_2$. If these unit elongations are designated by a, b, c, respectively, what are the strain components E_{11}, E_{22}, E_{12}.

3.23. (a) Do Problem 3.22 if the measured strains are 200×10^{-6}, 50×10^{-6}, 100×10^{-6}, respectively.
(b) If $E_{33} = E_{32} = E_{31} = 0$, find the principal strains and directions of part (a).
(c) How will the results of part (b) be altered if $E_{33} \neq 0$?

3.24. Repeat Problem 3.23, except that $a = b = c = 1000 \times 10^{-6}$.

3.25. The unit elongations at a certain point on the surface of a body are measured experimentally by means of strain gages that are arranged 60° apart (called the 60° strain rosette) in the directions $\mathbf{e}_1$, $\frac{1}{2}(\mathbf{e}_1 + \sqrt{3}\mathbf{e}_2)$, and $\frac{1}{2}(-\mathbf{e}_1 + \sqrt{3}\mathbf{e}_2)$. If these elongations are designated by a, b, c, respectively, what are the strain components E_{11}, E_{22}, E_{12}?

3.26. Do Problem 3.25 if the strain rosette measurements give $a = 2 \times 10^{-4}$, $b = 1 \times 10^{-4}$, $c = 1.5 \times 10^{-4}$.

3.27. Do Problem 3.26 except that $a = b = c = 2000 \times 10^{-6}$.

3.28. For the velocity field, $\mathbf{v} = (kx_2^2)\mathbf{e}_1$, (a) find the rate of deformation and spin tensors.
(b) Find the rate of extension of a material element $d\mathbf{x} = (ds)\mathbf{n}$, where

$$\mathbf{n} = (\sqrt{2}/2)(\mathbf{e}_1 + \mathbf{e}_2) \text{ at the position } \mathbf{x} = 5\mathbf{e}_1 + 3\mathbf{e}_2.$$

3.29. For the velocity field

$$\mathbf{v} = \left(\frac{t+k}{1+x_1}\right)\mathbf{e}_1,$$

find the rate of extension of material elements $d\mathbf{x}^1 = (ds_1)\mathbf{e}_1$ and $d\mathbf{x}^2 = (ds_2/\sqrt{2})(\mathbf{e}_1 + \mathbf{e}_2)$ as they pass the origin at $t = 1$.

3.30. (a) Find the rate of deformation and spin tensors for the velocity field $\mathbf{v} = (\cos t)(\sin \pi x_1)\mathbf{e}_2$.

(b) For the velocity field of part (a), find the rate of extension of elements $d\mathbf{x}^1 = (ds_1)\mathbf{e}_1$, $d\mathbf{x}^2 = (ds_2)\mathbf{e}_2$, $d\mathbf{x}^3 = (ds_3/\sqrt{2})(\mathbf{e}_1 + \mathbf{e}_2)$ at the origin at $t = 0$.

3.31. For the velocity field of Problem 3.9

(a) Find the rate of deformation and spin tensors.

(b) Find the rate of extension of a radial material line element.

3.32. A motion is said to be irrotational if the spin tensor vanishes. Show that the velocity field of Problem 3.10 describes an irrotational motion.

3.33. (a) Let $d\mathbf{x}^1 = (ds_1)\mathbf{n}$ and $d\mathbf{x}^2 = (ds_2)\mathbf{m}$ be two material elements that emanate from a particle P which at present has a rate of deformation $\mathbf{D}$. Consider $(D/Dt)(d\mathbf{x}^1 \cdot d\mathbf{x}^2)$ and show that

$$\left(\frac{1}{ds_1}\frac{D(ds_1)}{Dt} + \frac{1}{ds_2}\frac{D(ds_2)}{Dt}\right)\cos\theta - (\sin\theta)\frac{D\theta}{Dt} = 2\mathbf{m}\cdot\mathbf{D}\mathbf{n},$$

where θ is the angle between $\mathbf{m}$ and $\mathbf{n}$.

(b) Consider the special cases (i) $d\mathbf{x}^1 = d\mathbf{x}^2$, and (ii) $\theta = \pi/2$ and show how this expression reduces to the results of Section 3.8.

3.34. Let $\mathbf{e}_1, \mathbf{e}_2, \mathbf{e}_3$ and D_1, D_2, D_3 be the principal directions and values of a rate of deformation tensor $\mathbf{D}$. Further let $d\mathbf{x}^1 = (ds_1)\mathbf{e}_1$, $d\mathbf{x}^2 = (ds_2)\mathbf{e}_2$, $d\mathbf{x}^3 = (ds_3)\mathbf{e}_3$ be three material line elements. Consider the material derivative $(D/Dt)[d\mathbf{x}^1 \cdot (d\mathbf{x}^2 \times d\mathbf{x}^3)]$ and show that

$$\Delta = \frac{1}{dV}\frac{D(dV)}{Dt} = D_1 + D_2 + D_3,$$

where the infinitesimal volume $dV = (ds_1)(ds_2)(ds_3)$.

3.35. Consider a material element $d\mathbf{x} = ds\,\mathbf{n}$.

(a) Show that $(D/Dt)(\mathbf{n}) = \mathbf{D}\mathbf{n} + \mathbf{W}\mathbf{n} - (\mathbf{n}\cdot\mathbf{D}\mathbf{n})\mathbf{n}$.

(b) Show that if $\mathbf{n}$ is an eigenvector of $\mathbf{D}$ that $(D/Dt)(\mathbf{n}) = \mathbf{W}\mathbf{n} = \boldsymbol{\omega}\times\mathbf{n}$, where $\boldsymbol{\omega}$ is the dual vector of $\mathbf{W}$.

3.36. Given the following velocity field for an incompressible fluid:

$$v_1 = k(x_2 - 2)^2 x_3,$$

$$v_2 = -x_1 x_2,$$

$$v_3 = k x_1 x_3,$$

determine k such that the equation of mass conservation is satisfied.

3.37. Is the fluid motion described in (a) Problem 3.9 and (b) Problem 3.10 incompressible?

3.38. In a spatial description the density of an incompressible fluid is given by $\rho = kx_2$. Find the permissible form for the velocity field, with $v_3 = 0$, in order that the conservation of mass equation be satisfied.

3.39. Consider the velocity field

$$\mathbf{v} = \left(\frac{x_1}{1+t}\right)\mathbf{e}_1.$$

(a) Find the density if it is independent of spatial position, i.e., $\rho = \rho(t)$.
(b) Find the density if it is a function of x_1 alone.

3.40. Given the velocity field

$$\mathbf{v} = x_1 t\mathbf{e}_1 + x_2 t\mathbf{e}_2,$$

determine how the fluid density varies with time, if in a spatial description it is a function of time only.

3.41. Check whether or not the following distribution of the state of strain satisfies the compatibility conditions:

$$[\mathbf{E}] = \begin{bmatrix} X_1+X_2 & X_1 & X_2 \\ X_1 & X_2+X_3 & X_3 \\ X_2 & X_3 & X_1+X_3 \end{bmatrix}.$$

3.42. Check whether or not the following distribution of the state of strain satisfies the compatibility conditions:

$$[\mathbf{E}] = \begin{bmatrix} X_1^2 & X_2^2+X_3^2 & X_1X_3 \\ X_2^2+X_3^2 & 0 & X_1 \\ X_1X_3 & X_1 & X_2^2 \end{bmatrix}.$$

3.43. Does the displacement field

$$u_1 = \sin X_1, \quad u_2 = X_1^3X_2, \quad u_3 = \cos X_3$$

correspond to a compatible strain field?

3.44. Given the strain field

$$E_{12} = E_{21} = X_1X_2,$$

and all other $E_{ij} = 0$,
(a) does it satisfy the equations of compatibility?
(b) By attempting to integrate the strain field, show that it cannot correspond to a displacement field.

3.45. The strain components are given by

$$E_{11} = \frac{1}{\alpha}f(X_2, X_3), \quad E_{22} = E_{33} = -\frac{\nu}{\alpha}f(X_2, X_3),$$

$$E_{12} = E_{13} = E_{23} = 0.$$

Show that in order to be compatible $f(X_2, X_3)$ must be linear.

3.46. Using the result of Problem B50, obtain the acceleration components in cylindrical coordinates as

$$a_r = \frac{\partial v_r}{\partial t} + v_r \frac{\partial v_r}{\partial r} + \frac{v_\phi}{r}\frac{\partial v_r}{\partial \phi} + v_z \frac{\partial v_r}{\partial z} - \frac{v_\phi^2}{r},$$

$$a_\phi = \frac{\partial v_\phi}{\partial t} + v_r \frac{\partial v_\phi}{\partial r} + \frac{v_\phi}{r}\frac{\partial v_\phi}{\partial \phi} + v_z \frac{\partial v_\phi}{\partial z} + \frac{v_r v_\phi}{r},$$

$$a_z = \frac{\partial v_z}{\partial t} + v_r \frac{\partial v_z}{\partial r} + \frac{v_\phi}{r}\frac{\partial v_z}{\partial \phi} + v_z \frac{\partial v_z}{\partial z}.$$

3.47. Using the result of Problem B50, obtain the components of **E** (strain tensor) in cylindrical coordinates as, with u_r, u_ϕ, and u_z denoting displacement components

$$E_{rr} = \frac{\partial u_r}{\partial r}, \quad E_{\phi\phi} = \frac{1}{r}\frac{\partial u_\phi}{\partial \phi} + \frac{u_r}{r}, \quad E_{zz} = \frac{\partial u_z}{\partial z},$$

$$E_{r\phi} = \frac{1}{2}\left(\frac{1}{r}\frac{\partial u_r}{\partial \phi} - \frac{u_\phi}{r} + \frac{\partial u_\phi}{\partial r}\right), \quad E_{rz} = \frac{1}{2}\left(\frac{\partial u_r}{\partial z} + \frac{\partial u_z}{\partial r}\right),$$

and

$$E_{\phi z} = \frac{1}{2}\left(\frac{\partial u_\phi}{\partial z} + \frac{1}{r}\frac{\partial u_z}{\partial \phi}\right).$$

3.48. Using the result of Problem B50, obtain the components of **D** (rate of deformation tensor) in cylindrical coordinates as, with v_r, v_ϕ, v_z, denoting velocity components

$$D_{rr} = \frac{\partial v_r}{\partial r}, \quad D_{\phi\phi} = \frac{1}{r}\frac{\partial v_\phi}{\partial \phi} + \frac{v_r}{r}, \quad D_{zz} = \frac{\partial v_z}{\partial z},$$

$$D_{r\phi} = \frac{1}{2}\left(\frac{1}{r}\frac{\partial v_r}{\partial \phi} - \frac{v_\phi}{r} + \frac{\partial v_\phi}{\partial r}\right), \quad D_{rz} = \frac{1}{2}\left(\frac{\partial v_r}{\partial z} + \frac{\partial v_z}{\partial r}\right)$$

$$D_{\phi z} = \frac{1}{2}\left(\frac{\partial v_\phi}{\partial z} + \frac{1}{r}\frac{\partial v_z}{\partial \phi}\right).$$

3.49. Using the result of Problem B50, obtain the components of the spin tensor **W** in cylindrical coordinates as

$$W_{rr} = W_{\phi\phi} = W_{zz} = 0,$$

$$W_{r\phi} = \frac{1}{2}\left(\frac{1}{r}\frac{\partial v_r}{\partial \phi} - \frac{v_\phi}{r} - \frac{\partial v_\phi}{\partial r}\right) = -W_{\phi r}, \quad W_{rz} = \frac{1}{2}\left(\frac{\partial v_r}{\partial z} - \frac{\partial v_z}{\partial r}\right) = -W_{zr},$$

$$W_{\phi z} = \frac{1}{2}\left(\frac{\partial v_\phi}{\partial z} - \frac{1}{r}\frac{\partial v_z}{\partial \phi}\right) = -W_{z\phi}.$$

3.50. In cylindrical coordinates (r, ϕ, z), consider a differential volume bounded by the three pairs of faces $r = r_0$, $r = r_0 + dr$; $\phi = \phi_0$, $\phi = \phi_0 + d\phi$; and $z = z_0$,

$z = z_0 + dz$. The rate at which mass is flowing into the volume across the face $r = r_0$ is given by $(\rho v_r)(r_0 d\phi)(dz)$ and similar expressions for other faces. By demanding that the net rate of inflow of mass be equal to the rate of increase of mass inside the volume, obtain the equation of conservation of mass in cylindrical coordinates as

$$\frac{\partial \rho}{\partial t} + \frac{1}{r}\frac{\partial}{\partial r}(r\rho v_r) + \frac{1}{r}\frac{\partial}{\partial \phi}(\rho v_\phi) + \frac{\partial}{\partial z}(\rho v_z) = 0.$$

3.51. In invariant form the conservation of mass equation takes the form

$$\frac{\partial \rho}{\partial t} + (\nabla \rho)\cdot \mathbf{v} + \rho(\operatorname{tr}\mathbf{D}) = 0,$$

where $\operatorname{tr}\mathbf{D} = D_{ii}$.

Using the results of Problems B47 and 3.48, rederive the cylindrical coordinate equation of Problem 3.50.

4

Stress

In the previous chapter, we considered the purely kinematical description of the motion of a continuum without any consideration of the forces that cause the motion and deformation. In this chapter, we shall consider a means of describing the forces in the interior of a body. It is generally accepted that matter is formed of molecules which in turn consist of atoms and subatomic particles. Therefore, the internal forces in real matter are those between the above particles. In the classical continuum theory the internal forces are introduced through the concept of body forces and surface forces. Body forces are those that act throughout a volume (e.g., gravity, electrostatic, etc.) by a long-range interaction with matter or charge at a distance. Surface forces are those that act on a surface (real or imagined) separating parts of the body. We shall assume that it is adequate to describe the surface force at a point of a surface through the definition of a **stress vector**, discussed in Section 4.1, which pays no attention to the curvature of the surface at the point. Such an assumption is known as **Cauchy's stress principle** which is one of the basic axioms of classical continuum mechanics.

4.1 STRESS VECTOR

Let us consider a body depicted in Fig. 4.1. Imagine a plane such as S, which passes through an arbitrary internal point P and which has a unit normal vector $\mathbf{n}$. The plane cuts the body into two portions. One portion lies on the side of the arrow of $\mathbf{n}$ (designated by II in the figure) and the other portion on the tail of $\mathbf{n}$ (designated by I). Considering portion I as a

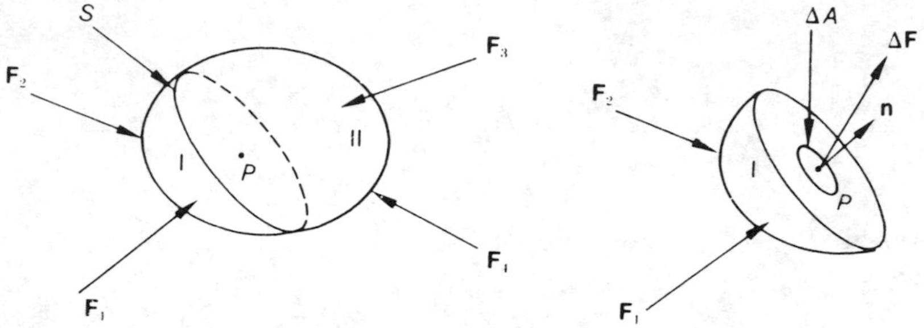

Fig. 4.1

free body, there will be on plane S a resultant force $\Delta\mathbf{F}$ acting on a small area ΔA containing P. We define the stress vector (from II to I) at the point P on the plane S as the limit of the ratio $\Delta\mathbf{F}/\Delta A$ as $\Delta A \to 0$. That is, with $\mathbf{t}_\mathbf{n}$ denoting the stress vector,

$$\mathbf{t}_\mathbf{n} = \lim_{\Delta A \to 0} \frac{\Delta \mathbf{F}}{\Delta A}. \tag{4.1}$$

If portion II is considered as a free body, then by Newton's law of action and reaction, we shall have a stress vector (from I to II), $\mathbf{t}_{-\mathbf{n}}$, at the same point on the same plane equal and opposite to that given by Eq. (4.1). That is,

$$\mathbf{t}_\mathbf{n} = -\mathbf{t}_{-\mathbf{n}}. \tag{4.2}$$

Next, let S be a surface (instead of a plane) passing the point P. Let $\Delta\mathbf{F}$ be the resultant force on a small area ΔS on the surface S. The **stress vector** at P on S is defined as

$$\mathbf{t} = \lim_{\Delta S \to 0} \frac{\Delta \mathbf{F}}{\Delta S}.$$

We now state the following principle, known as **Cauchy's stress principle**: The stress vector at any given place and time has a common value on all parts of material having a common tangent plane at P and lying on the same side of it. In other words, if $\mathbf{n}$ is the unit outward† normal to the tangent plane, then

$$\mathbf{t} = \mathbf{t}(\mathbf{x}, t, \mathbf{n}),$$

where the scalar t denotes time.

In the following section, we shall show that this dependence on $\mathbf{n}$ is, in fact, a linear one, i.e.,

$$\mathbf{t}(\mathbf{x}, t, \mathbf{n}) = \mathbf{T}(\mathbf{x}, t)\,\mathbf{n},$$

where T is a linear transformation

†That is, outward away from the material.

4.2 STRESS TENSOR

Let $\mathbf{T}$ be a transformation such that, if $\mathbf{n}$ is a unit normal vector to a plane, then the stress vector on the plane (from the material lying on the arrow side of $\mathbf{n}$) is given by

$$\mathbf{t_n} = \mathbf{Tn}. \tag{4.3}†$$

Through the use of Newton's second law, it will be shown in the following that $\mathbf{T}$ is in fact a linear transformation, i.e., a second-order tensor.

Let a small tetrahedron be isolated from the body with the point $\mathbf{P}$ as one of its vertices (*see* Fig. 4.2). The size of the tetrahedron will ultimately be made to approach zero so that, in the limit, the inclined plane will pass through the point P. From Eqs. (4.2) and (4.3), the stress vector on the face PAB, whose outward normal is in the direction of $-\mathbf{e}_1$, is given by

$$\mathbf{t}_{-\mathbf{e}_1} = -\mathbf{t}_{\mathbf{e}_1} = -\mathbf{Te}_1. \tag{i}$$

Similarly, the stress vectors on the faces PBC and PAC are

$$\mathbf{t}_{-\mathbf{e}_2} = -\mathbf{Te}_2 \tag{ii}$$

and

$$\mathbf{t}_{-\mathbf{e}_3} = -\mathbf{Te}_3 \tag{iii}$$

respectively. Thus, with ΔA_1, ΔA_2, ΔA_3, and ΔA_n denoting the areas of PAB, PBC, PAC, and ABC respectively, we have, from Newton's second law,

$$\Sigma\mathbf{F} = \mathbf{t}_{-\mathbf{e}_1}(\Delta A_1) + \mathbf{t}_{-\mathbf{e}_2}(\Delta A_2) + \mathbf{t}_{-\mathbf{e}_3}(\Delta A_3) + \mathbf{t_n}(\Delta A_n) = m\mathbf{a}. \tag{iv}$$

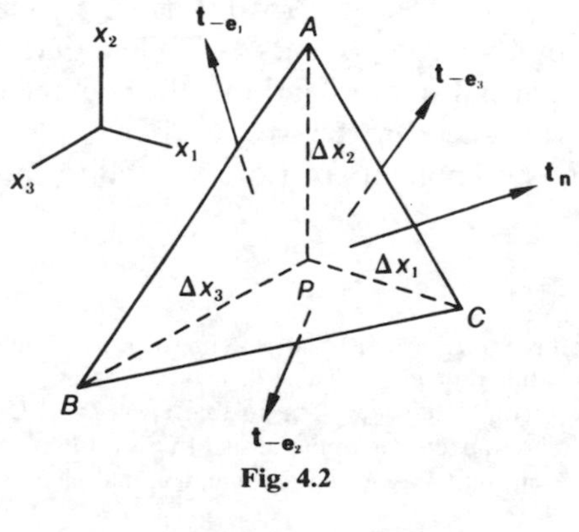

Fig. 4.2

†This equation is sometimes known as **Cauchy's fundamental lemma.**

Since the mass m = (density) (volume), and the volume of the tetrahedron is proportional to the product of three infinitesimal lengths, (in fact, $\frac{1}{6}\Delta x_1 \Delta x_2 \Delta x_3$), when the size of the tetrahedron approaches zero, the right-hand side of Eq. (iv) will approach zero faster than the terms on the left. Thus, in the limit, the acceleration term drops out exactly from Eq. (iv).†

Let the unit normal vector of the inclined plane ABC be

$$\mathbf{n} = n_1\mathbf{e}_1 + n_2\mathbf{e}_2 + n_3\mathbf{e}_3. \tag{v}$$

The areas ΔA_1, ΔA_2, and ΔA_3, being the projections of ΔA_n are related to ΔA_n by the relations

$$\Delta A_1 = n_1 \Delta A_n, \quad \Delta A_2 = n_2 \Delta A_n, \quad \Delta A_3 = n_3 \Delta A_n. \tag{vi}$$

Using Eqs. (i), (ii), (iii), (v), (vi), and (4.3), Eq. (iv) becomes

$$\mathbf{T}(n_1\mathbf{e}_1 + n_2\mathbf{e}_2 + n_3\mathbf{e}_3) = n_1(\mathbf{T}\mathbf{e}_1) + n_2(\mathbf{T}\mathbf{e}_2) + n_3(\mathbf{T}\mathbf{e}_3).$$

Thus, $\mathbf{T}$ is a linear transformation. It is called the **stress tensor**.

4.3 COMPONENTS OF STRESS TENSOR

From

$$\mathbf{t}_{\mathbf{e}_1} = \mathbf{T}\mathbf{e}_1 = T_{11}\mathbf{e}_1 + T_{21}\mathbf{e}_2 + T_{31}\mathbf{e}_3,$$

we see that T_{11} is the normal component or normal stress and T_{21} and T_{31} are the tangential components or shearing stress of the stress vector $\mathbf{t}_{\mathbf{e}1}$ on the plane whose normal is $\mathbf{e}_1$.‡ On this plane the magnitude of tangential or shearing stresses is $(T_{21}^2 + T_{31}^2)^{1/2}$. Similar interpretations can be given for the other components of $\mathbf{T}$. The diagonal elements T_{11}, T_{22}, and T_{33} are the **normal stresses** and the off-diagonal elements T_{21}, T_{31}, T_{12}, T_{32}, T_{13}, and T_{23} are **shearing stresses.**

From¶ $\mathbf{t} = \mathbf{T}\mathbf{n}$, the components of $\mathbf{t}$ are related to those of $\mathbf{T}$ and $\mathbf{n}$ by the equation

$$t_i = T_{ij} n_j \tag{4.4a}$$

†Note that any body force (e.g., weight) that is acting will be of the same order of magnitude as that of the acceleration term and will also drop out.

‡Some authors use the convention $\mathbf{t}_\mathbf{n} = \mathbf{T}^T\mathbf{n}$ so that $\mathbf{t}_{\mathbf{e}_i} = T_{ij}\mathbf{e}_j$. Under that convention, for example, T_{21} and T_{23} are tangential components of $\mathbf{t}_{\mathbf{e}_2}$ on the plane whose normal is $\mathbf{e}_2$, etc. These differences in meaning regarding the nondiagonal elements of $\mathbf{T}$ disappear for symmetric $\mathbf{T}$.

¶Whenever there is no confusion, we drop the subscript $\mathbf{n}$ for $\mathbf{t}_\mathbf{n}$.

or in a form more convenient for computations,

$$[\mathbf{t}] = [\mathbf{T}]\,[\mathbf{n}]. \tag{4.4b}$$

Thus, it is clear that if a matrix of **T** is known, the stress vector **t** on any inclined plane is uniquely determined from the above equation. In other words, the state of stress at a point is completely characterized by the stress tensor **T**. Also, since **T** is a second-order tensor, any one matrix of **T** determines the other matrices of **T** from Eq. (B29) of Chapter 2.

4.4 SYMMETRY OF STRESS TENSOR – PRINCIPLE OF MOMENT OF MOMENTUM

By the use of moment of momentum equation for a differential element, we shall now show that the stress tensor is generally a symmetric tensor.†

Consider the free-body diagram of a differential parallelepiped isolated from a body as shown in Fig. 4.3. Let us find the moment of all the forces about an axis passing through the center point A and parallel to the x_3-axis:

$$\Sigma M_A = T_{21}(\Delta x_2 \Delta x_3)\left(\frac{\Delta x_1}{2}\right) + (T_{21} + \Delta T_{21})\,(\Delta x_2 \Delta x_3)\left(\frac{\Delta x_1}{2}\right)$$

$$- T_{12}(\Delta x_1 \Delta x_3)\left(\frac{\Delta x_2}{2}\right) - (T_{12} + \Delta T_{12})\,(\Delta x_1 \Delta x_3)\left(\frac{\Delta x_2}{2}\right).$$

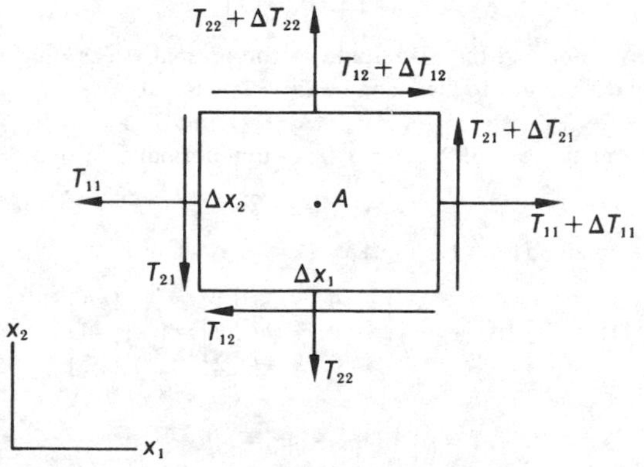

Fig. 4.3

†*See* Problem 4.20 for a case where the stress tensor is not symmetric.

Dropping out the terms containing small quantities of higher order, we obtain

$$\Sigma M_A = (T_{21} - T_{12})\,\Delta x_1 \Delta x_2 \Delta x_3.$$

Now, whether the element is in static equilibrium or not, ΣM_A is equal to zero because the angular acceleration term is proportional to the product of the volume and the square of a length† and is therefore a small quantity of higher order than the right side of the above equation. Thus $T_{12} = T_{21}$. Similar proofs give $T_{13} = T_{31}$ and $T_{23} = T_{32}$. That is,

$$T_{12} = T_{21}, \quad T_{13} = T_{31}, \quad T_{23} = T_{32}, \tag{4.5}$$

which means **T** is symmetric.

EXAMPLE 4.1

The state of stress at a certain point is $\mathbf{T} = -p\mathbf{I}$, where p is a scalar. Show that there is no shearing stress on any plane containing this point.

Solution. The stress vector on any plane passing through the point with normal $\mathbf{n}$ is

$$\mathbf{t} = \mathbf{Tn} = -p\mathbf{In} = -p\mathbf{n}.$$

Therefore, it is normal to the plane. This simple stress state is called a state of hydrostatic pressure.

EXAMPLE 4.2

With reference to an x-, y-, z-coordinate system, the matrix of a state of stress at a certain point of a body is given by:

$$[\mathbf{T}] = \begin{bmatrix} 2 & 4 & 3 \\ 4 & 0 & 0 \\ 3 & 0 & -1 \end{bmatrix} \text{MPa}$$

(a) Find the stress vector and the magnitude of the normal stress on a plane that passes through the point and is parallel to the plane $x + 2y + 2z - 6 = 0$.
(b) If $\mathbf{e}_1' = (\frac{1}{3})(2\mathbf{e}_1 + 2\mathbf{e}_2 + \mathbf{e}_3)$ and $\mathbf{e}_2' = (1/\sqrt{2})(\mathbf{e}_1 - \mathbf{e}_2)$, find T_{12}'.

Solution. (a) The plane $x + 2y + 2z - 6 = 0$ has a unit normal $\mathbf{n}$ given by

$$\mathbf{n} = \tfrac{1}{3}(\mathbf{e}_1 + 2\mathbf{e}_2 + 2\mathbf{e}_3).$$

The stress vector is obtained from Eq. (4.4) as

$$[\mathbf{t}] = [\mathbf{T}][\mathbf{n}] = \left(\frac{1}{3}\right)\begin{bmatrix} 2 & 4 & 3 \\ 4 & 0 & 0 \\ 3 & 0 & -1 \end{bmatrix}\begin{bmatrix} 1 \\ 2 \\ 2 \end{bmatrix} = \left(\frac{1}{3}\right)\begin{bmatrix} 16 \\ 4 \\ 1 \end{bmatrix}$$

or

$$\mathbf{t} = \left(\frac{1}{3}\right)(16\mathbf{e}_1 + 4\mathbf{e}_2 + \mathbf{e}_3).$$

†The angular acceleration term is $(I_A)_{33}\alpha_3$, where $(I_A)_{33} =$ (density)$(\Delta x_1 \Delta x_2 \Delta x_3)\cdot[(\Delta x_1)^2 + (\Delta x_2)^2]$, α_3 is the x_3-component of angular acceleration.

The magnitude of the normal stress is simply, with $T_n \equiv T_{(n)(n)}$,

$$T_n = \mathbf{t} \cdot \mathbf{n} = \left(\frac{1}{9}\right)(16+8+2) = 2.89 \text{ MPa}.$$

(b) To find the primed components of the stress we have,

$$T'_{12} = \mathbf{e}'_1 \cdot \mathbf{T}\mathbf{e}'_2 = \frac{1}{3\sqrt{2}}[2, 2, 1]\begin{bmatrix} 2 & 4 & 3 \\ 4 & 0 & 0 \\ 3 & 0 & -1 \end{bmatrix}\begin{bmatrix} 1 \\ -1 \\ 0 \end{bmatrix}.$$

Therefore,

$$T'_{12} = \frac{700}{3\sqrt{2}} = 1.65 \text{ MPa}.$$

EXAMPLE 4.3

The distribution of stress inside a body is given by the matrix

$$[\mathbf{T}] = \begin{bmatrix} -p+\rho g y & 0 & 0 \\ 0 & -p+\rho g y & 0 \\ 0 & 0 & -p+\rho g y \end{bmatrix},$$

where p, ρ, and g are constants. Figure 4.4 shows a rectangular block inside the body.

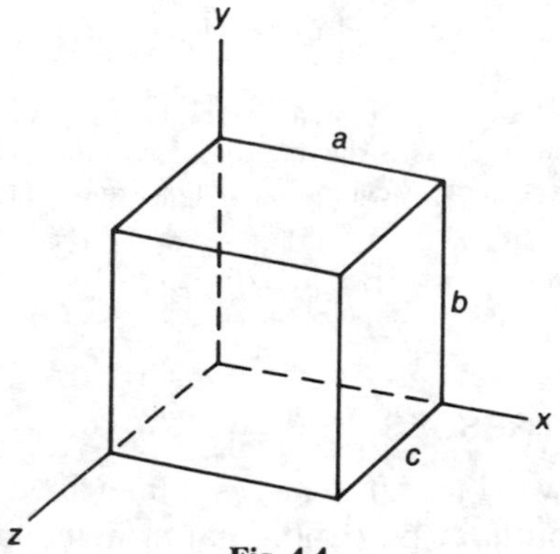

Fig. 4.4

(a) What is the distribution of the stress vector on the six faces of the block?
(b) Find the total resultant force acting on the faces $y = 0$ and $x = 0$.

Solution. (a) We have, from $\mathbf{t} = \mathbf{Tn}$,

$$\begin{aligned}
&\text{on } x = 0, \quad [\mathbf{n}] = [-1, \ 0, \ 0], \quad [\mathbf{t}] = [p - \rho g y, 0, 0]; \\
&\text{on } x = a, \quad [\mathbf{n}] = [+1, \ 0, \ 0], \quad [\mathbf{t}] = [-p + \rho g y, 0, 0]; \\
&\text{on } y = 0, \quad [\mathbf{n}] = [\ 0, -1, \ 0], \quad [\mathbf{t}] = [0, p, 0]; \\
&\text{on } y = b, \quad [\mathbf{n}] = [\ 0, +1, \ 0], \quad [\mathbf{t}] = [0, -p + \rho g b, 0]; \\
&\text{on } z = 0, \quad [\mathbf{n}] = [\ 0, \ 0, -1], \quad [\mathbf{t}] = [0, 0, p - \rho g y]; \\
&\text{on } z = c, \quad [\mathbf{n}] = [\ 0, \ 0, +1], \quad [\mathbf{t}] = [0, 0, -p + \rho g y].
\end{aligned}$$

A section of the distribution of the stress vector is shown in Fig. 4.5.

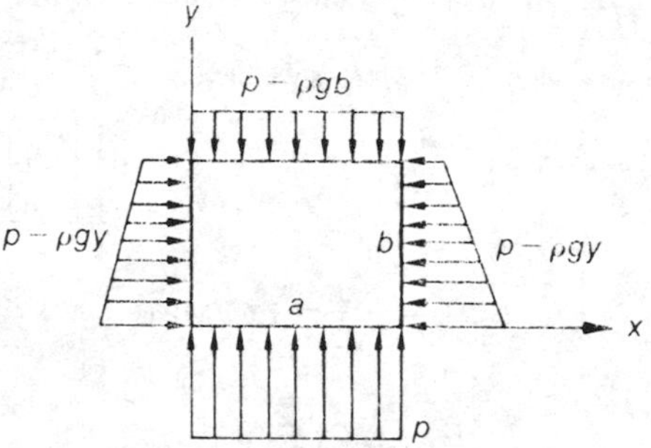

Fig. 4.5

(b) On the face $y = 0$ the total force is

$$\mathbf{F}_1 = \int \mathbf{t}\,dA = \left[p \int dA\right]\mathbf{e}_2 = p(ac)\mathbf{e}_2.$$

On the face $x = 0$ the total force is

$$\mathbf{F}_2 = \left[\int (p - \rho g y)\,dA\right]\mathbf{e}_1 = \left[p \int dA - \rho g \int y\,dA\right]\mathbf{e}_1.$$

The second integral can be evaluated directly by replacing (dA) by $(c\,dy)$ and integrating from $y = 0$ to $y = b$. Or, since $\int y\,dA$ is the first moment of the face area about the z-axis, it is therefore equal to the product of the centroidal distance and the total area. Thus,

$$\mathbf{F}_2 = \left[p(bc) - \frac{\rho g b^2 c}{2}\right]\mathbf{e}_1.$$

4.5 PRINCIPAL STRESSES

From Chapter 2, we know for any symmetric stress tensor **T**, there exists at least three mutually perpendicular principal directions (the eigenvectors of **T**). The planes having these directions as their normals are known as the **principal planes**. On these planes, the stress vector is normal to the plane (i.e., no shearing stresses) and the normal stresses are known as the **principal stresses**. Thus, the principal stresses (eigenvalues of **T**) include the maximum and the minimum values of normal stresses among all planes passing through a given point.

The principal stresses are to be obtained from the characteristic equation of **T**, which may be written

$$\lambda^3 - I_1\lambda^2 + I_2\lambda - I_3 = 0,$$

where

$$I_1 = T_{11} + T_{22} + T_{33},$$

$$I_2 = \begin{vmatrix} T_{11} & T_{12} \\ T_{21} & T_{22} \end{vmatrix} + \begin{vmatrix} T_{11} & T_{13} \\ T_{31} & T_{33} \end{vmatrix} + \begin{vmatrix} T_{22} & T_{23} \\ T_{32} & T_{33} \end{vmatrix}, \tag{4.6}$$

and

$$I_3 = \det [T_{ij}]$$

are the three scalar invariants of the stress tensor. For the computations of the principal directions, we refer the reader to Chapter 2.

4.6 MAXIMUM SHEARING STRESS

In this section, we show that the maximum shearing stress is equal to one-half the difference between the maximum and the minimum principal stresses and acts on the plane that bisects the angle between the directions of the maximum and minimum principal stresses.

Let $\mathbf{e}_1$, $\mathbf{e}_2$, and $\mathbf{e}_3$ be the principal directions of $\mathbf{T}$ and let T_1, T_2, T_3 be the principal stresses. If $\mathbf{n} = n_1\mathbf{e}_1 + n_2\mathbf{e}_2 + n_3\mathbf{e}_3$ is the unit normal to a plane, the components of the stress vector on the plane is given by

$$\begin{bmatrix} t_1 \\ t_2 \\ t_3 \end{bmatrix} = \begin{bmatrix} T_1 & 0 & 0 \\ 0 & T_2 & 0 \\ 0 & 0 & T_3 \end{bmatrix} \begin{bmatrix} n_1 \\ n_2 \\ n_3 \end{bmatrix} = \begin{bmatrix} n_1T_1 \\ n_2T_2 \\ n_3T_3 \end{bmatrix},$$

i.e.,

$$\mathbf{t} = n_1T_1\mathbf{e}_1 + n_2T_2\mathbf{e}_2 + n_3T_3\mathbf{e}_3. \tag{4.7}$$

and the normal stress on the same plane is given by

$$T_n = \mathbf{n} \cdot \mathbf{t} = n_1^2T_1 + n_2^2T_2 + n_3^2T_3. \tag{4.8}$$

Thus, if T_s denotes the magnitude of the total shearing stress on the plane, we have (*see* Fig. 4.6)

$$T_s^2 = |\mathbf{t}|^2 - T_n^2, \tag{4.9}$$

i.e.,

$$T_s^2 = T_1^2n_1^2 + T_2^2n_2^2 + T_3^2n_3^2 - (T_1n_1^2 + T_2n_2^2 + T_3n_3^2)^2. \tag{4.10}$$

To obtain the triple (n_1, n_2, n_3) for which T_s^2 is a maximum or minimum, we have to obtain all triples (n_1, n_2, n_3) for which T_s^2 is stationary in the sense that the total differential $d(T_s^2)$ is zero. That is,

$$d(T_s^2) = \frac{\partial(T_s^2)}{\partial n_1} dn_1 + \frac{\partial(T_s^2)}{\partial n_2} dn_2 + \frac{\partial(T_s^2)}{\partial n_3} dn_3 = 0. \tag{i}$$

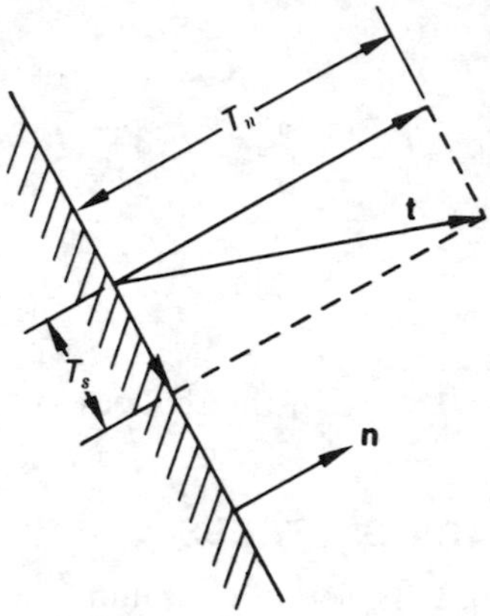

Fig. 4.6

If dn_1, dn_2, and dn_3 can vary independently of one another, then (i) gives the familiar conditions

$$\frac{\partial(T_s^2)}{\partial n_1}=0, \quad \frac{\partial(T_s^2)}{\partial n_2}=0, \quad \text{and} \quad \frac{\partial(T_s^2)}{\partial n_3}=0.$$

But the n_i's are related by

$$n_1^2+n_2^2+n_3^2=1, \tag{ii}$$

therefore

$$n_1dn_1+n_2dn_2+n_3dn_3=0. \tag{iii}$$

That is, the dn_i's cannot be varied independently.

Multiplying Eq. (iii) by an arbitrary multiplier λ and then subtracting it from Eq. (i), we get

$$d(T_s^2)=\left(\frac{\partial(T_s^2)}{\partial n_1}-\lambda n_1\right)dn_1+\left(\frac{\partial(T_s^2)}{\partial n_2}-\lambda n_2\right)dn_2+\left(\frac{\partial(T_s^2)}{\partial n_3}-\lambda n_3\right)dn_3=0.$$

We now choose λ to equal the value of

$$\frac{1}{n_1}\frac{\partial(T_s^2)}{\partial n_1}$$

at a stationary value of T_s^2, so that the first term in this equation vanishes for any dn_1, i.e.,

$$\frac{\partial(T_s^2)}{\partial n_1}=\lambda n_1. \tag{iv}$$

Now, dn_2 and dn_3 are arbitrary, therefore

$$\frac{\partial(T_s^2)}{\partial n_2} = \lambda n_2, \tag{v}$$

$$\frac{\partial(T_s^2)}{\partial n_3} = \lambda n_3. \tag{vi}$$

Equations (iv), (v), (vi), and (ii) determine the values of n_1, n_2, n_3, and λ that correspond to a stationary value of T_s^2.

Computing the partial derivatives from Eq. (4.10), Eqs. (iv), (v), (vi), and (ii) become

$$2n_1[T_1^2 - 2(T_1n_1^2 + T_2n_2^2 + T_3n_3^2)T_1] = n_1\lambda, \tag{a}$$

$$2n_2[T_2^2 - 2(T_1n_1^2 + T_2n_2^2 + T_3n_3^2)T_2] = n_2\lambda, \tag{b}$$

$$2n_3[T_3^2 - 2(T_1n_1^2 + T_2n_2^2 + T_3n_3^2)T_3] = n_3\lambda, \tag{c}$$

and

$$n_1^2 + n_2^2 + n_3^2 = 1. \tag{d}$$

From Eqs. (a), (b), (c), and (d), the following stationary points (n_1, n_2, n_3) are obtained:

$$(1, 0, 0) \quad (0, 1, 0) \quad (0, 0, 1) \tag{e}$$

$$\left(\frac{1}{\sqrt{2}}, \pm\frac{1}{\sqrt{2}}, 0\right), \quad \left(\frac{1}{\sqrt{2}}, 0, \pm\frac{1}{\sqrt{2}}\right), \quad \left(0, \frac{1}{\sqrt{2}}, \pm\frac{1}{\sqrt{2}}\right). \tag{f}$$

The planes determined by solutions (e) are nothing but the principal planes, on which $T_s = 0$. Thus, on these planes the value of T_s^2 is a minimum (in fact, zero).

The values of T_s^2 on the planes given by solutions (f) are easily obtained from Eq. (4.10) to be the following:

$$\text{for } \mathbf{n} = \frac{1}{\sqrt{2}}\mathbf{e}_1 \pm \frac{1}{\sqrt{2}}\mathbf{e}_2, \quad T_s^2 = \frac{(T_1 - T_2)^2}{4}, \tag{4.11a}$$

$$\text{for } \mathbf{n} = \frac{1}{\sqrt{2}}\mathbf{e}_1 \pm \frac{1}{\sqrt{2}}\mathbf{e}_3, \quad T_s^2 = \frac{(T_1 - T_3)^2}{4}, \tag{4.11b}$$

and

$$\text{for } \mathbf{n} = \frac{1}{\sqrt{2}}\mathbf{e}_2 \pm \frac{1}{\sqrt{2}}\mathbf{e}_3, \quad T_s^2 = \frac{(T_2 - T_3)^2}{4}. \tag{4.11c}$$

Thus, the maximum magnitude of the shearing stress is given by the largest of the three values

$$\frac{|T_1 - T_2|}{2}, \quad \frac{|T_1 - T_3|}{2}, \quad \text{and} \quad \frac{|T_2 - T_3|}{2}.$$

In other words,

$$(T_s)_{\max} = \frac{(T_n)_{\max} - (T_n)_{\min}}{2}. \tag{4.12}$$

It can also be shown that on the plane of maximum shearing stress, the normal stress is $T_n = [(T_n)_{\max} + (T_n)_{\min}]/2$.

EXAMPLE 4.4

If the state of stress is such that the components T_{13}, T_{23}, T_{33} are equal to zero, then it is called a state of plane stress.

(a) For plane stress find the principal values and corresponding principal directions.

(b) Determine the maximum shearing stress.

Solution. (a) For the stress matrix

$$[\mathbf{T}] = \begin{bmatrix} T_{11} & T_{12} & 0 \\ T_{12} & T_{22} & 0 \\ 0 & 0 & 0 \end{bmatrix},$$

the characteristic equation has the form

$$\lambda[\lambda^2 - (T_{11} + T_{22})\lambda + (T_{11}T_{22} - T_{12}^2)] = 0.$$

Therefore $\lambda = 0$ is an eigenvalue and its direction is obviously $\mathbf{n} = \mathbf{e}_3$. The remaining eigenvalues are

$$\left.\begin{matrix} T_1 \\ T_2 \end{matrix}\right\} = \frac{T_{11} + T_{22} \pm \sqrt{(T_{11} - T_{22})^2 + 4T_{12}^2}}{2}. \tag{4.13}$$

To find the corresponding eigenvectors, we set $(T_{ij} - \lambda\delta_{ij})n_j = 0$ and obtain for either $\lambda = T_1$ or T_2,

$$\begin{aligned} (T_{11} - \lambda)n_1 + T_{12}n_2 &= 0, \\ T_{12}n_1 + (T_{22} - \lambda)n_2 &= 0, \\ -\lambda n_3 &= 0. \end{aligned}$$

The third equation gives $n_3 = 0$. Let the eigenvector $\mathbf{n} = \cos\theta\mathbf{e}_1 + \sin\theta\mathbf{e}_2$ (*see* Fig. 4.7). Then, from the first equation

$$\tan\theta = \frac{n_2}{n_1} = -\frac{T_{11} - \lambda}{T_{12}}. \tag{4.14}$$

(b) Since the third eigenvalue is always zero, the maximum shearing stress will be the greatest of the values

$$|T_1/2|, \quad |T_2/2|$$

and

$$\left|\frac{T_1 - T_2}{2}\right| = \frac{\sqrt{(T_{11} - T_{22})^2 + 4T_{12}^2}}{2}. \tag{4.15}$$

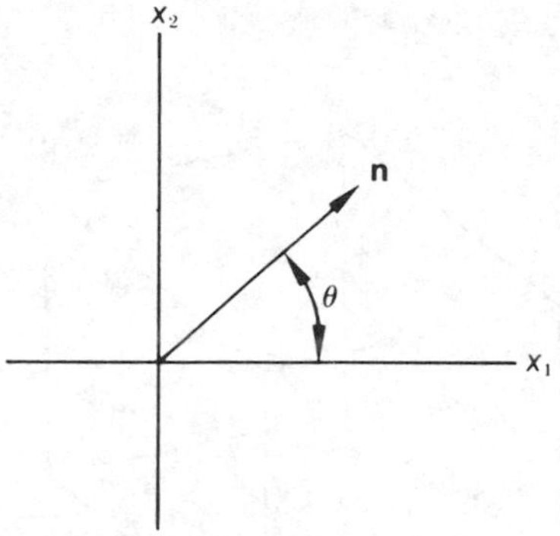

Fig. 4.7

EXAMPLE 4.5

Do Example 4.4 for the following state of stress: $T_{12} = T_{21} = 1000$ psi, all other T_{ij} are zero.

Solution. From Eq. (4.13), we have

$$\left.\begin{matrix} T_1 \\ T_2 \end{matrix}\right\} = \pm \frac{\sqrt{(4)(1000)^2}}{2} = \pm 1000 \text{ psi}.$$

Corresponding to the maximum normal stress $T_1 = 1000$ psi, Eq. (4.14) gives

$$\tan\theta_1 = -\frac{0-1000}{1000} = +1, \quad \text{i.e.,} \quad \theta_1 = 45°$$

and corresponding to the minimum normal stress $T_2 = -1000$ psi (i.e., maximum compressive stress),

$$\tan\theta_2 = -\frac{0-(-1000)}{1000} = -1, \quad \text{i.e.,} \quad \theta_2 = -45°.$$

The maximum shearing stress is given by

$$(T_s)_{\max} = \frac{1000-(-1000)}{2} = 1000 \text{ psi},$$

which acts on the planes bisecting the planes of maximum and minimum normal stresses, i.e., the $\mathbf{e}_1$-plane and $\mathbf{e}_2$-plane in this problem.

4.7 EQUATIONS OF MOTION – PRINCIPLE OF LINEAR MOMENTUM

In this section, we derive the differential equations of motion for any continuum in motion. The basic postulate is that each particle of the continuum must satisfy Newton's law of motion.

Figure 4.8 shows the stress vectors that are acting on the six faces of

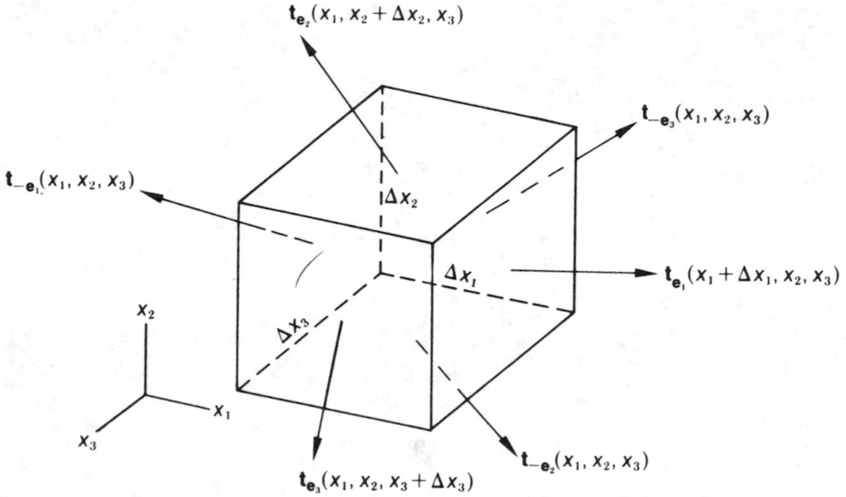

Fig. 4.8

a small rectangular element that is isolated from the continuum in the neighborhood of the position designated by x_i.

Let $\mathbf{B} = B_i\mathbf{e}_i$ be the body force (such as weight) per unit mass, ρ be the mass density at x_i, and $\mathbf{a}$ the acceleration of a particle currently at the position x_i; then Newton's law of motion takes the form, valid in rectangular Cartesian coordinate systems

$$\left[\left(\frac{\mathbf{t}_{\mathbf{e}_1}(x_1+\Delta x_1, x_2, x_3)-\mathbf{t}_{\mathbf{e}_1}(x_1, x_2, x_3)}{\Delta x_1}\right)\right.$$
$$+\left(\frac{\mathbf{t}_{\mathbf{e}_2}(x_1, x_2+\Delta x_2, x_3)-\mathbf{t}_{\mathbf{e}_2}(x_1, x_2, x_3)}{\Delta x_2}\right)$$
$$\left.+\left(\frac{\mathbf{t}_{\mathbf{e}_3}(x_1, x_2, x_3+\Delta x_3)-\mathbf{t}_{\mathbf{e}_3}(x_1, x_2, x_3)}{\Delta x_3}\right)\right]\Delta x_1\Delta x_2\Delta x_3+\rho\mathbf{B}\Delta x_1\Delta x_2\Delta x_3$$
$$=(\rho\mathbf{a})(\Delta x_1)(\Delta x_2)(\Delta x_3).$$

Dividing by $\Delta x_1\Delta x_2\Delta x_3$ and letting $\Delta x_i \to 0$ we have

$$\frac{\partial \mathbf{t}_{\mathbf{e}_1}}{\partial x_1}+\frac{\partial \mathbf{t}_{\mathbf{e}_2}}{\partial x_2}+\frac{\partial \mathbf{t}_{\mathbf{e}_3}}{\partial x_3}+\rho\mathbf{B}=\rho\mathbf{a}.$$

Since $\mathbf{t}_{\mathbf{e}_j} = \mathbf{T}\mathbf{e}_j = T_{ij}\mathbf{e}_i$, therefore we have (noting that $\mathbf{e}_i$ are of fixed direction)

$$\frac{\partial T_{ij}}{\partial x_j}\mathbf{e}_i+\rho B_i\mathbf{e}_i=\rho a_i\mathbf{e}_i.$$

In invariant form, the above equation can be written

$$\operatorname{div} \mathbf{T} + \rho \mathbf{B} = \rho \mathbf{a} \tag{4.16a†}$$

and, in component form

$$\frac{\partial T_{ij}}{\partial x_j} + \rho B_i = \rho a_i. \tag{4.16b}$$

These are the equations that must be satisfied for any continuum, whether it be solid or fluid, in motion. They are called "**Cauchy's equations of motion**." If the acceleration vanishes, then Eq. (4.16) reduces to the **equilibrium equations**

$$\frac{\partial T_{ij}}{\partial x_j} + \rho B_i = 0. \tag{4.17a}$$

Or, in invariant form

$$\operatorname{div} \mathbf{T} + \rho \mathbf{B} = 0. \tag{4.17b}$$

EXAMPLE 4.6

In the absence of body forces, does the stress distribution

$$T_{11} = x_2^2 + \nu(x_1^2 - x_2^2), \quad T_{12} = -2\nu x_1 x_2,$$
$$T_{22} = x_1^2 + \nu(x_2^2 - x_1^2), \quad T_{23} = T_{13} = 0,$$
$$T_{33} = \nu(x_1^2 + x_2^2)$$

satisfy the equations of equilibrium?

Solution. Writing the first ($i = 1$) equilibrium equation, we have

$$\frac{\partial T_{1j}}{\partial x_j} = \frac{\partial T_{11}}{\partial x_1} + \frac{\partial T_{12}}{\partial x_2} + \frac{\partial T_{13}}{\partial x_3} = 2\nu x_1 - 2\nu x_1 + 0 = 0.$$

Similarly, for $i = 2$ we have

$$\frac{\partial T_{2j}}{\partial x_j} = -2\nu x_2 + 2\nu x_2 + 0 = 0$$

and for $i = 3$, we have

$$\frac{\partial T_{3j}}{\partial x_j} = 0 + 0 + 0 = 0.$$

Therefore, the given stress distribution does satisfy the equilibrium equations.

EXAMPLE 4.7

Write the equations of motion if the stress components have the form $T_{ij} = -p\delta_{ij}$, where $p = p(x_1, x_2, x_3, t)$.

†By definition, div **T** is a vector whose Cartesian components are $\partial T_{ij}/\partial x_j$ (*see* Section 2B20).

Solution. Substituting the given stress distribution in the first term on the left-hand side of Eq. (4.16), we obtain

$$\frac{\partial T_{ij}}{\partial x_j} = -\frac{\partial p}{\partial x_j}\delta_{ij} = -\frac{\partial p}{\partial x_i}.$$

Therefore,

$$-\frac{\partial p}{\partial x_i} + \rho B_i = \rho a_i$$

or

$$-\nabla p + \rho \mathbf{B} = \rho \mathbf{a}.$$

4.8 BOUNDARY CONDITION FOR THE STRESS TENSOR

If on the actual boundary of some body there are applied distributive forces, we call them surface tractions. We wish to find the relation between the surface tractions and the stress field that is defined within the body.

If we consider an infinitesimal tetrahedron cut from the boundary of a body with its inclined face coinciding with the boundary surface (Fig. 4.9), then as in Section 4.1, we obtain

$$\mathbf{t} = \mathbf{T}\mathbf{n}, \tag{4.18}$$

where $\mathbf{n}$ is the unit vector normal to the boundary, $\mathbf{T}$ is the stress tensor evaluated at the boundary, and $\mathbf{t}$ is the force vector per unit area of surface traction. Equation (4.18) is called the stress boundary condition.

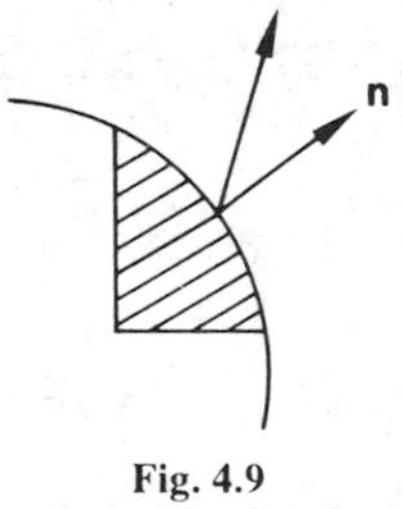

Fig. 4.9

PROBLEMS

4.1. The state of stress at a certain point of a body is given by

$$[\mathrm{T}] = \begin{bmatrix} 1 & 2 & 3 \\ 2 & 4 & 5 \\ 3 & 5 & 0 \end{bmatrix} \mathrm{MPa}$$

Determine on which of the three coordinate planes (normals $\mathbf{e}_1, \mathbf{e}_2, \mathbf{e}_3$) the (a) normal stress and the (b) shearing stress is greatest.

4.2. The state of stress at a certain point of a body is given by

$$[\mathrm{T}] = \begin{bmatrix} 2 & -1 & 3 \\ -1 & 4 & 0 \\ 3 & 0 & -1 \end{bmatrix} \text{MPa}$$

(a) Find the stress vector at the point on the plane whose normal is in the direction $2\mathbf{e}_1 + 2\mathbf{e}_2 + \mathbf{e}_3$.

(b) Determine the magnitude of the normal and shearing stress on this plane.

4.3. Do the previous problem for a plane which is parallel to the plane $x_1 - 2x_2 + 3x_3 = 4$.

4.4. The stress distribution in a certain body is given by

$$[\mathbf{T}] = \begin{bmatrix} 0 & 100x_1 & -100x_2 \\ 100x_1 & 0 & 0 \\ -100x_2 & 0 & 0 \end{bmatrix}.$$

Find the stress vector acting on a plane which passes through the point $(\frac{1}{2}, \sqrt{3}/2, 3)$ and is tangent to the circular cylindrical surface $x_1^2 + x_2^2 = 1$ at that point.

4.5. For the following state of stress

$$[\mathrm{T}] = \begin{bmatrix} 10 & 50 & -50 \\ 50 & 0 & 0 \\ -50 & 0 & 0 \end{bmatrix} \text{MPa}$$

find T'_{11} and T'_{13}, where x'_1 is in the direction of $\mathbf{e}_1 + 2\mathbf{e}_2 + 3\mathbf{e}_3$ and x'_2 is in the direction of $\mathbf{e}_1 + \mathbf{e}_2 - \mathbf{e}_3$.

4.6. Consider the following stress distribution

$$[\mathbf{T}] = \begin{bmatrix} \alpha x_2 & \beta & 0 \\ \beta & 0 & 0 \\ 0 & 0 & 0 \end{bmatrix},$$

where α and β are constants.

(a) Determine and sketch the distribution of the stress vector acting on the square in the $x_1 = 0$ plane, with vertices located at $(0, 1, 1)$, $(0, -1, 1)$, $(0, 1, -1)$, $(0, -1, -1)$.

(b) Find the total resultant force and moment about the origin of the stress vectors acting on the square of part (a).

4.7. Do the previous problem if the stress distribution is given by

$$T_{11} = \alpha x_2^2$$

and all other $T_{ij} = 0$.

4.8. Do Problem 4.6 for the stress distribution

$$T_{11} = \alpha, \quad T_{21} = T_{12} = \alpha x_3$$

and all other $T_{ij} = 0$.

4.9. Consider the following stress distribution for a certain circular cylindrical bar

$$[\mathbf{T}] = \begin{bmatrix} 0 & -\alpha x_3 & +\alpha x_2 \\ -\alpha x_3 & 0 & 0 \\ \alpha x_2 & 0 & 0 \end{bmatrix}.$$

(a) What is the distribution of the stress vector on the surfaces defined by $x_2^2 + x_3^2 = 4$, $x_1 = 0$, and $x_1 = l$?

(b) Find the total resultant force and moment on the end face $x_1 = l$.

4.10. For any stress state **T**, we define the deviatoric stress $\mathbf{T}^0$ to be

$$\mathbf{T}^0 = \mathbf{T} - \left(\frac{T_{kk}}{3}\right)\mathbf{I},$$

where T_{kk} is the first invariant of the stress tensor **T**.

(a) Show that the first invariant of the deviatoric stress vanishes.

(b) Given the stress tensor

$$[\mathbf{T}] = 100\begin{bmatrix} 6 & 5 & -2 \\ 5 & 3 & 4 \\ -2 & 4 & 9 \end{bmatrix} \text{kPa},$$

evaluate $\mathbf{T}^0$.

(c) Show that the principal directions of the stress and the deviatoric stress coincide.

(d) Find a relation between the principal values of the stress and the deviatoric stress.

4.11. An octahedral stress plane is defined to make equal angles with each of the principal axes of stress.

(a) How many independent octahedral planes are there at each point?

(b) Show that the normal stress on an octahedral plane is given by one-third the first stress invariant.

(c) Show that the shearing stress on the octahedral plane is given by

$$T_s = \tfrac{1}{3}[(T_1 - T_2)^2 + (T_2 - T_3)^2 + (T_1 - T_3)^2]^{1/2},$$

where T_1, T_2, T_3 are the principal values of the stress tensor.

4.12. (a) Let **m** and **n** be two unit vectors that define two planes M and N that pass through a point P. For an arbitrary state of stress defined at the point P, show that the component of the stress vector $\mathbf{t_m}$ in the **n**-direction is equal to the component of the stress vector $\mathbf{t_n}$ in the **m**-direction.

(b) If $\mathbf{m} = \mathbf{e}_1$ and $\mathbf{n} = \mathbf{e}_2$, what does the result of part (a) reduce to?

4.13. Let **m** be a unit vector that defines a plane M passing through a point P. Show that the stress vector on any plane that contains the stress traction $\mathbf{t}_m$ lies in the M-plane.

4.14. Let $\mathbf{t}_m$ and $\mathbf{t}_n$ be stress vectors on planes defined by the unit vectors **m** and **n** and that pass through the point P. Show that if **k** is a unit vector that determines a plane that contains $\mathbf{t}_m$ and $\mathbf{t}_n$, then $\mathbf{t}_k$ is perpendicular to **m** and **n**.

4.15. Why can the following two matrices not represent the same stress tensor?

$$\begin{bmatrix} 100 & 200 & 40 \\ 200 & 0 & -30 \\ 40 & -30 & -50 \end{bmatrix}, \quad \begin{bmatrix} 40 & 100 & 60 \\ 100 & 100 & 0 \\ 60 & 0 & 20 \end{bmatrix}$$

4.16. The principal values of a stress tensor **T** are $T_1 = 10$ MPa, $T_2 = -10$ MPa, and $T_3 = 30$ MPa. If a matrix of the stress is given by

$$[\mathbf{T}] = \begin{bmatrix} T_{11} & 0 & 0 \\ 0 & 1 & 2 \\ 0 & 2 & T_{33} \end{bmatrix} \times 10 \text{ MPa}$$

find the value of T_{11} and T_{33}.

4.17. If the state of stress at a point is

$$[\mathbf{T}] = \begin{bmatrix} 300 & 0 & 0 \\ 0 & -200 & 0 \\ 0 & 0 & 400 \end{bmatrix} \text{kPa}$$

find (a) the magnitude of the shearing stress on the plane whose normal is in the direction of $2\mathbf{e}_1 + 2\mathbf{e}_2 + \mathbf{e}_3$, and (b) the maximum shearing stress.

4.18. The stress state in which the only nonvanishing stress components are a single pair of shearing stresses is called simple shear. Take $T_{12} = T_{21} = \tau$ and all other $T_{ij} = 0$.

(a) Find the principal values and principal directions of this stress state.

(b) Find the maximum shearing stress and the plane on which it acts.

4.19. The stress state in which only the three normal stress components do not vanish is called triaxial normal stress. Take $T_{11} = \sigma_1$, $T_{22} = \sigma_2$, $T_{33} = \sigma_3$ with $\sigma_1 > \sigma_2 > \sigma_3$ and all other $T_{ij} = 0$. Find the maximum shearing stress and the plane on which it acts.

4.20. Show that the symmetry of the stress tensor is not valid if there are body moments per unit volume, as in the case of a polarized anisotropic dielectric solid.

4.21. Given the following stress distribution

$$[\mathbf{T}] = \begin{bmatrix} x_1 + x_2 & T_{12}(x_1, x_2) & 0 \\ T_{12}(x_1, x_2) & x_1 - 2x_2 & 0 \\ 0 & 0 & x_2 \end{bmatrix}$$

find T_{12} in order that the stress distribution is in equilibrium with zero body force, and that the stress vector on $x_1 = 1$ is given by $\mathbf{t} = (1+x_2)\mathbf{e}_1 + (5-x_2)\mathbf{e}_2$.

4.22. Suppose the body force vector is $\mathbf{B} = -g\mathbf{e}_3$, where g is a constant. Consider the following stress tensor

$$[\mathbf{T}] = \alpha\begin{bmatrix} x_2 & -x_3 & 0 \\ -x_3 & 0 & -x_2 \\ 0 & -x_2 & T_{33} \end{bmatrix}.$$

and find an expression for T_{33} such that $\mathbf{T}$ satisfies the equations of equilibrium.

4.23. In the absence of body forces, do the stress components

$$T_{11} = \alpha[x_2^2 + \nu(x_1^2 - x_2^2)], \quad T_{22} = \alpha[x_1^2 + \nu(x_2^2 - x_1^2)],$$

$$T_{33} = \alpha\nu(x_1^2 + x_2^2), \quad T_{12} = -2\alpha\nu x_1 x_2 \quad T_{13} = T_{23} = 0,$$

satisfy the equations of equilibrium?

4.24. Repeat Problem 4.23 for the stress distribution

$$[\mathbf{T}] = \alpha\begin{bmatrix} x_1 + x_2 & 2x_1 - x_2 & 0 \\ 2x_1 - x_2 & x_1 - 3x_2 & 0 \\ 0 & 0 & x_1 \end{bmatrix}.$$

4.25. Suppose that the stress distribution has the form (called plane stress)

$$[\mathbf{T}] = \begin{bmatrix} T_{11}(x_1, x_2) & T_{12}(x_1, x_2) & 0 \\ T_{12}(x_1, x_2) & T_{22}(x_1, x_2) & 0 \\ 0 & 0 & 0 \end{bmatrix}.$$

(a) What are the equilibrium equations in this special case?

(b) If we introduce a function $\phi(x_1, x_2)$ such that

$$T_{11} = \frac{\partial^2\phi}{\partial x_2^2}, \quad T_{22} = \frac{\partial^2\phi}{\partial x_1^2}, \quad T_{12} = -\frac{\partial^2\phi}{\partial x_1 \partial x_2},$$

will this stress distribution be in equilibrium with zero body force?

4.26. In cylindrical coordinates (r, ϕ, z), consider a differential volume of material bounded by the three pairs of faces $r = r_0$, $r = r_0 + dr$; $\phi = \phi_0$, $\phi = \phi_0 + d\phi$; and $z = z_0$, $z = z_0 + dz$ (*see* Fig. P4.1).

Derive the following equations of motion in cylindrical coordinates

$$\rho a_r = \frac{\partial T_{rr}}{\partial r} + \frac{1}{r}\frac{\partial T_{r\phi}}{\partial \phi} + \frac{T_{rr} - T_{\phi\phi}}{r} + \frac{\partial T_{rz}}{\partial z} + \rho B_r,$$

$$\rho a_\phi = \frac{\partial T_{r\phi}}{\partial r} + \frac{\partial T_{\phi\phi}}{r\partial \phi} + \frac{2T_{r\phi}}{r} + \frac{\partial T_{\phi z}}{\partial z} + \rho B_\phi,$$

$$\rho a_z = \frac{\partial (rT_{rz})}{r\partial r} + \frac{\partial T_{\phi z}}{r\partial \phi} + \frac{\partial T_{zz}}{\partial z} + \rho B_z.$$

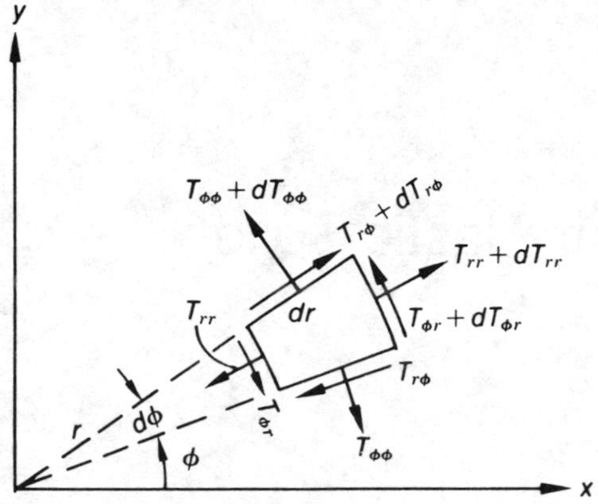
y
$T_{\phi\phi} + dT_{\phi\phi}$
$T_{r\phi} + dT_{r\phi}$
$T_{rr} + dT_{rr}$
T_{rr}
dr
$T_{\phi r} + dT_{\phi r}$
$T_{r\phi}$
$d\phi$
r
$T_{\phi r}$
$T_{\phi\phi}$
ϕ
x

Fig. P4.1

5

The Linear Elastic Solid

So far we have studied the kinematics of deformation, the description of the state of stress and three basic principles of continuum physics, the principle of conservation of mass [Eq. (3.29)], the principle of linear momentum [Eq. (4.16)], and the principle of moment of momentum [Eq. (4.5)]. All these relations† are valid for every continuum, indeed no mention was made of any material in the derivations.

These equations are however not sufficient to describe the response of a specific material due to a given loading. We know from experience that under the same loading conditions, the response of steel is different from that of water. Furthermore, for a given material, it varies with different loading conditions. For example, for moderate loadings, the deformation in steel caused by the application of loads disappears with the removal of the loads. This aspect of the material behavior is known as elasticity. Whereas beyond a certain loading, there will be permanent deformations, or even fracture exhibiting behavior quite different from that of elasticity. In this chapter, we shall study an idealized material which models the linear elastic behavior of a real solid. Such an idealized material is described by a linear stress-strain relationship (or, more generally, the constitutive equation for linear elastic solid). Using the constitutive equation, we then study some dynamic and static problems of such a solid.

†The principle of conservation of energy is treated in Chapter 6.

5.1 MECHANICAL PROPERTIES

We want to establish some appreciation of the mechanical behavior of solid materials. To do this, we perform some thought experiments modeled after real laboratory experiments.

Suppose we have a block of material, and we cut out a slender cylindrical test specimen of cross-sectional area A. The bar is now statically tensed by an axially applied load P, and the elongation Δl, over some axial gage length l, is measured. A typical plot of tensile force against elongation is shown in Fig. 5.1. Within the linear portion OA (sometimes called the proportional range), if the load had been removed, then the line OA is retraced and the specimen has exhibited an elasticity. Applying a load that is greater than A and then unloading, we typically traverse

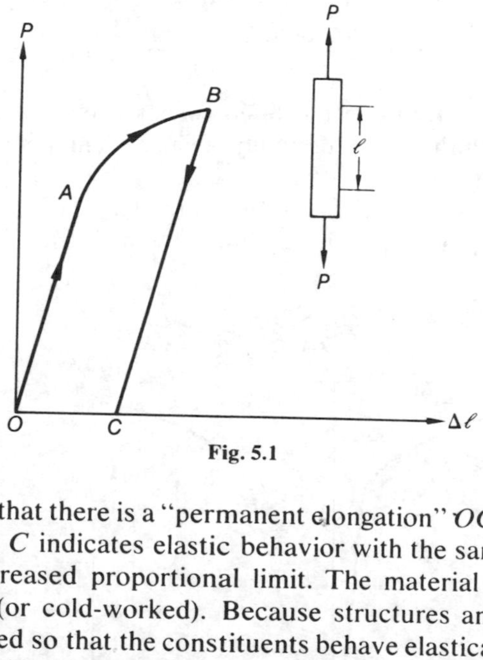

Fig. 5.1

$OABC$ and find that there is a "permanent elongation" OC. Reapplication of the load from C indicates elastic behavior with the same slope as OA, but with an increased proportional limit. The material is said to have work-hardened (or cold-worked). Because structures and machines are generally designed so that the constituents behave elastically and because the elastic behavior of materials is one of the simplest to model, we will restrict ourselves to linear elasticity. Restricting ourselves to the linear range we find that if the specimen is pulled at different rates of loading, we maintain the same proportionality between load and elongation. We therefore are justified in assuming that rate of loading has no effect on the linear elastic behavior.

The load-elongation diagram in Fig. 5.1 depends in the present form on the cross-section of the specimen and the axial gage length l. However, we want to have a representation of material behavior which is independent of specimen size and any variables introduced by the experimental setup. We accordingly plot the so-called stress P/A versus the axial strain or unit elongation $\epsilon_a = \Delta l/l$ as shown in Fig. 5.2. Now, the test results appear in a form which is not dependent on the specimen dimensions. The slope of the line OA will therefore be a material coefficient which is called **Young's modulus** (or, **modulus of elasticity**)

$$E_Y = \sigma/\epsilon_a.$$

The numerical value of E_Y for steel is around 207 GPa (30×10^6 psi). This means, for a steel bar of cross-sectional area 32.3 cm² (5 in²) that carries a load of 667,200 N (150,000 lbs), the axial strain is

$$\epsilon_a = \frac{667{,}200/(32.3 \times 10^{-4})}{207 \times 10^9} \approx 10^{-3}.$$

As expected, the strains in the linear elastic range of metals are quite small and we will, therefore, use infinitesimal strain theory to describe the deformation of metals.

In the tension test, we can also measure changes in the lateral dimension. If the bar was a circular cylinder, with diameter d, it can, under

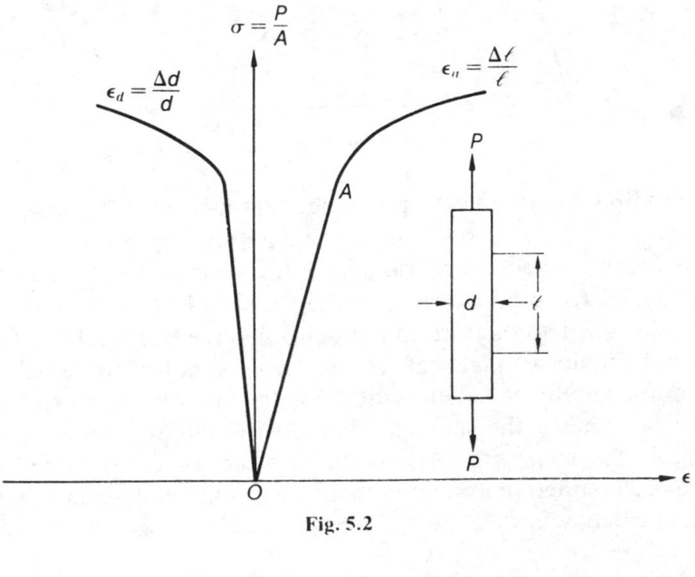

Fig. 5.2

certain conditions remain circular decreasing in diameter as the tensile load is increased. Letting ϵ_d be the lateral strain (equal to $\Delta d/d$), we find that the ratio $-\epsilon_d/\epsilon_a$ is a constant. We call this constant **Poisson's ratio** and denote it by ν. A typical value of ν for steel is 0.3.

So far we have only been considering a single specimen out of the block of material. It is conceivable that the modulus of elasticity E_Y as well as Poisson's ratio may depend on the orientation of the specimen relative to the block. In this case, the material is said to be **anisotropic** with respect to its elastic properties. Anisotropic properties are usually exhibited by materials with a definite internal structure such as wood or a rolled steel plate. If the specimens cut at different orientations at a sufficiently small neighborhood show the same σ-ϵ diagram, we can conclude that the material is **isotropic** with respect to its elastic properties.

In addition to a possible dependence on orientation of the elastic properties, we may also find that they may vary from one neighborhood to the other. In this case, we call the material **inhomogeneous**. If there is no change in the test results for specimens at different neighborhoods, we say the material is **homogeneous**.

Previously, we assumed that the circular cross-section of our test bar remained circular upon deformation. This will be the case if the material is homogeneous and isotropic with respect to its elastic properties.

Other characteristic tests with an elastic material are also possible. In one case, we may be interested in the change of volume of a block of homogeneous, isotropic material under uniform pressure, p, for which the stress state is

$$T_{ij} = -p\,\delta_{ij}.$$

In a suitable experiment, we measure the relation between p, the applied pressure, and e, the change in volume per initial volume. For an elastic material, linear relation is found and we define the **bulk modulus** k, as

$$k = \frac{-p}{e}.$$

A typical value of k for steel is 138 GPa (20×10^6 psi).

A torsion experiment yields another elastic constant. A cylindrical bar of circular cross-section of radius r is subjected to a torsional moment M_t along the cylinder axis. The bar has length l and will twist by an angle θ upon the application of the moment M_t.† For an elastic material, a

†Further discussion of a cylindrical bar under torsion is given later in the chapter, *see* Section 5.8B.

linear relation between angle of twist θ and the applied moment is obtained. We define a shear modulus μ

$$\mu = \frac{M_t l}{I_p \theta},$$

where $I_p = \pi r^4/2$ (the polar area moment of inertia).

A typical value of μ for steel is 76 GPa (11×10^6 psi).

Four different constants for the isotropic linear elastic material were introduced by three different experiments. We wonder whether these constants are independent, or how many we need to describe a linear isotropic elastic material.

5.2 LINEAR ELASTIC SOLID

Within certain limits, the experiments cited in Section 5.1 had the following features in common:

(a) The relation between the applied loading and a quantity measuring the deformation were linear.

(b) The rate of load application did not affect the linear relationship of (a).

(c) Upon removal of the loading, the deformations disappeared completely.

(d) The deformations which were observed in the experiments were very small.

The characteristics (a) through (d) are now used to formulate the constitutive equation of an ideal material, the linear elastic or Hookean elastic solid. The constitutive equation relates the stress to relevant quantities of deformation. In this case, deformations are small and the rate of load application has no effect. We therefore can write

$$\mathbf{T} = \mathbf{T}(\mathbf{E}), \tag{5.1}$$

where $\mathbf{T}(\mathbf{E})$ is a single-valued function of $\mathbf{E}$ with $\mathbf{T}(\mathbf{0}) = \mathbf{0}$. If, in addition the function is to be linear, then we have, in component form

$$T_{11} = C_{1111}E_{11} + C_{1112}E_{12} + \cdots + C_{1133}E_{33}$$
$$T_{12} = C_{1211}E_{11} + C_{1212}E_{12} + \cdots + C_{1233}E_{33}$$
$$\cdots\cdots\cdots\cdots\cdots\cdots\cdots\cdots\cdots\cdots\cdots\cdots$$
$$T_{33} = C_{3311}E_{11} + C_{3312}E_{12} + \cdots + C_{3333}E_{33}.$$

The above nine equations can be written as

$$T_{ij} = C_{ijkl}E_{kl}. \tag{5.2}$$

Since T_{ij} and E_{ij} are components of second-order tensors, from Example 2.8, we see that C_{ijkl} are components of a fourth-order tensor. It is known as the **elasticity tensor**. It is this tensor which characterizes the mechanical properties of a particular anisotropic Hookean elastic solid. The anisotropy of the material is represented by the fact that the components of C_{ijkl} are in general different for different basis as seen in Example 2.8. If the body is homogeneous, that is, the mechanical properties are the same for every particle in the body, then C_{ijkl} are constants (i.e., independent of position). We shall be concerned only with homogeneous bodies.

The tensor **E** is symmetric and we consider only the case where **T** is also symmetric, therefore, the above nine equations [Eq. (5.2)] reduce to only six equations relating the six independent stress components (T_{11}, T_{22}, T_{33}, T_{12}, T_{13}, T_{23}) to the six independent strain components (E_{11}, E_{22}, E_{33}, E_{12}, E_{13}, E_{23}). Thus, for a linear anisotropic elastic solid, we need no more than 36 material constants to specify its mechanical properties.

5.3 LINEAR ISOTROPIC ELASTIC SOLID

A material is said to be **isotropic** if its mechanical properties can be described without reference to direction. When this is not true the material is said to be **anisotropic**. Many structural metals such as steel and aluminum can be regarded as isotropic without appreciable error.

We had, for a linear elastic solid,

$$T_{ij} = C_{ijkl}E_{kl}.$$

If it is isotropic, then the components of the elasticity tensor, i.e., C_{ijkl} must remain the same regardless of how the rectangular basis are rotated and reflected. In other words,

$$C_{ijkl} = C'_{ijkl} \tag{5.3}$$

under all orthogonal transformation of basis. A tensor having the same components with respect to every rectangular unit basis is known as an **isotropic tensor**. For example, the identity tensor **I** is an isotropic tensor since its components δ_{ij} are the same for any Cartesian basis. It is obvious that the following three fourth-order tensors are isotropic: $A_{ijkl} \equiv \delta_{ij}\delta_{kl}$, $B_{ijkl} \equiv \delta_{ik}\delta_{jl}$ and $H_{ijkl} \equiv \delta_{il}\delta_{jk}$.† In fact, it can be shown that any isotropic

†Note

$$A_{1111} = 1, \quad A_{1212} = 0, \quad A_{1221} = 0,$$
$$B_{1111} = 1, \quad B_{1212} = 1, \quad B_{1221} = 0,$$
$$H_{1111} = 1, \quad H_{1212} = 0, \quad H_{1221} = 1.$$

fourth-order tensor can be represented as a linear combination of the above three tensors (we omit the proof). Thus, for an isotropic, linearly-elastic material, the elasticity tensor C_{ijkl} can be written

$$C_{ijkl} = \lambda A_{ijkl} + \alpha B_{ijkl} + \beta H_{ijkl}. \tag{5.4}$$

Substituting Eq. (5.4) into Eq. (5.2) and since

$$A_{ijkl}E_{kl} = \delta_{ij}\delta_{kl}E_{kl} = \delta_{ij}E_{kk} = \delta_{ij}e,$$

$$B_{ijkl}E_{kl} = \delta_{ik}\delta_{jl}E_{kl} = E_{ij},$$

$$H_{ijkl}E_{kl} = \delta_{il}\delta_{jk}E_{kl} = E_{ji} = E_{ij},$$

we have

$$T_{ij} = C_{ijkl}E_{kl} = \lambda e\delta_{ij} + (\alpha + \beta)E_{ij}.$$

Or, denoting $\alpha + \beta$ by 2μ, we have

$$T_{ij} = \lambda e\delta_{ij} + 2\mu E_{ij} \tag{5.5a}$$

or

$$\mathbf{T} = \lambda e\mathbf{I} + 2\mu\mathbf{E}, \tag{5.5b}$$

where $e = E_{kk}$ = first scalar invariant of **E**. In long form, Eq. (5.5) reads

$$T_{11} = \lambda(E_{11} + E_{22} + E_{33}) + 2\mu E_{11}, \tag{5.6a}$$

$$T_{22} = \lambda(E_{11} + E_{22} + E_{33}) + 2\mu E_{22}, \tag{5.6b}$$

$$T_{33} = \lambda(E_{11} + E_{22} + E_{33}) + 2\mu E_{33}, \tag{5.6c}$$

$$T_{12} = 2\mu E_{12}, \tag{5.6d}$$

$$T_{13} = 2\mu E_{13}, \tag{5.6e}$$

$$T_{23} = 2\mu E_{23}. \tag{5.6f}$$

Equations (5.5) are the constitutive equations for a linear isotropic elastic solid. The two material constants λ and μ are known as **Lamé's constants.** Since E_{ij} are dimensionless, λ and μ are of the same dimension as the stress tensor, force per unit area. For a given real material, the values of Lamé's constants are to be determined from suitable experiments. We shall have more to say about this later.

EXAMPLE 5.1

Find the components of stress at a point if the strain matrix is

$$[\mathbf{E}] = 10^{-6}\begin{bmatrix} 30 & 50 & 20 \\ 50 & 40 & 0 \\ 20 & 0 & 30 \end{bmatrix}$$

and the material is steel with $\lambda = 119.2$ GPa (17.3×10^6 psi) and $\mu = 79.2$ GPa (11.5×10^6 psi).

Solution. We use Hooke's Law $T_{ij} = 2\mu E_{ij} + \lambda e \delta_{ij}$ by first evaluating the dilatation, $e = 100 \times 10^{-6}$. The stress components can now be obtained:

$$\begin{aligned}
T_{11} &= 2\mu E_{11} + \lambda e = 1.67 \times 10^{-2}\,\text{GPa}, \\
T_{22} &= 2\mu E_{22} + \lambda e = 1.83 \times 10^{-2}\,\text{GPa}, \\
T_{33} &= 2\mu E_{33} + \lambda e = 1.67 \times 10^{-2}\,\text{GPa}, \\
T_{12} &= T_{21} = 2\mu E_{12} = 7.92 \times 10^{-2}\,\text{GPa}, \\
T_{13} &= T_{31} = 2\mu E_{13} = 3.17 \times 10^{-3}\,\text{GPa}, \\
T_{23} &= T_{32} = 0.
\end{aligned}$$

EXAMPLE 5.2

(a) For an isotropic Hookean material, show that the principal directions of stress and strain coincide.

(b) Find a relation between the principal values of stress and strain.

Solution. (a) Let $\mathbf{n}_1$ be an eigenvector of the strain tensor $\mathbf{E}$ (i.e., $\mathbf{E}\mathbf{n}_1 = E_1\mathbf{n}_1$). Then, by Hooke's Law we have

$$\mathbf{T}\mathbf{n}_1 = 2\mu\mathbf{E}\mathbf{n}_1 + \lambda e\mathbf{I}\mathbf{n}_1 = (2\mu E_1 + \lambda e)\mathbf{n}_1. \tag{i}$$

Therefore, $\mathbf{n}_1$ is also an eigenvector of the tensor $\mathbf{T}$.

(b) Let E_1, E_2, E_3 be the eigenvalues of $\mathbf{E}$, then $e = E_1 + E_2 + E_3$ and from Eq. (i),

$$T_1 = 2\mu E_1 + \lambda(E_1 + E_2 + E_3).$$

In a similar fashion,

$$T_2 = 2\mu E_2 + \lambda(E_1 + E_2 + E_3),$$

$$T_3 = 2\mu E_3 + \lambda(E_1 + E_2 + E_3).$$

EXAMPLE 5.3

For an isotropic elastic material

(a) Find a relation between the first invariants of stress and strain.

(b) Use the result of part (a) to invert Hooke's law, so that strain is a function of stress.

Solution. (a) By adding Eqs. (5.6a, b, c) we have

$$T_{kk} = (2\mu + 3\lambda)E_{kk} = (2\mu + 3\lambda)e.$$

(b) We now invert Eq. (5.5b) as

$$\mathbf{E} = \frac{1}{2\mu}\mathbf{T} - \frac{\lambda}{2\mu}e\mathbf{I} = \frac{1}{2\mu}\mathbf{T} - \frac{\lambda T_{kk}}{2\mu(2\mu + 3\lambda)}\mathbf{I}.$$

5.4 YOUNG'S MODULUS, POISSON'S RATIO, SHEAR MODULUS, AND BULK MODULUS

Equations (5.6) express the stress components in terms of the strain components. These equations can be inverted, as in Example 5.3, to give

$$E_{ij} = \frac{1}{2\mu}\left[T_{ij} - \frac{\lambda}{3\lambda + 2\mu}(T_{kk})\delta_{ij}\right]. \tag{5.7a}$$

We also have

$$e = \left(\frac{1}{2\mu + 3\lambda}\right) T_{kk}. \tag{5.7b}$$

We will refer to the stress state as **uniaxial tension** (or compression), if only one normal stress component is not zero. In fact, this is a good approximation to the actual state of stress in the cylindrical bar used in the tensile test† and described in Section 5.1. In particular, we take the $\mathbf{e}_1$-direction to be axial and $T_{11} \neq 0$. In this case, Eq. (5.7a) reduces to

$$E_{11} = \frac{1}{2\mu}\left[T_{11} - \frac{\lambda}{3\lambda + 2\mu} T_{11}\right] = \frac{\lambda + \mu}{\mu(3\lambda + 2\mu)} T_{11},$$

$$E_{33} = E_{22} = -\frac{\lambda}{2\mu(3\lambda + 2\mu)} T_{11} = -\frac{\lambda}{2(\lambda + \mu)} E_{11},$$

$$E_{12} = E_{13} = E_{23} = 0.$$

The ratio of T_{11}/E_{11} corresponds to the σ/ϵ_a ratio of the tensile test (*see* Section 5.1) and we have

$$\frac{\sigma}{\epsilon_a} = \frac{T_{11}}{E_{11}} = E_Y = \frac{\mu(3\lambda + 2\mu)}{\lambda + \mu}, \tag{5.8}$$

where E_Y is the **Young's modulus** or **modulus of elasticity.**

Poisson's ratio ν, which is equal to the negative ratio of the transverse strain (E_{22}, E_{33}) to the axial strain is found to be

$$-\frac{E_{33}}{E_{11}} = -\frac{E_{22}}{E_{11}} = \nu = \frac{\lambda}{2(\lambda + \mu)} = -\frac{\epsilon_d}{\epsilon_a}. \tag{5.9}$$

Using Eqs. (5.8) and (5.9) we write Eq. (5.7) in the frequently used engineering form:

$$E_{11} = \frac{1}{E_Y}[T_{11} - \nu(T_{22} + T_{33})],$$

†*See* Section 5.8A for a more detailed discussion.

$$
\begin{aligned}
E_{22} &= \frac{1}{E_Y}[T_{22} - \nu(T_{11} + T_{33})],\\
E_{33} &= \frac{1}{E_Y}[T_{33} - \nu(T_{11} + T_{22})],\\
E_{12} &= \frac{1}{2\mu}T_{12},\\
E_{13} &= \frac{1}{2\mu}T_{13},\\
E_{23} &= \frac{1}{2\mu}T_{23}.
\end{aligned}
\tag{5.10}
$$

Even though there are three material constants in Eq. (5.10), it is important to remember that only two of them are independent for the isotropic material. In fact, by eliminating λ from Eqs. (5.8) and (5.9), we have the useful relation

$$
\mu = \frac{E_Y}{2(1+\nu)}. \tag{5.11}
$$

A second stress state, called **simple shear**, corresponds with one which has all the stress components except one pair of off-diagonal elements vanishing. In particular, we choose $T_{12} = T_{21} \neq 0$, and Eq. (5.7a) reduces to

$$
E_{12} = E_{21} = T_{12}/2\mu.
$$

Defining the shear modulus as the ratio of the shearing stress in simple shear to the decrease in angle between elements that are initially in the $\mathbf{e}_1$- and $\mathbf{e}_2$-directions, we have

$$
\frac{T_{12}}{2E_{12}} \equiv \mu.
$$

Therefore, Lamé's constant μ is also called the shear modulus.

A third stress state, called **hydrostatic stress**, is defined by the stress tensor $\mathbf{T} = \sigma\mathbf{I}$. In this case, Eq. (5.7b) gives

$$
e = \frac{3\sigma}{2\mu + 3\lambda}.
$$

As mentioned in Section 5.1 the bulk modulus k, is defined as the ratio of the hydrostatic normal stress σ, to the unit volume change, we have

$$
k = \frac{\sigma}{e} = \frac{2\mu + 3\lambda}{3}. \tag{5.12}
$$

From Eqs. (5.8), (5.9), (5.11), and (5.12) we see that the elastic constants are interrelated. Equation (5.5a) shows only two constants are necessary to describe a linear, elastic isotropic material. Depending on the application, two basic pairs are frequently used. They are either E_Y and ν or μ and λ. Table 5.1 expresses the various elastic constants in terms of the basic pairs. Table 5.2 gives some numerical values for some common materials.

Table 5.1 Conversion of constants for an isotropic elastic material.

	Basic Pair	
	λ, μ	E_Y, ν
λ	λ	$\dfrac{\nu E_Y}{(1+\nu)(1-2\nu)}$
μ	μ	$\dfrac{E_Y}{2(1+\nu)}$
k	$\lambda+\frac{2}{3}\mu$	$\dfrac{E_Y}{3(1-2\nu)}$
E_Y	$\dfrac{\mu(3\lambda+2\mu)}{\lambda+\mu}$	E_Y
ν	$\dfrac{\lambda}{2(\lambda+\mu)}$	ν

EXAMPLE 5.4

(a) If for a specific material the ratio of the bulk modulus to Young's modulus is very large, find the approximate value of Poisson's ratio.
(b) Indicate why the material of part (a) can be called incompressible.

Solution. (a) In terms of Lamé's constants, we have

$$\frac{k}{E_Y}=\frac{1}{3}\left(\frac{\lambda}{\mu}+1\right) \quad \text{and} \quad \frac{\lambda}{\mu}=\frac{2\nu}{1-2\nu}.$$

Combining these two equations gives

$$\frac{k}{E_Y}=\frac{1}{3(1-2\nu)}.$$

Therefore, if $k/E_Y \to \infty$, then Poisson's ratio $\nu \to \frac{1}{2}$.
(b) For an arbitrary stress state the dilatation or unit volume change is given by

$$e=\frac{T_{ii}}{3k}=\left(\frac{1-2\nu}{E_Y}\right)T_{ii}.$$

If $\nu \to \frac{1}{2}$, then $e \to 0$. That is, the material is incompressible. It has never been observed in real material that hydrostatic compression results in an increase of volume; therefore the upper limit of Poisson's ratio is $\nu = \frac{1}{2}$.

Table 5.2 Elastic constants for isotropic materials at room temperature‡.

Material	Composition	Modulus of Elasticity E_Y		Poisson's Ratio ν	Sheer Modulus μ		Lamé Constant λ		Bulk Modulus k	
		10^6 psi	GPa		10^6 psi	GPa	10^6 psi	GPa	10^6 psi	GPa
Aluminum	Pure and alloy	9.9–11.4	68.2–78.5	0.32–0.34	3.7–3.85	25.5–26.53	6.7–9.1	46.2–62.7	9.2–11.7	63.4–80.6
Brass	60–70% Cu, 40–30% Zn	14.5–15.9	99.9–109.6	0.33–0.36	5.3–6.0	36.5–41.3	10.6–15.0	73.0–103.4	14.1–19.0	97.1–130.9
Copper		17–18	117–124	0.33–0.36	5.8–6.7	40.0–46.2	12.4–19.0	85.4–130.9	16.3–21.5	112.3–148.1
Iron, cast	2.7–3.6% C	13–21	90–145	0.21–0.30	5.2–8.2	35.8–56.5	3.9–12.1	26.9–83.4	7.4–17.6	51.0–121.3
Steel	Carbon and low alloy	28–32	193–220	0.26–0.29	11.0–11.9	75.8–82.0	12.0–17.1	82.7–117.8	19.3–25.0	133.0–172.3
Stainless steel	18% Cr, 8% Ni	28–30	193–207	0.30	10.6	73.0	16.2–17.3	111.6–119.2	23.2–24.4	160.5–168.1
Titanium	Pure and alloy	15.4–16.6	106.1–114.4	0.34	6.0	41.3	12.2–13.2	84.1–90.9	16.2–17.2	111.6–118.5
Glass	Various	7.2–11.5	49.6–79.2	0.21–0.27	3.8–4.7	26.2–32.4	2.2–5.3	15.2–36.5	4.7–8.4	32.4–57.9
Methyl methacrylate		0.35–0.5	2.41–3.45	—	—	—	—	—	—	—
Polyethylene		0.02–0.055	0.14–0.38	—	—	—	—	—	—	—
Rubber		0.00011–0.00060	0.00076–0.00413	0.50	0.00004–0.00020	0.00028–0.00138	∞†	∞†	∞†	∞†

†As ν approaches 0.5 the ratio of k/E_Y and $\lambda/\mu \to \infty$ as shown in Example 5.4. The actual value of k and λ for some rubbers may be close to the values of steel.

‡Partly from "An Introduction to the Mechanics of Solids," S. H. Crandall and N. C. Dahl, (Eds.), McGraw-Hill, 1959. (Used with permission of McGraw-Hill Book Company.)

5.5 EQUATIONS OF THE INFINITESIMAL THEORY OF ELASTICITY

In Section 4.7, we derived the Cauchy's equations of motion, to be satisfied by any continuum, in the following form

$$\rho a_i = \rho B_i + \frac{\partial T_{ij}}{\partial x_j}, \tag{5.13}$$

where ρ is the density, a_i the acceleration component, ρB_i the component of body force per unit volume and T_{ij} the stress components. All terms in the equation are quantities associated with a particle which is currently at the position (x_1, x_2, x_3).

We shall consider only the case of small motions, that is, motions such that every particle is always in a small neighborhood of the natural state.†. More specifically, if X_i denote the position in the natural state of a typical particle, we assume that $x_i \approx X_i$ and that the magnitude of the components of the displacement gradient $\partial u_i/\partial X_j$, is very much smaller than unity.

Since

$$x_1 = X_1 + u_1,$$

$$x_2 = X_2 + u_2,$$

and

$$x_3 = X_3 + u_3,$$

we have

$$\frac{\partial T_{ij}}{\partial X_1} = \frac{\partial T_{ij}}{\partial x_1}\frac{\partial x_1}{\partial X_1} + \frac{\partial T_{ij}}{\partial x_2}\frac{\partial x_2}{\partial X_1} + \frac{\partial T_{ij}}{\partial x_3}\frac{\partial x_3}{\partial X_1}$$

$$= \frac{\partial T_{ij}}{\partial x_1}\left(1 + \frac{\partial u_1}{\partial X_1}\right) + \frac{\partial T_{ij}}{\partial x_2}\frac{\partial u_2}{\partial X_1} + \frac{\partial T_{ij}}{\partial x_3}\frac{\partial u_3}{\partial X_1}.$$

Thus, for small motion

$$\left(\frac{\partial T_{ij}}{\partial X_1}\right) \approx \left(\frac{\partial T_{ij}}{\partial x_1}\right).$$

Similarly,

$$\frac{\partial T_{ij}}{\partial X_2} \approx \frac{\partial T_{ij}}{\partial x_2} \quad \text{and} \quad \frac{\partial T_{ij}}{\partial X_3} \approx \frac{\partial T_{ij}}{\partial x_3}.$$

Therefore,

$$\frac{\partial T_{ij}}{\partial X_j} \approx \frac{\partial T_{ij}}{\partial x_j}.$$

†We assume the existence of a state, called **natural state**, in which the body is unstressed.

Since

$$a_i = \left.\frac{\partial^2 x_i}{\partial t^2}\right|_{X_i\text{-fixed}} = \left.\frac{\partial^2 u_i}{\partial t^2}\right|_{X_i\text{-fixed}}$$

and (*see* Problem 3.21b)

$$\rho = \rho_0(1 - E_{kk}),$$

so that

$$\rho \frac{\partial^2 u_i}{\partial t^2} \approx \rho_0 \frac{\partial^2 u_i}{\partial t^2},$$

where ρ_0 is the density in the natural state, we finally have the equations of motion governing small motions of elastic bodies

$$\rho_0 \frac{\partial^2 u_i}{\partial t^2} = \rho B_i + \frac{\partial T_{ij}}{\partial X_j}. \tag{5.14}$$

A displacement field u_i is said to describe a possible motion in an elastic medium if it satisfies Eq. (5.14). When a displacement field $u_i = u_i(X_1, X_2, X_3, t)$ is given, to make sure that it is a possible motion, we can first compute the strain field E_{ij} from

$$E_{ij} = \frac{1}{2}\left(\frac{\partial u_i}{\partial X_j} + \frac{\partial u_j}{\partial X_i}\right) \tag{5.15}$$

and then the corresponding elastic stress field T_{ij} from

$$T_{ij} = \lambda e \delta_{ij} + 2\mu E_{ij}. \tag{5.16}$$

The substitution of u_i and T_{ij} in Eq. (5.14) will then verify whether or not the given motion is possible. If the motion is found to be possible, the surface tractions, on the boundary of the body, needed to maintain the motion are given by

$$t_i = T_{ij} n_j. \tag{5.17}$$

On the other hand, if the boundary conditions are prescribed (e.g., certain boundaries of the body must remain fixed at all times and other boundaries must remain traction-free at all times, etc.) then, in order that u_i be the solution to the problem, it must meet the prescribed conditions on the boundaries.†

EXAMPLE 5.5

Combine Eqs. (5.14), (5.15), and (5.16) to obtain the Navier's equations of motion in terms of the displacement components only.

†More generally, there are also initial conditions to be met throughout the body.

Solution. From

$$T_{ij} = \lambda e \delta_{ij} + 2\mu E_{ij} = \lambda e \delta_{ij} + \mu \left(\frac{\partial u_i}{\partial X_j} + \frac{\partial u_j}{\partial X_i} \right)$$

we have

$$\frac{\partial T_{ij}}{\partial X_j} = \lambda \frac{\partial e}{\partial X_j} \delta_{ij} + \mu \left(\frac{\partial^2 u_i}{\partial X_j \partial X_j} + \frac{\partial^2 u_j}{\partial X_j \partial X_i} \right).$$

Now

$$\frac{\partial e}{\partial X_j} \delta_{ij} = \frac{\partial e}{\partial X_i},$$

$$\frac{\partial^2 u_j}{\partial X_j \partial X_i} = \frac{\partial}{\partial X_i} \left(\frac{\partial u_j}{\partial X_j} \right) = \frac{\partial e}{\partial X_i}.$$

Therefore the equation of motion becomes

$$\rho_0 \frac{\partial^2 u_i}{\partial t^2} = \rho_0 B_i + (\lambda + \mu) \frac{\partial e}{\partial X_i} + \mu \frac{\partial^2 u_i}{\partial X_j \partial X_j}, \tag{5.18}$$

where

$$e = \frac{\partial u_j}{\partial X_j} = \frac{\partial u_1}{\partial X_1} + \frac{\partial u_2}{\partial X_2} + \frac{\partial u_3}{\partial X_3}.$$

For $i = 1$, we have

$$\rho_0 \frac{\partial^2 u_1}{\partial t^2} = \rho B_1 + (\lambda + \mu) \frac{\partial e}{\partial X_1} + \mu \left(\frac{\partial^2}{\partial X_1{}^2} + \frac{\partial^2}{\partial X_2{}^2} + \frac{\partial^2}{\partial X_3{}^2} \right) u_1.$$

For $i = 2$ and 3, the equations are of similar form.

5.6 PRINCIPLE OF SUPERPOSITION

Let $u_i^{(1)}$ and $u_i^{(2)}$ be two possible displacement fields corresponding to two body force fields $B_i^{(1)}$ and $B_i^{(2)}$. Let $T_{ij}^{(1)}$ and $T_{ij}^{(2)}$ be the corresponding stress fields. Then

$$\rho_0 \frac{\partial^2 u_i^{(1)}}{\partial t^2} = \rho_0 B_i^{(1)} + \frac{\partial T_{ij}^{(1)}}{\partial X_j}$$

and

$$\rho_0 \frac{\partial^2 u_i^{(2)}}{\partial t^2} = \rho_0 B_i^{(2)} + \frac{\partial T_{ij}^{(2)}}{\partial X_j}.$$

Adding the two equations, we get

$$\rho_0 \frac{\partial^2}{\partial t^2} (u_i^{(1)} + u_i^{(2)}) = \rho_0 (B_i^{(1)} + B_i^{(2)}) + \frac{\partial}{\partial X_j} (T_{ij}^{(1)} + T_{ij}^{(2)}).$$

It is clear from the linearity of Eqs. (5.15) and (5.16) that $T_{ij}^{(1)} + T_{ij}^{(2)}$ is the stress field corresponding to the displacement field $u_i^{(1)} + u_i^{(2)}$. Thus, $u_i^{(1)} + u_i^{(2)}$ is also a possible motion under the body force field $\rho B_i^{(1)} + \rho B_i^{(2)}$. The corresponding stress fields are given by $T_{ij}^{(1)} + T_{ij}^{(2)}$, and the surface tractions needed to maintain the total motion are given by $t_i^{(1)} +$

$t_i^{(2)}$. This is the principle of superposition. One application of this principle is that in a given problem, we shall often assume that the body force is absent having in mind that its effect, if not negligible, can always be obtained separately and then superposed onto the solution of vanishing body force.

5.7 SOME EXAMPLES OF ELASTODYNAMICS

A. Plane Irrotational Wave

Consider the motion

$$u_1 = \epsilon \sin \frac{2\pi}{l} (X_1 - c_L t), \quad u_2 = 0, \quad u_3 = 0, \tag{i}$$

representing an infinite train of sinusoidal plane waves. In this motion, every particle executes simple harmonic oscillations of small amplitude ϵ around its natural state, the motion being always parallel to the $\mathbf{e}_1$-direction. All particles on a plane perpendicular to $\mathbf{e}_1$ are at the same phase of the harmonic motion at any one time [i.e., the same value of $2\pi(X_1 - c_L t)/l$]. A particle denoted by $X_1 + dX_1$, acquires at $t + dt$, the same phase of the particle denoted by X_1 at time t if $(X_1 + dX_1) - c_L(t + dt) = X_1 - c_L t$, i.e., $dX_1/dt = c_L$. Thus c_L is known as the **phase velocity** (the velocity with which the phase is being transmitted). It is also the velocity with which the sinusoidal disturbance of wavelength l is moving in the $\mathbf{e}_1$-direction. Since the motions of the particles are parallel to the direction of the propagation of wave, it is a longitudinal wave.

We shall now consider if this wave is a possible motion in an elastic medium.

The strain components corresponding to the u_i given in Eq. (i) are

$$E_{11} = \epsilon \left(\frac{2\pi}{l}\right) \cos \frac{2\pi}{l} (X_1 - c_L t),$$

$$E_{22} = E_{33} = E_{12} = E_{13} = E_{23} = 0.$$

The stress components are (note $e = E_{11} + 0 + 0$)

$$T_{11} = (\lambda + 2\mu) E_{11},$$

$$T_{22} = \lambda E_{11},$$

$$T_{33} = \lambda E_{11},$$

$$T_{12} = T_{13} = T_{23} = 0.$$

Substituting T_{ij} and u_i into the equation of motion†

$$\rho_0 \frac{\partial^2 u_i}{\partial t^2} = \frac{\partial T_{ij}}{\partial X_j}$$

we easily see that the second and third equations of motion are automatically satisfied $(0=0)$ and the first equation demands that

$$\rho_0 \frac{\partial^2 u_1}{\partial t^2} = \frac{\partial T_{11}}{\partial X_1}$$

or

$$-\rho_0 \epsilon \left(\frac{2\pi}{l}\right)^2 c_L^2 \sin \frac{2\pi}{l}(X_1 - c_L t) = -(\lambda + 2\mu)\, \epsilon \left(\frac{2\pi}{l}\right)^2 \sin \frac{2\pi}{l}(X_1 - c_L t)$$

so that the phase velocity c_L is obtained to be

$$c_L = \sqrt{\frac{\lambda + 2\mu}{\rho_0}}. \tag{5.19}$$

Thus we see that with c_L given by Eq. (5.19), the wave motion considered is a possible one. Since, for this motion, the components of the rotation tensor

$$\frac{1}{2}\left(\frac{\partial u_i}{\partial X_j} - \frac{\partial u_j}{\partial X_i}\right)$$

are zero at all times, it is known as a plane **irrotational wave**. As a particle oscillates, its volume also changes harmonically [the dilatation $e = E_{11} = \epsilon(2\pi/l) \cos (2\pi/l)(X_1 - c_L t)$], the wave is thus also known as a dilatational wave.

From Eq. (5.19), we see that for the plane wave discussed, the phase velocity c_L depends only on the material properties and not on the wavelength l. Thus any disturbance represented by the superposition of any number of one-dimensional plane irrotational wave trains of different wavelengths propagates, without changing the form of the disturbance (no longer sinusoidal), with the velocity equal to the phase velocity c_L. In fact, it can be easily seen [from Eq. (5.18)] that, any irrotational disturbance given by

$$u_1 = u_1(X_1, t), \quad u_2 = u_3 = 0 \tag{5.20}$$

is a possible motion in the absence of body forces provided that $u_1(X_1, t)$ is a solution of the simple wave equation

$$\frac{\partial^2 u_1}{\partial t^2} = c_L^2 \frac{\partial^2 u_1}{\partial X_1^2}. \tag{5.21}$$

†Body forces B_i are neglected.

It can be easily verified that $u_1 = f(s)$, where $s = X_1 \pm c_L t$ satisfies the above equation for any function f, so that disturbances of any form given by $f(s)$ propagate without changing its form with wave speed c_L. In other words, the phase velocity is also the rate of advance of a finite train of waves, or, of any arbitrary disturbance, into an undisturbed region.

EXAMPLE 5.6

Consider a displacement field

$$u_1 = \alpha \sin \frac{2\pi}{l}(X_1 - c_L t) + \beta \cos \frac{2\pi}{l}(X_1 - c_L t),$$

$$u_2 = u_3 = 0$$

for a material half-space that lies to the right of the plane $X_1 = 0$.
(a) Determine α, β, and l if the applied displacement on the plane $X_1 = 0$ is given by $\mathbf{u} = (b \sin \omega t)\mathbf{e}_1$.
(b) Determine α, β, and l if the applied surface traction on $X_1 = 0$ is given by $\mathbf{t} = (d \sin \omega t)\mathbf{e}_1$.

Solution. The given displacement field is the superposition of two longitudinal elastic waves having the same velocity of propagation c_L in the positive X_1-direction and is therefore a possible elastic solution.
(a) To satisfy the displacement boundary condition, one simply sets

$$u_1(0, t) = b \sin \omega t$$

or

$$-\alpha \sin\left(\frac{2\pi c_L}{l}\right)t + \beta \cos\left(\frac{2\pi c_L}{l}\right)t = b \sin \omega t.$$

Since this relation must be satisfied for all time t, we have

$$\beta = 0, \quad \alpha = -b, \quad l = \frac{2\pi c_L}{\omega}$$

and the elastic wave has the form

$$u_1 = -b \sin \frac{\omega}{c_L}(X_1 - c_L t).$$

Note that the wavelength is inversely proportional to the forcing frequency ω. That is, the higher the forcing frequency the smaller the wavelength of the elastic wave.
(b) To satisfy the traction boundary condition on $X_1 = 0$, one requires that

$$\mathbf{t} = \mathbf{T}(-\mathbf{e}_1) = -(T_{11}\mathbf{e}_1 + T_{21}\mathbf{e}_2 + T_{31}\mathbf{e}_3) = (d \sin \omega t)\mathbf{e}_1,$$

that is, at $X_1 = 0$, $T_{11} = -d \sin \omega t$, $T_{21} = T_{31} = 0$. For the assumed displacement field

$$T_{11} = (2\mu + \lambda)\frac{\partial u_1}{\partial X_1}, \quad T_{21} = T_{31} = 0,$$

therefore,

$$-d \sin \omega t = (2\mu + \lambda)\left[\alpha\left(\frac{2\pi}{l}\right)\cos \frac{2\pi}{l}(X_1 - c_L t) - \beta\left(\frac{2\pi}{l}\right)\sin \frac{2\pi}{l}(X_1 - c_L t)\right]_{X_1=0},$$

i.e.,

$$-d \sin \omega t = (2\mu + \lambda)\frac{2\pi}{l}\left[\alpha \cos \frac{2\pi}{l} c_L t + \beta \sin \frac{2\pi}{l} c_L t\right].$$

To satisfy this relation for all time t, we have

$$\alpha = 0, \quad \beta = \frac{-d}{2\mu+\lambda}\left(\frac{l}{2\pi}\right), \quad \omega = \frac{2\pi c_L}{l}$$

or

$$\alpha = 0, \quad \beta = \frac{-dc_L}{(2\mu+\lambda)\omega}, \quad l = \frac{2\pi c_L}{\omega},$$

and the resulting wave has the form,

$$u_1 = \frac{-dc_L}{(2\mu+\lambda)\omega}\cos\frac{\omega}{c_L}(X_1 - c_L t).$$

We note, that not only the wavelength but the amplitude of the resulting wave is inversely proportional to the forcing frequency.

B. Plane Equivoluminal Wave

Consider the motion

$$u_1 = 0, \quad u_2 = \epsilon \sin\frac{2\pi}{l}(X_1 - c_T t), \quad u_3 = 0. \tag{ii}$$

This infinite train of plane harmonic wave differs from that discussed in Section 5.7A in that it is a transverse wave: the particle motion is parallel to $\mathbf{e}_2$-direction, whereas the disturbance is propagated in the $\mathbf{e}_1$-direction.

For this motion, the strain components are

$$E_{11} = E_{22} = E_{33} = E_{13} = E_{23} = 0$$

and

$$E_{12} = \frac{1}{2}\epsilon\left(\frac{2\pi}{l}\right)\cos\frac{2\pi}{l}(X_1 - c_T t),$$

and the stress components are

$$T_{11} = T_{22} = T_{33} = T_{13} = T_{23} = 0$$

and

$$T_{12} = \mu\epsilon\left(\frac{2\pi}{l}\right)\cos\frac{2\pi}{l}(X_1 - c_T t).$$

Substitution of T_{ij} and u_i in the equations of motion, neglecting body force, gives the phase velocity c_T to be

$$c_T = \sqrt{\frac{\mu}{\rho_0}}. \tag{5.22}$$

Since, in this motion, the dilatation ϵ is zero at all times, it is known as an "equivoluminal wave." It is also called "shear wave."

Here again the phase velocity c_T is independent of the wavelength l, so that it again has the additional significance of being the wave velocity of a finite train of equivoluminal waves, or of any arbitrary equivoluminal disturbance into an undisturbed region.

The ratio of the two phase velocities c_L and c_T is

$$\frac{c_L}{c_T}=\sqrt{\frac{\lambda+2\mu}{\mu}}.$$

Since $\lambda=2\mu\nu/(1-2\nu)$, the ratio is found to depend only on ν, in fact

$$\frac{c_L}{c_T}=\sqrt{\frac{2(1-\nu)}{1-2\nu}}=\sqrt{\left(1+\frac{1}{1-2\nu}\right)}. \tag{5.23}$$

For steel with $\nu=0.3$, $c_L/c_T=\sqrt{7/2}=1.87$. Since $\nu<\frac{1}{2}$, c_L is always greater than c_T.

EXAMPLE 5.7

Consider a displacement field

$$u_2=\alpha\sin\frac{2\pi}{l}(X_1-c_Tt)+\beta\cos\frac{2\pi}{l}(X_1-c_Tt),$$

$$u_1=u_3=0$$

for a material half-space that lies to the right of the plane $X_1=0$.

(a) Determine α, β, and l if the applied displacement on $X_1=0$ is given by $\mathbf{u}=(b\sin\omega t)\mathbf{e}_2$.

(b) Determine α, β, and l if the applied surface traction on $X_1=0$ is $\mathbf{t}=(d\sin\omega t)\mathbf{e}_2$.

Solution. The problem is analogous to that of the previous example.

(a) Using $u_2(0,t)=b\sin\omega t$, we have

$$\beta=0,\quad \alpha=-b,\quad l=\frac{2\pi c_T}{\omega}$$

and

$$u_2=-b\sin\frac{\omega}{c_T}(X_1-c_Tt).$$

(b) Using $\mathbf{t}=-T_{21}\mathbf{e}_2=(d\sin\omega t)\mathbf{e}_2$ gives

$$\alpha=0,\quad \beta=\frac{-dc_T}{\mu\omega},\quad l=\frac{2\pi c_T}{\omega},$$

and

$$u_2=\frac{-dc_T}{\mu\omega}\cos\frac{\omega}{c_T}(X_1-c_Lt).$$

EXAMPLE 5.8

Consider the displacement field

$$u_3=\alpha\cos pX_2\cos\frac{2\pi}{l}(X_1-ct),$$

$$u_2=u_1=0.$$

(a) Show that this is an equivoluminal motion.
(b) From the equation of motion, determine the phase velocity c in terms of p, l, ρ_0 and μ (assuming no body forces).
(c) This displacement field is used to describe a type of waveguide that is bounded by the planes $X_2 = \pm h$. Find the phase velocity c if these planes are traction-free.

Solution. (a) Since

$$\operatorname{div}\mathbf{u} = \frac{\partial u_1}{\partial X_1} + \frac{\partial u_2}{\partial X_2} + \frac{\partial u_3}{\partial X_3} = 0+0+0 = 0,$$

thus, there is no change of volume at any time.
(b) For convenience, let $k \equiv 2\pi/l$ and $\omega \equiv kc = 2\pi c/l$, then

$$u_3 = \alpha \cos pX_2 \cos (kX_1 - \omega t),$$

where k is known as the wave number and ω is the circular frequency. The only nonzero stresses are given by (note: $u_1 = u_2 = 0$)

$$T_{13} = T_{31} = \mu \frac{\partial u_3}{\partial X_1} = \alpha\mu k[-\cos pX_2 \sin (kX_1 - \omega t)],$$

$$T_{23} = T_{32} = \mu \frac{\partial u_3}{\partial X_2} = \alpha\mu p[-\sin pX_2 \cos (kX_1 - \omega t)].$$

The substitution of the stress components into the third equation of motion yields (the first two equations are trivially satisfied)

$$\frac{\partial T_{31}}{\partial X_1} + \frac{\partial T_{32}}{\partial X_2} = (\mu k^2 + \mu p^2)\,(-u_3) = \rho_0 \frac{\partial^2 u_3}{\partial t^2} = \rho_0\omega^2(-u_3).$$

Therefore, with $c_T^2 \equiv \mu/\rho_0$,

$$k^2 = \frac{\rho_0}{\mu}\omega^2 - p^2 = \left(\frac{\omega}{c_T}\right)^2 - p^2.$$

Since $k = 2\pi/l$, $\omega = 2\pi c/l$, therefore

$$c = c_T\left[\left(\frac{lp}{2\pi}\right)^2 + 1\right]^{1/2}.$$

(c) To satisfy the traction-free boundary condition at $X_2 = \pm h$, we require that

$$\mathbf{t} = \pm \mathbf{T}\mathbf{e}_2 = \pm\,(T_{12}\mathbf{e}_1 + T_{22}\mathbf{e}_2 + T_{32}\mathbf{e}_3) = \pm\, T_{32}\mathbf{e}_3 = 0,$$

therefore, $T_{32}\,]_{X_2=\pm h} = \mp \mu p\alpha \sin ph \cos (kX_1 - \omega t) = 0$. In order for this relation to be satisfied for all X_1 and t, we must have

$$\sin ph = 0.$$

Thus,

$$p = \frac{n\pi}{h}, \qquad n = 0, 1, 2, \ldots.$$

Each value of n determines a possible displacement field, and the phase velocity c corresponding to each mode is given by

$$c = c_T\left[\left(\frac{ln}{2h}\right)^2 + 1\right]^{1/2}.$$

This result indicates that the equivoluminal wave is propagating with a speed c greater than the speed of a plane equivoluminal wave c_T. Note that when $p = 0$, $c = c_T$ as expected.

EXAMPLE 5.9

An infinite train of harmonic plane waves propagates in the direction of the unit vector $\mathbf{e}_n$. Express the displacement field in vector form for (a) a longitudinal wave, (b) a transverse wave.

Solution. Let $\mathbf{X}$ be the position vector of any point on a plane whose normal is $\mathbf{e}_n$ and whose distance from the origin is d (Fig. 5.3). Then $\mathbf{X}\cdot\mathbf{e}_n = d$. Thus, in order that the particles on the plane be at the same phase of the harmonic oscillation at any one time, the argument of sine (or cosine) must be of the form $(2\pi/l)\,(\mathbf{X}\cdot\mathbf{e}_n - ct - \eta)$, where η is an arbitrary constant.

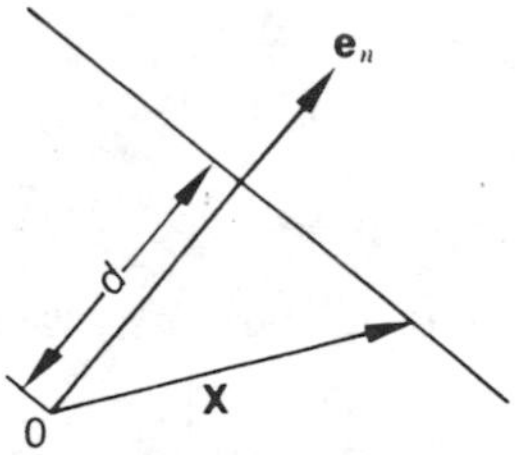

Fig. 5.3

(a) For longitudinal waves, $\mathbf{u}$ is parallel to $\mathbf{e}_n$, thus

$$\mathbf{u} = \epsilon\left[\sin\frac{2\pi}{l}\,(\mathbf{X}\cdot\mathbf{e}_n - c_L t - \eta)\right]\mathbf{e}_n. \tag{5.24}$$

In particular, if $\mathbf{e}_n = \mathbf{e}_1$, then $u_1 = \epsilon \sin (2\pi/l)\,(X_1 - c_L t - \eta)$, $u_2 = u_3 = 0$.

(b) For transverse waves, $\mathbf{u}$ is perpendicular to $\mathbf{e}_n$. Let $\mathbf{e}_t$ be a unit vector perpendicular to $\mathbf{e}_n$. Then

$$\mathbf{u} = \epsilon\left[\sin\frac{2\pi}{l}\,(\mathbf{X}\cdot\mathbf{e}_n - c_T t - \eta)\right]\mathbf{e}_t. \tag{5.25}$$

The plane of $\mathbf{e}_t$ and $\mathbf{e}_n$ is known as the plane of polarization. In particular, if $\mathbf{e}_n = \mathbf{e}_1$, $\mathbf{e}_t = \mathbf{e}_2$ then

$$u_1 = 0, \quad u_2 = \epsilon \sin\frac{2\pi}{l}\,(X_1 - c_T t - \eta), \quad u_3 = 0.$$

EXAMPLE 5.10

In Fig. 5.4 all three unit vectors $\mathbf{e}_{n_1}$, $\mathbf{e}_{n_2}$, and $\mathbf{e}_{n_3}$ lie in the X_1X_2 plane. Express the displacement components with respect to the X_i-coordinates of plane harmonic waves for (a) a transverse wave of amplitude ϵ_1, wavelength l_1 polarized in the X_1X_2-plane and propagating in the direction of e_{n_1}, (b) a transverse wave of amplitude ϵ_2, wavelength l_2, polarized in the X_1X_2-plane and propagating in the direction of $\mathbf{e}_{n_2}$, (c) a longitudinal wave of amplitude ϵ_3, wavelength l_3 propagating in the direction of $\mathbf{e}_{n_3}$.

Solution. Using the results of Example 5.9, we have

(a) $\mathbf{e}_{n_1} = \sin\alpha_1\mathbf{e}_1 - \cos\alpha_1\mathbf{e}_2$, $\quad \mathbf{X}\cdot\mathbf{e}_{n_1} = X_1\sin\alpha_1 - X_2\cos\alpha_1$,

$\mathbf{e}_{t_1} = \pm(\cos\alpha_1\mathbf{e}_1 + \sin\alpha_1\mathbf{e}_2)$.

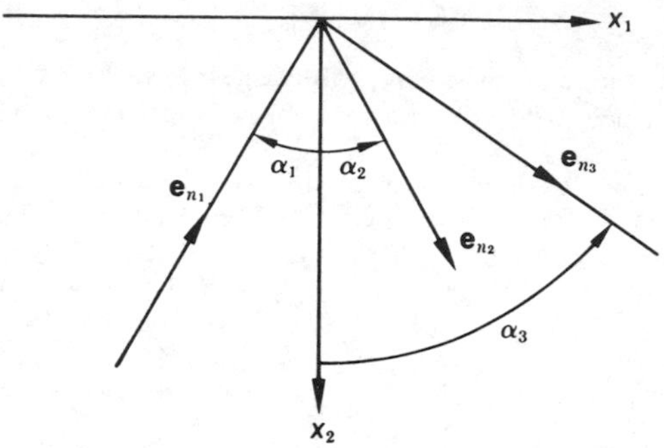

Fig. 5.4

Thus†

$$\left.\begin{aligned} u_1 &= \cos\alpha_1 \\ u_2 &= \sin\alpha_1, \end{aligned}\right\} \epsilon_1 \sin\frac{2\pi}{l_1}(X_1\sin\alpha_1 - X_2\cos\alpha_1 - c_T t - \eta_1).$$
$$u_3 = 0.$$

(b) $\mathbf{e}_{n_2} = \sin\alpha_2\mathbf{e}_1 + \cos\alpha_2\mathbf{e}_2, \quad \mathbf{X}\cdot\mathbf{e}_{n_2} = X_1\sin\alpha_2 + X_2\cos\alpha_2,$
$\mathbf{e}_{t_2} = \pm(\cos\alpha_2\mathbf{e}_1 - \sin\alpha_2\mathbf{e}_2).$

Thus

$$\left.\begin{aligned} u_1 &= \cos\alpha_2 \\ u_2 &= -\sin\alpha_2, \end{aligned}\right\} \epsilon_2 \sin\frac{2\pi}{l_2}(X_1\sin\alpha_2 + X_2\cos\alpha_2 - c_T t - \eta_2),$$
$$u_3 = 0.$$

(c) $\mathbf{e}_{n_3} = \sin\alpha_3\mathbf{e}_1 + \cos\alpha_3\mathbf{e}_2, \quad \mathbf{X}\cdot\mathbf{e}_{n_3} = X_1\sin\alpha_3 + X_2\cos\alpha_3.$

Thus

$$\left.\begin{aligned} u_1 &= \sin\alpha_3 \\ u_2 &= \cos\alpha_3, \end{aligned}\right\} \epsilon_3 \sin\frac{2\pi}{l_3}(X_1\sin\alpha_3 + X_2\cos\alpha_3 - c_L t - \eta_3),$$
$$u_3 = 0.$$

C. Reflection of Plane Elastic Waves

In Fig. 5.5, the plane $X_2 = 0$ is the free boundary of an elastic medium, occupying the lower half-space $X_2 \geq 0$. We wish to study how an incident plane wave is reflected by the boundary.

Consider an incident transverse wave of wavelength l_1, polarized in the plane of incidence with an incident angle α_1 (*see* Fig. 5.5). Since $X_2 = 0$ is a free boundary, the surface traction on the plane is zero at all times. Thus, the boundary will generate reflection waves in such a way that when they are superposed on the incident wave, the stress vector on the boundary vanishes at all times.

Let us superpose on the incident transverse wave two reflection waves

†The sign of $\mathbf{e}_{t_1}$ can be absorbed in the arbitrary constant η_1.

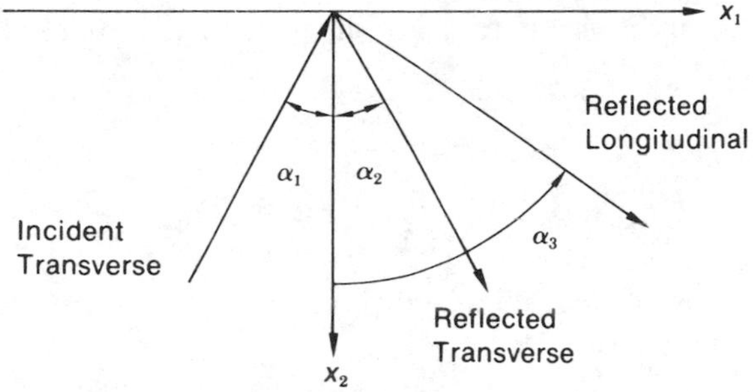

Fig. 5.5

(*see* Fig. 5.5), one transverse, the other longitudinal, both oscillating in the plane of incidence. The reason for superposing not only a reflected transverse wave but also a longitudinal one is that if only one is superposed, the stress-free condition on the boundary in general cannot be met, as will become obvious in the following derivation.

Let u_i denote the displacement components of the superposition of the three waves, then from the results of Example 5.10, we have

$$\begin{matrix} u_1 = \\ u_2 = \end{matrix} \begin{Bmatrix} \cos\alpha_1 \\ \sin\alpha_1 \end{Bmatrix} \epsilon_1 \sin\phi_1 + \begin{Bmatrix} \cos\alpha_2 \\ -\sin\alpha_2 \end{Bmatrix} \epsilon_2 \sin\phi_2 + \begin{Bmatrix} \sin\alpha_3 \\ \cos\alpha_3 \end{Bmatrix} \epsilon_3 \sin\phi_3,$$
$$u_3 = 0, \tag{A1}$$

where

$$\phi_1 = \frac{2\pi}{l_1}(X_1 \sin\alpha_1 - X_2 \cos\alpha_1 - c_T t - \eta_1),$$

$$\phi_2 = \frac{2\pi}{l_2}(X_1 \sin\alpha_2 + X_2 \cos\alpha_2 - c_T t - \eta_2), \tag{A2}$$

$$\phi_3 = \frac{2\pi}{l_3}(X_1 \sin\alpha_3 + X_2 \cos\alpha_3 - c_L t - \eta_3).$$

On the free boundary $(X_2 = 0)$, where $\mathbf{n} = -\mathbf{e}_2$, the condition $\mathbf{t} = \mathbf{0}$ leads to

$$\mathbf{Te}_2 = \mathbf{0},$$

i.e.,

$$T_{12} = T_{22} = T_{32} = 0.$$

Using Hooke's law, and noting that $u_3 = 0$, and u_1 and u_2 do not depend on X_3, we easily see that the condition $T_{32} = 0$ is automatically satisfied.

The other two conditions, in terms of displacement components, are

$$\frac{\partial u_1}{\partial X_2}+\frac{\partial u_2}{\partial X_1}=0 \qquad \text{on } X_2=0, \tag{B1}$$

$$(\lambda+2\mu)\frac{\partial u_2}{\partial X_2}+\lambda\frac{\partial u_1}{\partial X_1}=0. \tag{B2}$$

Performing the required differentiation, we obtain from Eqs. (B1) and (B2)

$$\frac{\epsilon_1}{l_1}(\sin^2\alpha_1-\cos^2\alpha_1)\cos\phi_1+\frac{\epsilon_2}{l_2}(\cos^2\alpha_2-\sin^2\alpha_2)\cos\phi_2$$
$$+\frac{\epsilon_3}{l_3}(\sin 2\alpha_3)\cos\phi_3=0, \tag{b1}$$

$$\frac{\epsilon_1}{l_1}\mu\sin 2\alpha_1\cos\phi_1+\frac{\epsilon_2}{l_2}\mu\sin 2\alpha_2\cos\phi_2-\frac{\epsilon_3}{l_3}(\lambda+2\mu\cos^2\alpha_3)\cos\phi_3=0. \tag{b2}$$

Since these equations are to be satisfied on $X_2=0$ for whatever values of X_1 and t, we must have

$$\cos\phi_1=\cos\phi_2=\cos\phi_3 \qquad \text{on } X_2=0$$

so that they drop out from Eq. (b1) and (b2). Thus at $X_2=0$, $\phi_1=\phi_2\pm 2p\pi=\phi_3\pm 2q\pi$, p and q are integers, i.e.,

$$\frac{2\pi}{l_1}(X_1\sin\alpha_1-c_Tt-\eta_1)=\frac{2\pi}{l_2}(X_1\sin\alpha_2-c_Tt-\eta_2')$$
$$=\frac{2\pi}{l_3}(X_1\sin\alpha_3-c_Lt-\eta_3'), \tag{C}$$

where $\eta_2'=\eta_2\mp pl_2$ and $\eta_3'=\eta_3\mp ql_3$.

Equation (C) can be satisfied for whatever values of X_1 and t only if

$$\frac{\sin\alpha_1}{l_1}=\frac{\sin\alpha_2}{l_2}=\frac{\sin\alpha_3}{l_3}, \tag{D1}$$

$$\frac{c_T}{l_1}=\frac{c_T}{l_2}=\frac{c_L}{l_3}, \tag{D2}$$

and

$$\frac{\eta_1}{l_1}=\frac{\eta_2'}{l_2}=\frac{\eta_3'}{l_3}. \tag{D3}$$

Thus

$$l_2=l_1, \tag{5.26a}$$

$$nl_3 = l_1, \qquad \text{where } \frac{1}{n} = \frac{c_L}{c_T} = \sqrt{\frac{\lambda + 2\mu}{\mu}}, \tag{5.26b}$$

$$\alpha_1 = \alpha_2 \tag{5.26c}$$

$$n \sin \alpha_3 = \sin \alpha_1, \tag{5.26d}$$

$$\eta_2' = \eta_1, n\eta_3' = \eta_1. \tag{5.26e}$$

That is, the reflected transverse wave has the same wavelength as that of the incident transverse wave and the angle of reflection is the same as the incident angle, the longitudinal wave has a different wavelength and a different reflection angle depending on the so-called "refraction index n."

With $\cos \phi_i$ dropped out, and in view of Eqs. (5.26), the boundary conditions (b1) and (b2) now become

$$\epsilon_1(\sin^2\alpha_1 - \cos^2\alpha_1) + \epsilon_2(\cos^2\alpha_1 - \sin^2\alpha_1) + \epsilon_3 n \sin 2\alpha_3 = 0, \tag{F1}$$

$$\epsilon_1(\mu \sin 2\alpha_1) + \epsilon_2(\mu \sin 2\alpha_1) - \epsilon_3 n(2\mu \cos^2\alpha_3 + \lambda) = 0. \tag{F2}$$

These two equations uniquely determine the amplitudes of the reflected waves in terms of the incident amplitude (which is arbitrary). In fact,

$$\epsilon_3 = \frac{n \sin 4\alpha_1}{\cos^2 2\alpha_1 + n^2 \sin 2\alpha_1 \sin 2\alpha_3}\epsilon_1, \tag{5.27a}$$

$$\epsilon_2 = \frac{\cos^2 2\alpha_1 - n^2 \sin 2\alpha_1 \sin 2\alpha_3}{\cos^2 2\alpha_1 + n^2 \sin 2\alpha_1 \sin 2\alpha_3}\epsilon_1. \tag{5.27b}$$

Thus, the problem of the reflection of a transverse wave polarized in the plane of incidence is solved. We mention that if the incident transverse wave is polarized normal to the plane of incidence, no longitudinal component occurs. Also, when an incident longitudinal wave is reflected, in addition to the regularly reflected longitudinal wave, there is also a transverse wave polarized in the plane of incidence.

Equation (5.26d) is analogous to the optical Snell's law, except here we have reflection instead of refraction. If $\sin \alpha_1 > n$, then $\sin \alpha_3 > 1$ and there is no longitudinal reflected wave but rather, waves of a more complicated nature will be generated. The angle $\alpha_i = \sin^{-1} n$ is called the critical angle.

D. Vibration of an Infinite Plate

Consider an infinite plate bounded by the planes $X_1 = 0$, and $X_1 = l$. These plane faces may have either a prescribed motion (fixed corresponding to no motion) or a prescribed surface traction (free corresponding to no surface traction).

The presence of these two boundaries indicates the possibility of a vibration (a standing wave). We begin by assuming the vibration is of the form

$$u_1 = u_1(X_1, t), \quad u_2 = u_3 = 0,$$

and, just as for longitudinal waves, the displacement must satisfy the equation

$$c_L{}^2 \frac{\partial^2 u_1}{\partial X_1{}^2} = \frac{\partial^2 u_1}{\partial t^2}.$$

A steady-state vibration solution to this equation is of the form

$$u_1 = (A \cos \lambda X_1 + B \sin \lambda X_1)(C \cos c_L \lambda t + D \sin c_L \lambda t), \tag{5.28}$$

where the constants A, B, C, D, and λ are determined by the boundary conditions. This vibration mode is sometimes termed a "**thickness stretch**" vibration because the plate is being stretched through its thickness. It is analogous to acoustic vibration of organ pipes and to the longitudinal vibration of slender rods.

Another vibration mode can be obtained by assuming the displacement field

$$u_2 = u_2(X_1, t), \quad u_1 = u_3 = 0.$$

In this case, the displacement field must satisfy the equation

$$c_T{}^2 \frac{\partial^2 u_2}{\partial X_1{}^2} = \frac{\partial^2 u_2}{\partial t^2}$$

and the solution is of the same form as in the previous case.

This vibration is termed "**thickness-shear**" and it is analogous to the vibrating string.

EXAMPLE 5.11

(a) Find the thickness-stretch vibration of a plate, where the left face ($X_1 = 0$) is subjected to a forced displacement $\mathbf{u} = (\alpha \cos \omega t)\mathbf{e}_1$ and the right face ($X_1 = l$) is fixed.
(b) Determine the values of ω that give resonance.

Solution. (a) Using Eq. (5.28) and the first boundary condition, we have

$$\alpha \cos \omega t = u_1(0, t) = AC \cos c_L \lambda t + AD \sin c_L \lambda t.$$

Therefore,

$$AC = \alpha, \quad \lambda = \frac{\omega}{c_L}, \quad \text{and } D = 0.$$

The second boundary condition gives

$$0 = u_1(l, t) = \left(\alpha \cos \frac{\omega l}{c_L} + BC \sin \frac{\omega l}{c_L}\right) \cos \omega t.$$

Therefore,

$$BC = -\alpha \cot \frac{\omega l}{c_L}$$

and the vibration is given by

$$u_1(X_1, t) = \alpha \left[\cos \frac{\omega}{c_L} X_1 - \frac{\sin \frac{\omega}{c_L} X_1}{\tan \frac{\omega}{c_L} l} \right] \cos \omega t.$$

(b) Resonance is indicated by unbounded displacements. This occurs in part (a) for forcing frequencies corresponding to $\tan \omega l / c_L = 0$, that is, when†

$$\omega = \frac{n\pi c_L}{l}, \qquad n = 1, 2, 3 \ldots .$$

EXAMPLE 5.12

(a) Find the thickness-shear vibration of an infinite plate which has an applied surface traction $\mathbf{t} = -(\beta \cos \omega t)\mathbf{e}_2$ on the plane $X_1 = 0$ and is fixed at the plane $X_1 = l$.
(b) Determine the resonant frequencies.

Solution. (a) The traction on $X_1 = 0$ determines the stress $T_{12}]_{X_1=0} = \beta \cos \omega t$. This shearing stress forces a vibration of the form

$$u_2 = (A \cos \lambda X_1 + B \sin \lambda X_1)\,(C \cos c_T\lambda t + D \sin c_T\lambda t).$$

Using Hooke's law, we have

$$T_{12}\Big]_{X_1=0} = \mu \frac{\partial u_2}{\partial X_1}\Big]_{X_1=0} = \beta \cos \omega t$$

or

$$\frac{\beta}{\mu} \cos \omega t = \lambda BC \cos c_T\lambda t + \lambda BD \sin c_T\lambda t.$$

Thus,

$$\lambda = \frac{\omega}{c_T}, \quad D = 0, \quad \text{and } BC = \frac{\beta c_T}{\omega \mu}.$$

The boundary condition at $X_1 = l$ gives

$$u_2(l, t) = 0 = \left(AC \cos \frac{\omega l}{c_T} + \frac{\beta c_T}{\omega \mu} \sin \frac{\omega l}{c_T} \right) \cos \omega t.$$

Thus,

$$AC = -\frac{\beta c_T}{\omega \mu} \tan \frac{\omega l}{c_T}$$

and

$$u_2(X_1, t) = \frac{\beta c_T}{\omega \mu} \left(\sin \frac{\omega}{c_T} X_1 - \tan \frac{\omega l}{c_T} \cos \frac{\omega}{c_T} X_1 \right) \cos \omega t.$$

†Note: These values of ω correspond to the natural free-vibration frequencies with both faces fixed.

(b) Resonance occurs for

$$\tan\frac{\omega l}{c_T} = \infty$$

or†

$$\omega = \frac{n\pi c_T}{2l}, \qquad n = 1, 3, 5 \ldots.$$

5.8 SOME EXAMPLES OF ELASTOSTATIC PROBLEMS

A. Simple Extension

A cylindrical elastic bar of arbitrary cross-section (Fig. 5.6) is under the action of equal opposite normal traction σ at its end faces. Its lateral surface is free from any surface traction and body forces are assumed to be absent.

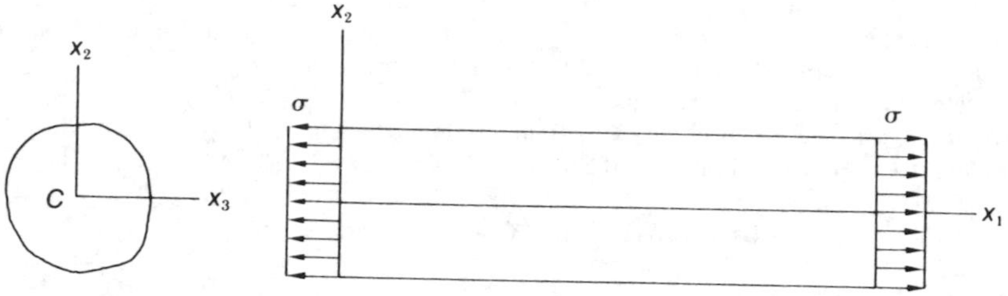

Fig. 5.6

Intuitively, one expects that the state of stress at any point will depend neither on the length of the bar nor on its lateral dimension‡. In other words, the state of stress in the bar is the same everywhere. By considering points on the boundary surface, one then concludes that the stress components are

$$T_{11} = \sigma, \quad T_{22} = T_{33} = T_{12} = T_{13} = T_{23} = 0. \tag{5.29}$$

We now show that this result is a possible solution.

(i) The differential equation of equilibrium $\partial T_{ij}/\partial X_j = 0$ is identically satisfied.

(ii) The boundary condition on the end faces is obviously satisfied. On the lateral surface,

$$\mathbf{n} = 0\mathbf{e}_1 + n_2\mathbf{e}_2 + n_3\mathbf{e}_3$$

†Note: These values of ω correspond to free-vibration natural frequencies with one face traction-free and one face fixed.

‡In the case of $\sigma < 0$, it is assumed that buckling does not take place.

and

$$\mathbf{t} = \mathbf{Tn} = n_2(\mathbf{Te}_2) + n_3(\mathbf{Te}_3) = n_2(\mathbf{0}) + n_3(\mathbf{0}) = \mathbf{0}.$$

Thus, the traction-free condition is also satisfied.

(iii) The stress components being constants, the strain components are also constant. The conditions of compatibility are automatically satisfied. In fact,

$$E_{11} = \frac{1}{E_Y}[T_{11} - \nu(T_{22} + T_{33})] = \frac{\sigma}{E_Y}, \tag{5.30a}$$

$$E_{22} = \frac{1}{E_Y}[T_{22} - \nu(T_{11} + T_{33})] = -\nu\frac{\sigma}{E_Y}, \tag{5.30b}$$

$$E_{33} = \frac{1}{E_Y}[T_{33} - \nu(T_{22} + T_{11})] = -\nu\frac{\sigma}{E_Y}, \tag{5.30c}$$

$$E_{12} = E_{13} = E_{23} = 0,$$

and it is easily verified that the following single-valued continuous displacement field corresponds to the strain field of Eq. (5.30):

$$u_1 = \left(\frac{\sigma}{E_Y}\right)X_1, \quad u_2 = -\left(\frac{\nu\sigma}{E_Y}\right)X_2, \quad u_3 = -\left(\frac{\nu\sigma}{E_Y}\right)X_3. \tag{5.31}$$

Thus, we have completed the solution of the problem of simple extension or compression.

If the constant cross-sectional area of the bar is A, the surface traction σ on either end face gives rise to a resultant force of magnitude,

$$P = \sigma A \tag{5.32}$$

passing through the centroid of the area.† Thus, in terms of P and A, the stress components in the bar are

$$[\mathrm{T}] = \begin{bmatrix} \frac{P}{A} & 0 & 0 \\ 0 & 0 & 0 \\ 0 & 0 & 0 \end{bmatrix}. \tag{5.33}$$

Since the matrix is diagonal, we know from Chapter 2 that the principal stresses are P/A, 0, 0. Thus, the maximum normal stress is

$$(T_n)_{\max} = \frac{P}{A}, \tag{5.34}$$

†If (y_P, z_P) denote the line of action of P, then $y_P = (\int y\sigma dA)/P = (\int y dA)/A = y_c$ and similarly, $z_P = z_c$.

acting on cross-sectional planes, and the maximum shearing stress is

$$(T_s)_{\max} = \frac{P}{2A}, \tag{5.35}$$

acting on planes making 45° with the normal cross-sectional plane.

Let the undeformed length of the bar be l and let Δl be its elongation. Then from the first equation of Eq. (5.31) and Eq. (5.32) we have

$$\Delta l = \frac{Pl}{AE_Y}. \tag{5.36}$$

Also, if d is the undeformed length of a line in the transverse direction, its elongation Δd is given by

$$\Delta d = -\frac{\nu Pd}{AE_Y}. \tag{5.37}$$

The minus sign indicates the expected contraction of the lateral dimension for a bar under tension.

In reality, when a bar is pulled, the exact nature of the distribution of surface traction is often not known, only the resultant force is known. The question naturally arises under what conditions can an elasticity solution such as the one we just obtained for simple extension be applicable to real problems. The answer to the question is given by the so-called **Saint-Venant's principle** which can be stated as follows:

> If some distribution of forces acting on a portion of the surface of a body is replaced by a different distribution of forces acting on the same portion of the body, then the effects of the two different distributions on the parts of the body sufficiently far removed from the region of application of the forces are essentially the same, provided that the two distributions of forces have the same resultant force and the same resultant couple.

The validity of the principle can be demonstrated in specific instances and a number of sufficient conditions have been established. We state only that in most cases the principle has been proven to be in close agreement with experiment.

By invoking Saint-Venant's principle, we now regard the solution we obtained for simple extension to be valid at least in most part of a slender bar, provided the resultant force on either end passes through the centroid of the cross-sectional area.

EXAMPLE 5.13

A steel circular bar, 2 ft (0.61 m) long, 1 in. (2.54 cm) radius is pulled by equal and opposite axial forces P at its ends. Find the maximum normal and shear stresses if $P =$ 10,000 lb (44.48 kN). $E_Y = 30 \times 10^6$ psi (207 GPa), $\nu = 0.3$.

Solution. The maximum normal stress is

$$(T_n)_{\max} = \frac{P}{A} = \frac{10{,}000}{\pi} = 3180 \text{ psi. } (21.93 \text{ MPa}).$$

The maximum shearing stress is

$$(T_s)_{\max} = \frac{3180}{2} = 1590 \text{ psi. } (10.96 \text{ MPa}).$$

and total elongation is

$$\Delta l = \frac{Pl}{AE_Y} = \frac{(10{,}000)\ (2\times 12)}{\pi(30\times 10^6)} = 2.54\times 10^{-3} \text{ in. } (64.5\ \mu\text{m}).$$

The diameter will contract by an amount

$$-\Delta d = \frac{\nu P}{EA} d = \frac{(0.3)\ (10{,}000)\ (2)}{(30\times 10^6)\ (\pi)} = 0.636\times 10^{-4} \text{ in. } (1.61\ \mu\text{m}).$$

EXAMPLE 5.14

A composite bar, formed by welding two slender bars of equal length and equal cross-sectional area, is loaded by an axial force P as shown in Fig. 5.7. If Young's moduli of the two portions are $E_Y{}^1$ and $E_Y{}^2$, find how the applied force is distributed between the two halves.

Solution. The single nontrivial equation of static equilibrium requires that

$$P = P_1 - P_2. \tag{a}$$

Statics alone does not determine the distribution of the load P (a statically indeterminate problem), so we must consider the deformation induced by the load P. In this problem there is no net elongation of the composite bar, therefore

$$\frac{P_1 l}{AE_Y{}^1} + \frac{P_2 l}{AE_Y{}^2} = 0. \tag{b}$$

Combining Eqs. (a) and (b) we obtain

$$P_1 = \frac{P}{1+(E_Y{}^2/E_Y{}^1)}, \quad P_2 = \frac{-P}{1+(E_Y{}^1/E_Y{}^2)}.$$

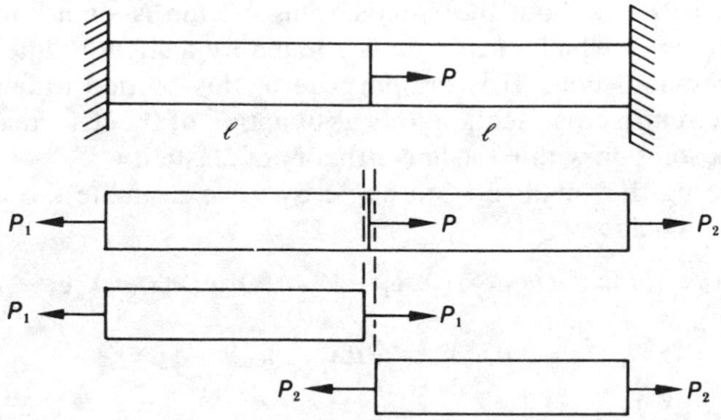

Fig. 5.7

If, in particular, Young's moduli are $E_{Y^1} = 207$ GPa (steel) and $E_{Y^2} = 69$ GPa (aluminum then

$$P_1 = \frac{3P}{4}, \quad P_2 = \frac{-P}{4}.$$

B. Torsion of a Circular Cylinder

Let us consider the elastic deformation of a cylindrical bar of circular cross-section (of radius a and length l), that is being twisted by equal and opposite end moments M_t (*see* Fig. 5.8). We choose the x_1-axis to coincide with the axis of the cylinder and the left-hand and right-hand faces to correspond to the planes $X_1 = 0$ and $X_1 = l$ respectively.

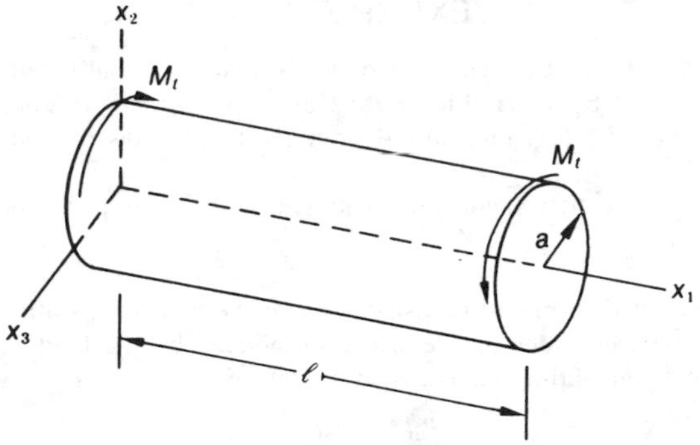

Fig. 5.8

By the symmetry of the problem, it is reasonable to assume that the motion of each cross-sectional plane induced by the end moments is a rigid body rotation about the x_1-axis. This motion is similar to that of a stack of coins in which each coin is rotated by a slightly different angle than the previous coin. It is the purpose of this section to demonstrate that for a circular cross-section, this assumption of the deformation leads to an exact solution within the linear theory of elasticity.

Designating the small rotation angle by θ we evaluate the associated displacement field as

$$\mathbf{u} = (\theta \mathbf{e}_1) \times \mathbf{R} = (\theta \mathbf{e}_1) \times (X_1 \mathbf{e}_1 + X_2 \mathbf{e}_2 + X_3 \mathbf{e}_3) = \theta (X_2 \mathbf{e}_3 - X_3 \mathbf{e}_2)$$

or

$$u_1 = 0, \quad u_2 = -\theta X_3, \quad u_3 = \theta X_2, \tag{5.38}$$

where $\theta = \theta(X_1)$.

Corresponding to this displacement field are the nonzero strains

$$E_{12} = E_{21} = -\frac{1}{2} X_3 \frac{d\theta}{dX_1}, \tag{5.39a}$$

$$E_{13} = E_{31} = \frac{1}{2} X_2 \frac{d\theta}{dX_1}. \tag{5.39b}$$

The nonzero stress components are

$$T_{12} = T_{21} = -\mu X_3 \frac{d\theta}{dX_1}, \tag{5.40a}$$

$$T_{13} = T_{31} = \mu X_2 \frac{d\theta}{dX_1}. \tag{5.40b}$$

To determine if this is a possible state of stress we check the equilibrium equations ($\partial T_{ij}/\partial X_j = 0$). The first equation is identically satisfied, whereas for the second and third equations we have

$$-\mu X_3 \left(\frac{d^2\theta}{dX_1{}^2}\right) = 0,$$

$$+\mu X_2 \left(\frac{d^2\theta}{dX_1{}^2}\right) = 0.$$

Thus,

$$\frac{d\theta}{dX_1} \equiv \theta' = \text{constant}. \tag{5.41}$$

Interpreted physically, we satisfy equilibrium if the increment in angular rotation (i.e., twist per unit length) is a constant. Now that the displacement field has been shown to generate a possible stress field, we must determine the surface tractions that correspond to the stress field.

On the lateral surface (*see* Fig. 5.9) we have a unit normal vector $\mathbf{n} = (1/a)\,(X_2\mathbf{e}_2 + X_3\mathbf{e}_3)$. Therefore, the surface traction on the lateral surface

$$[\mathbf{t}] = [\mathbf{T}]\,[\mathbf{n}] = \frac{1}{a}\begin{bmatrix} 0 & T_{12} & T_{13} \\ T_{21} & 0 & 0 \\ T_{31} & 0 & 0 \end{bmatrix}\begin{bmatrix} 0 \\ X_2 \\ X_3 \end{bmatrix} = \frac{1}{a}\begin{bmatrix} X_2T_{12} + X_3T_{13} \\ 0 \\ 0 \end{bmatrix}.$$

Substituting from Eqs. (5.40), we have

$$\mathbf{t} = \frac{\mu}{a}\,(-X_2X_3\theta' + X_2X_3\theta')\,\mathbf{e}_1 = 0.$$

Thus, in agreement with the bar being twisted by end moments only, the lateral surface is traction-free.

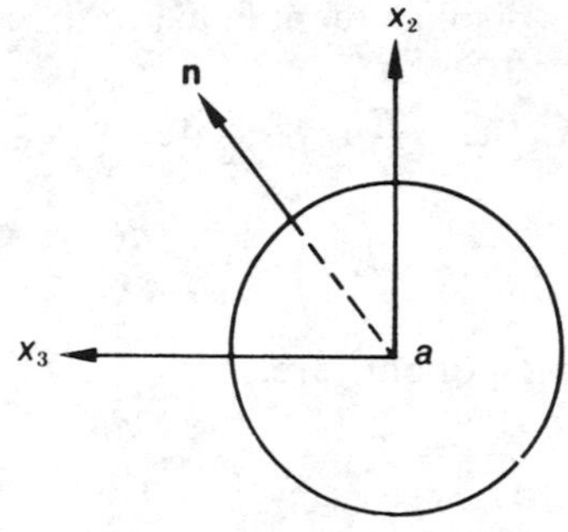

Fig. 5.9

On the face $X_1 = l$ we have a unit normal $\mathbf{n} = \mathbf{e}_1$ and a surface traction

$$\mathbf{t} = \mathbf{T}\mathbf{e}_1 = T_{21}\mathbf{e}_2 + T_{31}\mathbf{e}_3.$$

This distribution of surface traction on the end face gives rise to the following resultant (Fig. 5.10):

$$R_1 = \int T_{11} dA = 0,$$

$$R_2 = \int T_{21} dA = -\mu\theta' \int X_3 dA = 0,$$

$$R_3 = \int T_{31} dA = \mu\theta' \int X_2 dA = 0,$$

$$M_1 = \int (X_2 T_{31} - X_3 T_{21}) dA = \mu\theta' \int (X_2^2 + X_3^2) dA = \mu\theta' I_p,$$

$$M_2 = M_3 = 0,$$

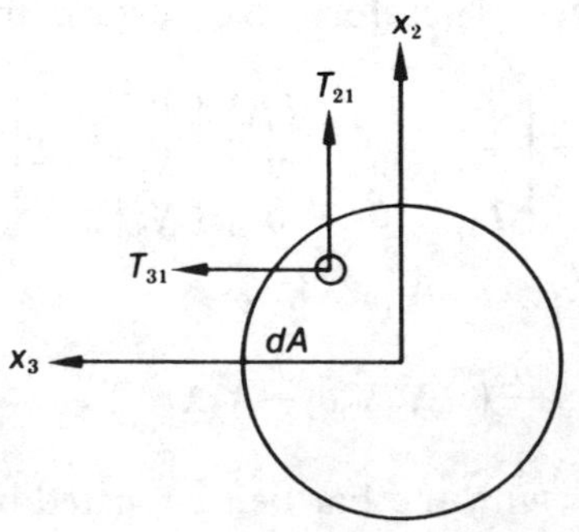

Fig. 5.10

where I_p is the polar moment of inertia of the cross-sectional area and is given by $I_p = \pi a^4/2$ for a circle of radius a.

The resultant force system on the face $X_1 = 0$ will similarly give rise to a counterbalancing couple $-\mu\theta' I_p$. Therefore, the resultant force system on either end face is a twisting couple $M_1 = M_t$ and it induces a twist per unit length given by

$$\theta' = \frac{M_t}{\mu I_p}. \tag{5.42}$$

This indicates that we can, as indicated in Section 5.1, determine the shear modulus from a simple torsion experiment.

In terms of the twisting couple M_t, the stress tensor becomes

$$[\mathbf{T}] = \begin{bmatrix} 0 & -\dfrac{M_t X_3}{I_p} & +\dfrac{M_t X_2}{I_p} \\ -\dfrac{M_t X_3}{I_p} & 0 & 0 \\ +\dfrac{M_t X_2}{I_p} & 0 & 0 \end{bmatrix}. \tag{5.43}$$

In reality, when a bar is twisted the exact distribution of the applied forces is rarely, if ever, known. We shall assume that as long as the resultants of the applied forces on the two ends of a slender bar are equal and opposite couples of strength M_t, the state of stress inside the bar is given by Eq. (5.43)†.

EXAMPLE 5.15

For a circular bar of radius a in torsion (a) find the magnitude and location of the greatest normal and shearing stresses throughout the bar; (b) find the principal direction at the position $X_2 = 0, X_3 = a$.

Solution. (a) We first evaluate the principal stresses as a function of position by solving the characteristic equation:

$$\lambda^3 - \lambda \left(\frac{M_t}{I_p}\right)^2 (X_2^2 + X_3^2) = 0.$$

Thus, the principal values at any point are

$$\lambda = 0 \quad \text{and} \quad \lambda = \pm \frac{M_t}{I_p}\sqrt{(X_2^2 + X_3^2)} = \pm \frac{M_t r}{I_p},$$

where r is the distance from the axis of the bar.

In this case, the magnitude of the maximum shearing and normal stress at any point are equal and are proportional to the distance r. Therefore, the greatest shearing and normal

† Saint-Venant's Principle is invoked here.

stress both occur on the boundary, $r = a$, with

$$(T_n)_{\max} = (T_s)_{\max} = \frac{M_t a}{I_p}. \tag{5.44}$$

(b) For the principal value $\lambda = M_t a/I_p$ at the boundary points $(X_1, 0, a)$ the eigenvalue equation becomes

$$-\frac{M_t a}{I_p} n_1 - \frac{M_t a}{I_p} n_2 = 0,$$

$$-\frac{M_t a}{I_p} n_1 - \frac{M_t a}{I_p} n_2 = 0,$$

$$-\left(\frac{M_t a}{I_p}\right) n_3 = 0.$$

Therefore, the eigenvector is given by $\mathbf{n} = (\sqrt{2}/2)(\mathbf{e}_1 - \mathbf{e}_2)$. This normal determines a plane perpendicular to the lateral face which makes a 45° angle with the axis. This is why a crack along a helix inclined at 45° to the axis of a circular cast iron torsion shaft, which is weaker in tension than in shear, often occurs.

EXAMPLE 5.16

In Fig. 5.11, a twisting torque M_t is applied to the rigid disc A. Find the twisting moments transmitted to the circular shafts on either side of the disc.

Solution. Let M_1 be the twisting moment transmitted to the left shaft and M_2 that to the right shaft. Then the equilibrium of the disc demands that

$$M_1 + M_2 = M_t. \tag{a}$$

In addition, because the disc is assumed to be rigid, the angle of twist of the left and right shaft must be equal:

$$\frac{M_1 l_1}{\mu I_p} = \frac{M_2 l_2}{\mu I_p}.$$

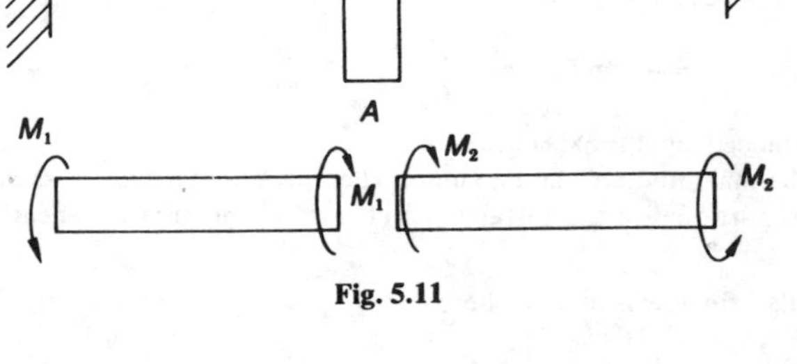

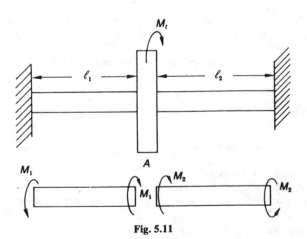

Fig. 5.11

Thus

$$M_1 l_1 = M_2 l_2. \tag{b}$$

From Eqs. (a) and (b), we then obtain

$$M_1 = \left(\frac{l_2}{l_1 + l_2}\right) M_t, \quad M_2 = \left(\frac{l_1}{l_1 + l_2}\right) M_t.$$

EXAMPLE 5.17

Consider the angle of twist for a circular cylinder under torsion to be a function of X_1 and time t, i.e., $\theta = \theta(X_1, t)$.
(a) Determine the differential equation that θ must satisfy for it to be a possible solution in the absence of body forces. What are the boundary conditions that θ must satisfy if: (b) the plane $X_1 = 0$ is a fixed end; (c) the plane $X_1 = 0$ is a free end.

Solution. (a) From the displacements

$$u_1 = 0, \quad u_2 = -\theta X_3, \quad u_3 = \theta X_2,$$

we find the stresses to be

$$T_{12} = T_{21} = 2\mu E_{12} = 2\mu E_{21} = -\mu X_3 \frac{\partial \theta}{\partial X_1},$$

$$T_{13} = T_{31} = 2\mu E_{13} = 2\mu E_{31} = \mu X_2 \frac{\partial \theta}{\partial X_1},$$

and

$$T_{11} = T_{22} = T_{33} = T_{23} = 0.$$

The second and third equations of motion give

$$-\mu X_3 \frac{\partial^2 \theta}{\partial X_1^2} = -\rho_0 X_3 \frac{\partial^2 \theta}{\partial t^2},$$

$$\mu X_2 \frac{\partial^2 \theta}{\partial X_1^2} = \rho_0 X_2 \frac{\partial^2 \theta}{\partial t^2}.$$

Therefore, $\theta(X_1, t)$ must satisfy the equation

$$c_T^2 \frac{\partial^2 \theta}{\partial X_1^2} = \frac{\partial^2 \theta}{\partial t^2},$$

where $c_T = \sqrt{(\mu/\rho_0)}$.
(b) At the fixed end $X_1 = 0$ there is no displacement, therefore

$$\theta(0, t) = 0.$$

(c) At the traction-free end $X_1 = 0$, $\mathbf{t} = -\mathbf{T}\mathbf{e}_1 = 0$. Thus $T_{21}|_{X_1=0} = 0$, $T_{31}|_{X_1=0} = 0$, therefore

$$\frac{\partial \theta}{\partial X_1}(0, t) = 0.$$

EXAMPLE 5.18

A cylindrical bar of square cross-section (*see* Fig. 5.12) is twisted by end moments. Show that the displacement field of the torsion of the circular bar does not give a correct solution to this problem.

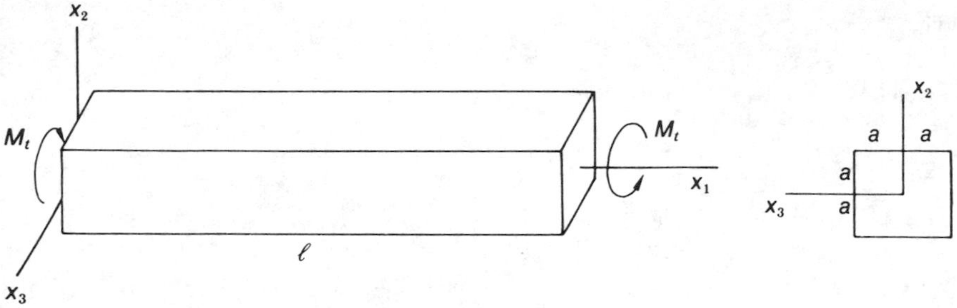

Fig. 5.12

Solution. The displacement field for the torsion of circular cylinders has already been shown to generate an equilibrium stress field. We therefore, check if the surface traction of the lateral surface vanishes. The unit vector on the plane $X_3 = a$ is $\mathbf{e}_3$, so that the surface traction for the stress tensor of Eq. (5.43) is given by

$$\mathbf{t} = \mathbf{T}\mathbf{e}_3 = T_{13}\mathbf{e}_1 = \frac{M_t X_2}{I_p}\mathbf{e}_1.$$

Similarly, there will be surface tractions in the $\mathbf{e}_1$-direction on the remainder of the lateral surface. Thus, the previously assumed displacement field must be altered. To obtain the actual solution for twisting by end moments only, we must somehow remove these axial surface tractions. As will be seen in the next section, this will cause the cross-sectional planes to warp.

C. Torsion of a Noncircular Cylinder

For cross-sections other than circular, the simple displacement field of Section 5.8B will not satisfy the tractionless lateral surface boundary condition (*see* Example 5.18). We wish to show that in order to satisfy this boundary condition, the cross-sections will not remain plane.

We begin by assuming a displacement field that still rotates each cross-section by a small angle θ, but in addition there may be a displacement in the axial direction. This warping of the cross-sectional plane will be defined by $u_1 = \phi(X_2, X_3)$. Our displacement field now has the form

$$u_1 = \phi(X_2, X_3), \quad u_2 = -X_3\theta(X_1), \quad u_3 = X_2\theta(X_1). \tag{5.45}$$

The associated nonzero strains and stresses are given by

$$T_{12} = T_{21} = 2\mu E_{12} = 2\mu E_{21} = -\mu X_3\theta' + \mu\frac{\partial\phi}{\partial X_2}, \tag{5.46a}$$

$$T_{13} = T_{31} = 2\mu E_{13} = 2\mu E_{31} = \mu X_2\theta' + \mu\frac{\partial\phi}{\partial X_3}. \tag{5.46b}$$

The second and third equilibrium equations are still satisfied if $\theta' =$ constant. However, the first equilibrium equation requires that

$$\frac{\partial^2 \phi}{\partial X_2{}^2} + \frac{\partial^2 \phi}{\partial X_3{}^2} = 0. \tag{5.47}$$

Therefore, the displacement field of Eq. (5.45) will generate a possible state of stress if ϕ satisfies Eq. (5.47). Now we compute the traction on the lateral surface. Since the bar is cylindrical, the unit normal to the lateral surface has the form $\mathbf{n} = n_2\mathbf{e}_2 + n_3\mathbf{e}_3$ and the associated surface traction is given by

$$\mathbf{t} = \mathbf{T}\mathbf{n} = \left\{\mu\theta'\left[-n_2X_3 + n_3X_2\right] + \mu\left[\frac{\partial \phi}{\partial X_2} n_2 + \frac{\partial \phi}{\partial X_3} n_3\right]\right\}\mathbf{e}_1.$$

We require that the lateral surface be traction-free, i.e., $\mathbf{t} = 0$, so that on the boundary the function ϕ must satisfy the condition

$$\frac{d\phi}{dn} \equiv (\nabla\phi)\cdot\mathbf{n} = \theta'[n_2X_3 - n_3X_2]. \tag{5.48}$$

Equations (5.47) and (5.48) define a boundary-value problem which is known to admit an exact solution for the function ϕ. Here, we will only consider the torsion of an elliptic cross-section by demonstrating that

$$\phi = AX_2X_3 \tag{5.49}$$

gives the correct solution.

Taking A as a constant, this choice of ϕ obviously satisfies the equilibrium equation [Eq. (5.47)]. To check the boundary condition we begin by defining the elliptic boundary by the equation

$$f(X_2, X_3) = \frac{X_2{}^2}{a^2} + \frac{X_3{}^2}{b^2} = 1.$$

The unit normal vector is given by

$$\mathbf{n} = \frac{\nabla f}{|\nabla f|} = \frac{2}{|\nabla f|}\left[\frac{X_2}{a^2}\mathbf{e}_2 + \frac{X_3}{b^2}\mathbf{e}_3\right]$$

and the boundary condition of Eq. (5.48) becomes

$$\left(\frac{\partial \phi}{\partial X_2}\right) b^2X_2 + \left(\frac{\partial \phi}{\partial X_3}\right) a^2X_3 = \theta' X_2X_3(b^2 - a^2).$$

Substituting our choice of ϕ into this equation, we find that

$$A = \theta' \left(\frac{b^2 - a^2}{a^2 + b^2} \right). \tag{5.50}$$

Because A does turn out to be a constant, we have satisfied both Eq. (5.47) and Eq. (5.48). Substituting the value of ϕ into Eq. (5.46), we obtain the associated stresses

$$T_{21} = T_{12} = -\left(\frac{2\mu a^2}{a^2 + b^2} \right) \theta' X_3,$$

$$T_{31} = T_{13} = \left(\frac{2\mu b^2}{a^2 + b^2} \right) \theta' X_2.$$

This distribution of stress gives a surface traction on the end face, $X_1 = l$,

$$\mathbf{t} = T_{21}\mathbf{e}_2 + T_{31}\mathbf{e}_3$$

and the following resultant force system

$$R_1 = R_2 = R_3 = M_2 = M_3 = 0,$$

$$M_1 = \int (X_2 T_{31} - X_3 T_{21})\, dA = \frac{2\mu\theta'}{a^2 + b^2} \left[a^2 \int X_3^2 dA + b^2 \int X_2^2 dA \right]$$

$$= \frac{2\mu\theta'}{a^2 + b^2} [a^2 I_{22} + b^2 I_{33}].$$

Denoting $M_1 = M_t$ and recalling that for an ellipse $I_{33} = \pi a^3 b/4$ and $I_{22} = \pi b^3 a/4$, we obtain

$$\theta' = \frac{(a^2 + b^2)}{\pi a^3 b^3 \mu} M_t.$$

Similarly, the resultant on the other end face $X_1 = 0$ will give rise to a counterbalancing couple.

In terms of the twisting moment, the stress tensor becomes

$$[\mathbf{T}] = \begin{bmatrix} 0 & \dfrac{-2M_t X_3}{\pi a b^3} & \dfrac{2M_t X_2}{\pi a^3 b} \\ \dfrac{-2M_t X_3}{\pi a b^3} & 0 & 0 \\ \dfrac{2M_t X_2}{\pi a^3 b} & 0 & 0 \end{bmatrix}. \tag{5.51}$$

EXAMPLE 5.19

For an elliptic cylindrical bar in torsion, (a) find the magnitude of the maximum normal and shearing stress at any point of the bar, and (b) find the ratio of the maximum shearing stresses at the extremities of the elliptic minor and major axes.

Solution. (a) As in Example 5.12 we first solve the characteristic equation

$$\lambda^3 - \lambda\left(\frac{2M_t}{\pi ab}\right)^2\left[\frac{X_2^2}{a^4}+\frac{X_3^2}{b^4}\right]=0.$$

The principal values are

$$\lambda = 0 \quad \text{and} \quad \lambda = \pm\frac{2M_t}{\pi ab}\sqrt{\frac{X_2^2}{a^4}+\frac{X_3^2}{b^4}},$$

which determines the maximum normal and shearing stresses:

$$(T_s)_{\max} = (T_n)_{\max} = \frac{2M_t}{\pi ab}\sqrt{\frac{X_2^2}{a^4}+\frac{X_3^2}{b^4}}.$$

(b) Supposing that $b > a$, we have at the end of the minor axis ($X_2 = a$, $X_3 = 0$),

$$(T_s)_{\max} = \left(\frac{2M_t}{\pi ab}\right)\left(\frac{1}{a}\right)$$

and at the end of the major axis ($X_2 = 0$, $X_3 = b$),

$$(T_s)_{\max} = \left(\frac{2M_t}{\pi ab}\right)\left(\frac{1}{b}\right).$$

The ratio of the maximum stresses is therefore b/a and the greater stress occurs at the end of the minor axis.

D. Pure Bending of a Beam

A beam is a bar acted on by forces or couples in an axial plane, which chiefly cause bending of the bar. When a beam or portion of a beam is acted on by end couples only, it is said to be in **pure** or **simple bending.** We shall consider the case of a cylindrical bar of arbitrary cross-section that is in pure bending.

Figure 5.13 shows a bar of uniform cross-section. We choose the x_1-axis to pass through the cross-sectional centroids and let $X_1 = 0$ and $X_1 = l$ correspond to the left- and right-hand faces of the bar.

To determine a linear elasticity solution to this problem we must specify a possible state of stress that corresponds to a tractionless lateral surface and some distribution of normal surface tractions on the end faces that is equipollent to bending couples $\mathbf{M}_R = M_2\mathbf{e}_2 + M_3\mathbf{e}_3$ and $\mathbf{M}_L = -\mathbf{M}_R$ (note that the M_1 component is absent because it is a twisting couple). Guided by the state of stress associated with simple extension, we tentatively assume that T_{11} is the only nonzero stress component and that it is an arbitrary function of X_i.

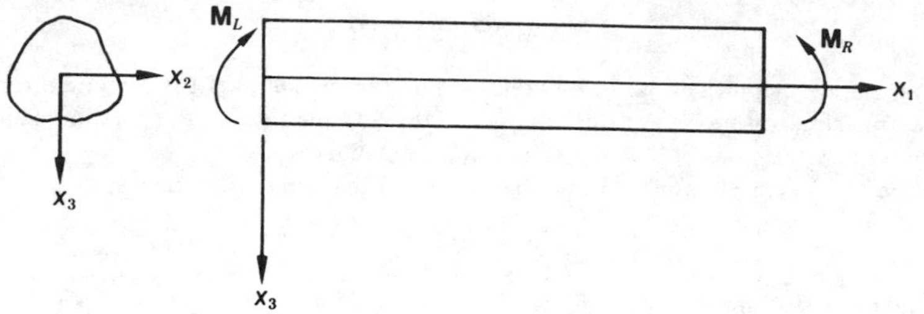

Fig. 5.13

To satisfy equilibrium, we require

$$\frac{\partial T_{11}}{\partial X_1} = 0,$$

i.e., $T_{11} = T_{11}(X_2, X_3)$.

The corresponding strains are

$$E_{11} = \frac{1}{E_Y} T_{11}, \quad E_{22} = E_{33} = -\frac{\nu}{E_Y} T_{11},$$

$$E_{12} = E_{13} = E_{23} = 0.$$

Since we have begun with an assumption on the state of stress, we must check whether these strains are compatible. Substituting the strains into the compatibility equations (Eq. 3.36) we obtain

$$\frac{\partial^2 T_{11}}{\partial X_2^2} = 0, \quad \frac{\partial^2 T_{11}}{\partial X_3^2} = 0, \quad \frac{\partial^2 T_{11}}{\partial X_3 \partial X_2} = 0,$$

which can be satisfied only if T_{11} is a linear function of the form

$$T_{11} = \alpha + \beta X_2 + \gamma X_3.$$

Now that we have a possible stress distribution, let us consider the nature of the boundary tractions. As is the case with simple extension, the lateral surface is obviously traction-free. On the end face $X_1 = l$, we have a surface traction

$$\mathbf{t} = \mathbf{T}\mathbf{e}_1 = T_{11}\mathbf{e}_1,$$

which gives a resultant force system†

†The integrals $\int X_2 dA$ and $\int X_3 dA$ are zero because each is the first moment of area about a centroidal axis.

$$R_1 = \int T_{11} dA = \alpha \int dA + \beta \int X_2 dA + \gamma \int X_3 dA = \alpha A,$$

$$R_2 = R_3 = 0,$$

$$M_1 = 0,$$

$$M_2 = \int X_3 T_{11} dA = \alpha \int X_3 dA + \beta \int X_2 X_3 dA + \gamma \int X_3^2 dA$$
$$= \beta I_{23} + \gamma I_{22},$$

$$M_3 = -\int X_2 T_{11} dA = -\alpha \int X_2 dA - \beta \int X_2^2 dA - \gamma \int X_2 X_3 dA$$
$$= -\beta I_{33} - \gamma I_{23},$$

where A is the cross-sectional area, I_{22}, I_{33}, and I_{23} are the moments and product of inertia of the cross-sectional area. On the face $X_1 = 0$, the resultant force system is equal and opposite to that given above.

Note that in the above equations, the resultant force $\mathbf{R} = R_1 \mathbf{e}_1$ passes through the centroid of the cross-section, its effect (simple extension) can be superposed onto that of bending, so we set $\alpha = 0$. Note that if a bar is loaded by an axial force that does not coincide with the centroid of the cross-section, it may be treated by replacing it by an equipollent bending couple and a force that passes through the centroid of the cross-section.

We now assume without any loss in generality that we have chosen the x_2- and x_3-axis to coincide with the principal axes of the cross-sectional area (e.g., along lines of symmetry) so that $I_{23} = 0$. In this case, the stress distribution for the cylindrical bar is given by

$$T_{11} = \frac{M_2}{I_{22}} X_3 - \frac{M_3}{I_{33}} X_2 \tag{5.52}$$

and all other $T_{ij} = 0$.

To investigate the nature of the deformation that is induced by bending moments, for simplicity we let $M_3 = 0$. The corresponding strains are

$$E_{11} = \frac{M_2}{I_{22} E_Y} X_3, \tag{5.53a}$$

$$E_{22} = E_{33} = -\frac{\nu M_2}{I_{22} E_Y} X_3. \tag{5.53b}$$

These equations can be integrated (we are assured that this is possible since the strains are compatible) to give the following displacement field:

$$u_1 = \frac{M_2}{E_Y I_{22}} X_1 X_3 - \alpha_3 X_2 + \alpha_2 X_3 + \alpha_4,$$

$$u_2 = -\frac{\nu M_2}{E_Y I_{22}} X_2 X_3 + \alpha_3 X_1 - \alpha_1 X_3 + \alpha_5,$$

$$u_3 = -\frac{M_2}{2E_Y I_{22}} [X_1^2 - \nu(X_2^2 - X_3^2)] - \alpha_2 X_1 + \alpha_1 X_2 + \alpha_6,$$

where α_i are constants of integration. In fact, α_4, α_5, α_6 define an overall rigid body translation of the bar and α_1, α_2, α_3 being constant parts of the antisymmetric part of the displacement gradient, define an overall small rigid body rotation. For convenience we let $\alpha_i = 0$ [note that this corresponds to requiring $\mathbf{u} = 0$ and $(\nabla \mathbf{u})^A = 0$ at the origin]. The displacements are therefore,

$$u_1 = \frac{M_2}{E_Y I_{22}} X_1 X_3, \quad u_2 = -\frac{\nu M_2}{E_Y I_{22}} X_2 X_3,$$

$$u_3 = -\frac{M_2}{2E_Y I_{22}} [X_1^2 - \nu(X_2^2 - X_3^2)]. \tag{5.54}$$

Considering the cross-sectional plane X_1 = constant, we note that the displacement perpendicular to the plane is given by

$$u_1 = \left(\frac{M_2 X_1}{E_Y I_{22}}\right) X_3.$$

Since u_1 is a linear function of X_3, the cross-sectional plane remains plane and is rotated about the x_2-axis (*see* Fig. 5.14) by an angle

$$\theta \approx \tan\theta = \frac{+u_1}{X_3} = \frac{M_2 X_1}{E_Y I_{22}}.$$

In addition, consider the displacement of the material that is initially along the x_1-axis ($X_2 = X_3 = 0$),

$$u_1 = u_2 = 0, \quad u_3 = -\frac{M_2 X_1^2}{2E_Y I_{22}} \tag{5.55}$$

The displacement of this material element (often called the neutral axis or neutral fiber) is frequently used to define the deflection of the beam. Note, that since

$$\left|\frac{du_3}{dX_1}\right| = \frac{M_2 X_1}{E_Y I_{22}} = \tan\theta$$

the cross-sectional planes remain perpendicular to the neutral axis. This is clearly a result of the absence of shearing stress in pure bending.

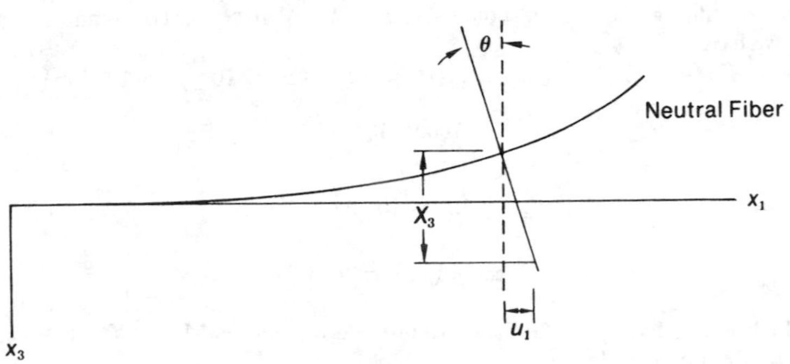

Fig. 5.14

EXAMPLE 5.20

In Fig. 5.15 a rectangular beam of width 15 cm and height 20 cm, is subjected to pure bending couples. The right-hand couple is given $\mathbf{M} = 7000\, \mathbf{e}_2$ N·m. Find the greatest normal and shearing stresses throughout the beam.

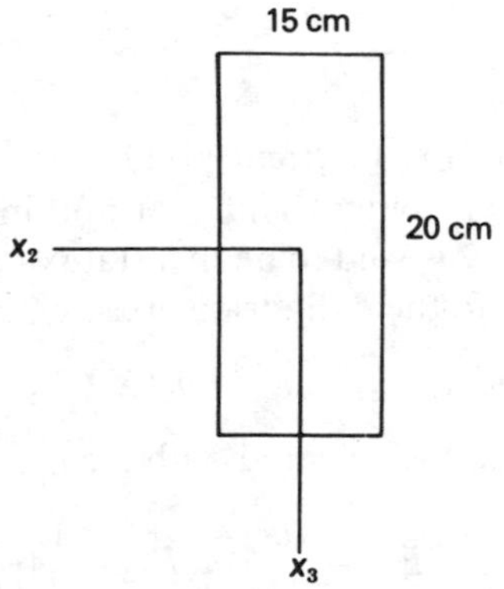

Fig. 5.15

Solution. We have

$$T_{11} = \frac{M_2}{I_{22}} X_3$$

and the remaining stress components vanish. Therefore, at any point,

$$(T_n)_{\max} = \frac{M_2}{I_{22}} X_3$$

and

$$(T_s)_{\max} = \frac{M_2 X_3}{2I_{22}}.$$

The greatest value will be at the boundary, i.e., $X_3 = 10^{-1}$ m. To obtain a numerical answer, we have

$$I_{22} = \tfrac{1}{12}(15 \times 10^{-2})(20 \times 10^{-2})^3 = 10^{-4}\ \text{m}^4$$

and the greatest stresses are

$$(T_n)_{\max} = \frac{(7000)(10^{-1})}{10^{-4}} = 7 \times 10^6\ \text{Pa}.$$

$$(T_s)_{\max} = 3.5 \times 10^6\ \text{Pa}$$

EXAMPLE 5.21

For the beam of Example 5.20, if the right-hand moment is $\mathbf{M} = 7000\,(\mathbf{e}_2 + \mathbf{e}_3)$ N·m, find the maximum normal stress.

Solution. We have

$$I_{33} = 0.563 \times 10^{-4}\ \text{m}^4, \qquad I_{22} = 10^{-4}\ \text{m}^4$$

$$T_{11} = \frac{M_2 X_3}{I_{22}} - \frac{M_3}{I_{33}} X_2 = (7X_3 - 12.4X_2) \times 10^1\ \text{Pa}$$

The maximum normal stress occurs at $X_2 = -7.5 \times 10^{-2}$ m and $X_3 = 10^{-1}$ m with

$$T_{11} = 16.3\ \text{MPa}.$$

E. Plane Strain

If the deformation of a cylindrical body is such that there is no axial components of the displacement and that nothing depends on the axial coordinate, then the body is said to be in a state of plane strain. Letting the $\mathbf{e}_3$-direction correspond to the cylindrical axis, we have

$$u_1 = u_1(X_1, X_2), \quad u_2 = u_2(X_1, X_2), \quad u_3 = 0.$$

These displacements are associated with the nonzero strain components

$$E_{11} = \frac{\partial u_1}{\partial X_1}, \quad E_{22} = \frac{\partial u_2}{\partial X_2}, \quad E_{12} = \frac{1}{2}\left(\frac{\partial u_1}{\partial X_2} + \frac{\partial u_2}{\partial X_1}\right).$$

The corresponding nonzero stress components are T_{11}, T_{22}, T_{12}, and T_{33}. Note that the normal stress T_{33} is given by $T_{33} = \nu(T_{11} + T_{22})$ in order to maintain $E_{33} = 0$, and is not a function of X_3.

Considering a static stress field with no body forces, the equilibrium equations reduce to

$$\frac{\partial T_{11}}{\partial X_1} + \frac{\partial T_{12}}{\partial X_2} = 0,$$

$$\frac{\partial T_{12}}{\partial X_1} + \frac{\partial T_{22}}{\partial X_2} = 0,$$

$$\frac{\partial T_{33}}{\partial X_3} = 0.$$

Because $T_{33} = T_{33}(X_1, X_2)$, the third equilibrium equation is trivially satisfied. We can conveniently reduce the number of equations from two to one by introducing the "**Airy stress function**" $\phi(X_1, X_2)$ such that

$$T_{11} = \frac{\partial^2 \phi}{\partial X_2^2}, \quad T_{12} = -\frac{\partial^2 \phi}{\partial X_1 \partial X_2},$$
$$T_{22} = \frac{\partial^2 \phi}{\partial X_1^2}, \quad T_{33} = \nu\left(\frac{\partial^2 \phi}{\partial X_1^2} + \frac{\partial^2 \phi}{\partial X_2^2}\right). \tag{5.56}$$

For arbitrary $\phi(X_1, X_2)$ it can easily be verified that the equilibrium equations are automatically satisfied. Since the problem is cast in terms of stress, we must check the compatibility equations. For plane strain, the only equation that is not identically satisfied is

$$\frac{\partial^2 E_{11}}{\partial X_2^2} + \frac{\partial^2 E_{22}}{\partial X_1^2} = 2\frac{\partial^2 E_{12}}{\partial X_1 \partial X_2}.$$

Determining the strains, by Hooke's law, in terms of ϕ and substituting into this equation, we find that ϕ must satisfy the differential equation

$$\frac{\partial^2 \phi}{\partial X_1^4} + 2\frac{\partial^4 \phi}{\partial X_1^2 \partial X_2^2} + \frac{\partial^4 \phi}{\partial X_2^4} = 0. \tag{5.57}$$

Thus, any $\phi(X_1, X_2)$ that satisfies Eq. (5.57) generates a possible elastic solution. In particular, any third degree polynomial (generating a linear stress and strain field) may be utilized.

EXAMPLE 5.22

Consider the state of stress given by

$$[\mathbf{T}] = \begin{bmatrix} 0 & 0 & 0 \\ 0 & 0 & 0 \\ 0 & 0 & G(X_1, X_2) \end{bmatrix}.$$

Show that the most general form of $G(X_1, X_2)$ which gives rise to a possible state of stress in the absence of body force is

$$G(X_1, X_2) = \alpha X_1 + \beta X_2 + \gamma.$$

Solution. The strain components are

$$E_{11} = -\frac{\nu}{E_y} G(X_1, X_2) = E_{22},$$

$$E_{33} = \frac{1}{E_y} G(X_1, X_2),$$

$$E_{12} = E_{13} = E_{23} = 0.$$

From the compatibility conditions [Eq. (3.36)], we have

$$\frac{\partial^2 G}{\partial X_2^2} = 0,$$

$$\frac{\partial^2 G}{\partial X_1^2} = 0,$$

$$\frac{\partial^2 G}{\partial X_1 \partial X_2} = 0.$$

Thus, $G(X_1, X_2) = \alpha X_1 + \beta X_2 + \gamma$. In the absence of body force, the equations of equilibrium are obviously satisfied.

EXAMPLE 5.23

Consider the stress function

$$\phi(X_1, X_2) = \frac{\beta}{6} X_2^3.$$

(a) Obtain the stresses for the state of plane strain: (b) if the stresses of part (a) are those inside a rectangular prism bounded by $X_1 = 0$, $X_1 = l$, $X_2 = \pm (h/2)$ and $X_3 = \pm (b/2)$, find the surface tractions on the boundaries; (c) if the boundary surfaces $X_3 = \pm (b/2)$ are traction-free, find the solution.

Solution. (a) From Eq. (5.56)

$$T_{11} = \beta X_2, \quad T_{22} = 0, \quad T_{33} = G(X_1, X_2) = \nu \beta X_2,$$

$$T_{12} = T_{13} = T_{23} = 0.$$

(b) On the face $X_1 = 0$, $\mathbf{t} = \mathbf{T}(-\mathbf{e}_1) = -\beta X_2 \mathbf{e}_1$.
On the face $X_1 = l$, $\mathbf{t} = \mathbf{T}(\mathbf{e}_1) = \beta X_2 \mathbf{e}_1$.
On the faces $X_2 = \pm (h/2)$, $\mathbf{t} = \mathbf{T}(\pm \mathbf{e}_2) = 0$.
On the faces $X_3 = \pm (b/2)$, $\mathbf{t} = \mathbf{T}(\pm \mathbf{e}_3) = \pm \nu \beta X_2 \mathbf{e}_3$.

(c) From the previous example, we see that the state of stress

$$[\mathbf{T}] = \begin{bmatrix} 0 & 0 & 0 \\ 0 & 0 & 0 \\ 0 & 0 & -G(X_1, X_2) \end{bmatrix},$$

where $G(X_1, X_2) = \nu \beta X_2$ (linear in X_1 and X_2), is a possible state of stress. Superposing this state of stress with that of part (a), we obtain

$$[\mathbf{T}] = \begin{bmatrix} \beta X_2 & 0 & 0 \\ 0 & 0 & 0 \\ 0 & 0 & 0 \end{bmatrix},$$

which requires no surface traction on $X_3 = \pm (b/2)$ to produce it. We note that this is the exact solution for pure bending with couple vectors parallel to the direction of $\mathbf{e}_3$.

A state of stress with $T_{33} = T_{31} = T_{32} = 0$, i.e.,

$$[\mathbf{T}] = \begin{bmatrix} T_{11} & T_{12} & 0 \\ T_{12} & T_{22} & 0 \\ 0 & 0 & 0 \end{bmatrix},$$

is called a state of **plane stress**. For the state of plane stress to be an exact solution to an elasticity problem, it is in general necessary to have T_{11}, T_{12}, and T_{22} dependent on X_3. If these stress components are independent of X_3, then the state of plane stress is possible only if $T_{11} + T_{22}$ is a linear function of X_1 and X_2 (*see* Examples 5.22 and 5.23).

EXAMPLE 5.24

Consider the stress function $\phi = \alpha X_1 X_2^3 + \beta X_1 X_2$.

(a) Is this an allowable stress function?

(b) Determine the associated stresses. (Plane strain case.)

(c) Determine α and β in order to solve the problem of a cantilevered beam with end load P (Fig. 5.16). Discuss both the plane strain and plane stress cases.

Solution. (a) Yes. Equation (5.57) is satisfied.

(b) $T_{11} = 6\alpha X_1 X_2$, $T_{22} = 0$,

$T_{12} = -\beta - 3\alpha X_2^2$, $T_{33} = 6\nu\alpha X_1 X_2$.

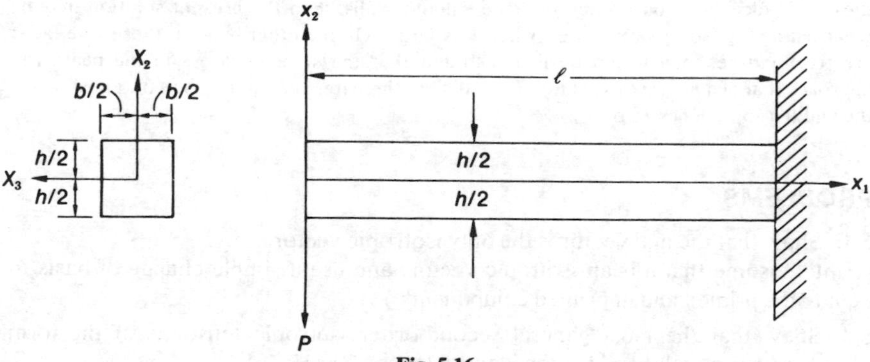

Fig. 5.16

(c) On the boundaries $X_2 = \pm (h/2)$ the tractions are

$$\mathbf{t} = \pm (T_{22}\mathbf{e}_2 + T_{12}\mathbf{e}_1) = \pm\left(-\beta - \frac{3\alpha h^2}{4}\right)\mathbf{e}_1.$$

But, we wish the lateral surface $(X_2 = \pm h/2)$ to be traction-free, therefore,

$$\beta = -\frac{3h^2}{4}\alpha.$$

On the boundary $X_1 = 0$,

$$\mathbf{t} = -\mathbf{T}\mathbf{e}_1 = (\beta + 3\alpha X_2^2)\mathbf{e}_2.$$

This shearing traction can be made equipollent to an applied load $-P\mathbf{e}_2$ by setting

$$-P = \beta \int dA + 3\alpha \int X_2^2\, dA,$$
$$= \beta A + 3\alpha I,$$

where $A = bh$ and $I = bh^3/12$. Substituting for β, we have

$$P = \alpha\left(\frac{3}{4}bh^3 - \frac{bh^3}{4}\right) = \left(\frac{bh^3}{2}\right)\alpha.$$

Therefore, $\alpha = 2P/bh^3$, $\beta = -3P/2bh$ and the stresses are

$$T_{11} = \frac{12P}{bh^3}X_1X_2 = \frac{PX_1X_2}{I},$$

$$T_{12} = \frac{3P}{2A} - \frac{P}{2I}X_2^2.$$

On the plane $X_1 = 0$ the maximum shearing stress, occurring at $X_2 = 0$, is equal to $3P/2A$. On the boundary $X_1 = l$, we have

$$\mathbf{t} = \left(\frac{Pl}{I}X_2\right)\mathbf{e}_1 + \left(\frac{3P}{2A} - \frac{P}{2I}X_2^2\right)\mathbf{e}_2.$$

The first term corresponds to the normal stress induced by a bending moment equal to Pl and the second term is exactly the same parabolic shearing stress distribution as at $X_1 = 0$.

In order that a state of plane strain is achieved, it is necessary to have normal tractions acting on the side faces $X_3 = \pm b/2$. The tractions are in fact $\mathbf{t} = \pm T_{33}\mathbf{e}_3 = \pm 6\nu\alpha X_1 X_2 \mathbf{e}_3$. If these side faces are constrained by fixed smooth walls, then the present solution may be regarded as a good approximation when b is large. On the other hand, if these side faces are traction-free, then it can be demonstrated that the state of stress in the beam will approach that of a plane stress state ($T_{33} = 0$, the other stress components as obtained) when the width b approaches zero.

PROBLEMS

5.1. Show that the null vector is the only isotropic vector.
(Hint: Assume that **a** is an isotropic vector, and use a simple change of basis to equate the primed and unprimed components.)

5.2. Show that the most general second-order isotropic tensor is of the form $\alpha\mathbf{I}$, where α is a scalar and $\mathbf{I}$ is the identity tensor [*see* Eq. (4.1)].

5.3. Show that for an anisotropic linear elastic material the principle directions of stress and strain are usually not coincident.

5.4. If the Lamé constants for a material are

$$\lambda = 119.2 \text{ GPa } (17.3 \times 10^6 \text{ psi}), \quad \mu = 79.2 \text{ GPa } (11.5 \times 10^6 \text{ psi}),$$

find Young's modulus, Poisson's ratio, and the bulk modulus.

5.5. Given Young's modulus $E_Y = 103$ GPa (15×10^6 psi) and Poisson's ratio $\nu = 0.34$, find the Lamé constants λ and μ. Also find the bulk modulus.

5.6. Given Young's modulus $E_Y = 193$ GPa (28×10^6 psi) and shear modulus $\mu = 76$ GPa (11×10^6 psi), find Poisson's ratio ν, Lamé's constant λ, and the bulk modulus k.

5.7. If the components of strain at a point of structural steel are

$$E_{11} = 36 \times 10^{-6}, \quad E_{22} = 40 \times 10^{-6}, \quad E_{33} = 25 \times 10^{-6},$$
$$E_{12} = 12 \times 10^{-6}, \quad E_{23} = 0, \quad E_{13} = 30 \times 10^{-6}.$$

find the stress components. $\lambda = 119.2$ GPa (17.3×10^6psi), $\mu = 79.2$ GPa (11.5×10^6psi)

5.8. Do Problem 5.7 if the strain components are

$$E_{11} = 100 \times 10^{-6}, \quad E_{22} = -50 \times 10^{-6}, \quad E_{33} = 200 \times 10^{-6},$$
$$E_{12} = -100 \times 10^{-6}, \quad E_{23} = 0, \quad E_{13} = 0.$$

5.9. (a) If the state of stress at a point of structural steel is

$$[\mathbf{T}] = \begin{bmatrix} 100 & 42 & 6 \\ 42 & -2 & 0 \\ 6 & 0 & 15 \end{bmatrix} \text{MPa}$$

what are the strain components? $E_Y = 207$ GPa (30×10^6 psi), $\mu = 79.2$ GPa (11.5×10^6 psi), $\nu = 0.30$.

(b) Suppose that a five-centimeter cube of structural steel has a constant state of stress given in part (a). Determine the total change in volume induced by this stress field.

5.10. (a) For the constant stress field given below, find the strain components

$$[\mathbf{T}] = \begin{bmatrix} 6 & 2 & 0 \\ 2 & -3 & 0 \\ 0 & 0 & 0 \end{bmatrix} \text{MPa}$$

(b) Suppose that a sphere of 5 cm radius is under the influence of this stress field, what will be the change in volume of the sphere? Use elastic constants of Prob 5.9.

5.11. Show that for an incompressible material ($\nu = \frac{1}{2}$, Example 5.4) that (a)

$$\mu = E_Y/3, \quad \lambda = \infty, \quad k = \infty,$$

and (b) Hooke's law becomes

$$\mathbf{T} = 2\mu\mathbf{E} + \tfrac{1}{3}(T_{kk})\mathbf{I}.$$

5.12. Given a function $f(a, b) = ab$ and a motion

$$x_1 = X_1 + k(X_1 + X_2),$$
$$x_2 = X_2 + k(X_1 - X_2),$$

where $k = 10^{-4}$.
(a) Show that $f(X_1, X_2) \approx f(x_1, x_2)$.
(b) Show that

$$\frac{\partial f(x_1, x_2)}{\partial x_1} \approx \frac{\partial f(X_1, X_2)}{\partial X_1}$$

and

$$\frac{\partial f(x_1, x_2)}{\partial x_2} \approx \frac{\partial f(X_1, X_2)}{\partial X_2}.$$

5.13. Do the previous problem for $f(a, b) = a^2 + b^2$.

5.14. Given the following displacement field

$$u_1 = kX_3X_2, \quad u_2 = kX_3X_1, \quad u_3 = k(X_1{}^2 - X_2{}^2), \qquad k = 10^{-4},$$

(a) Find the corresponding stress components.
(b) In the absence of body forces, is the state of stress a possible equilibrium stress field?

5.15. Repeat Problem 5.14, except that the displacement components are

$$u_1 = kX_2X_3, \quad u_2 = kX_1X_3, \quad u_3 = kX_1X_2, \qquad k = 10^{-4}.$$

5.16. Repeat Problem 5.14, except that the displacement components are:

$$u_1 = -kX_3X_2, \quad u_2 = kX_1X_3, \quad u_3 = k \sin X_2, \qquad k = 10^{-4}.$$

5.17. Calculate the ratio of c_L/c_T for Poisson's ratio equal to $\frac{1}{3}$, 0.49, 0.499.

5.18. Assume an arbitrary displacement field that depends only on the field variable X_2 and time t, determine what differential equations the displacement field must satisfy in order to be a possible motion (with zero body force).

5.19. Consider a linear elastic medium. Assume the following form for the displacement field

$$u_1 = \epsilon \{\sin \beta (X_3 - ct) + \alpha \sin \beta (X_3 + ct)\}, \quad u_2 = u_3 = 0.$$

(a) What is the nature of this elastic wave (longitudinal, transverse, direction of propagation)?
(b) Find the associated strains, stresses and determine under what conditions the equations of motion are satisfied with zero body force.
(c) Suppose that there is a boundary at $X_3 = 0$ that is traction-free. Under what conditions will the above motion satisfy this boundary condition for all time?
(d) Suppose that there is a boundary at $X_3 = l$ that is also traction-free. What further conditions will be imposed on the above motion to satisfy this boundary condition for all time?

5.20. Do the previous problem if the boundary $X_3 = 0$ is a fixed (no motion) and $X_3 = l$ is still traction-free.

5.21. Do Problem 5.20 if the boundaries $X_3 = 0$ and $X_3 = l$ are both rigidly fixed.

5.22. Do Problem 5.19 if the assumed displacement field is of the form

$$u_3 = \sin\beta(X_3 - ct) + \alpha\sin\beta(X_3 + ct),$$

$$u_1 = u_2 = 0.$$

5.23. Do Problem 5.22 if the boundary $X_3 = 0$ is fixed (no motion) and $X_3 = l$ is traction-free ($\mathbf{t} = 0$).

5.24. Do Problem 5.22 if the boundary $X_3 = 0$ and $X_3 = l$ are both rigidly fixed.

5.25. Consider an arbitrary displacement field $\mathbf{u} = \mathbf{u}(X_1, t)$.

(a) Show that if the motion is equivoluminal ($\partial u_i/\partial X_i = 0$) that $\mathbf{u}$ must satisfy the equation

$$\mu \frac{\partial^2 u_i}{\partial X_j \partial X_j} = \rho_0 \frac{\partial^2 u_i}{\partial t^2}.$$

(b) Show that if the motion is irrotational ($\partial u_i/\partial X_j = \partial u_j/\partial X_i$) that the dilatation $e = \partial u_i/\partial X_i$ must satisfy the equation

$$(2\mu + \lambda)\frac{\partial^2 e}{\partial X_i \partial X_i} = \rho_0 \frac{\partial^2 e}{\partial t^2}.$$

5.26. (a) Write a displacement field for an infinite train of longitudinal waves propagating in the direction $3\mathbf{e}_1 + 4\mathbf{e}_2$.

(b) Write a displacement field for an infinite train of transverse waves propagating in the direction $3\mathbf{e}_1 + 4\mathbf{e}_2$ and polarized in the X_1, X_2-plane.

5.27. Consider a material with Poisson's ratio equal to $\frac{1}{3}$, and a transverse elastic wave (as in Section 5.7C) of amplitude ϵ_1 and incident on a plane boundary at an angle α_1. Determine the amplitudes and angles of reflection of the reflected waves if

(a) $\alpha_1 = 0$,

(b) $\alpha_1 = 15°$.

5.28. Consider an incident transverse wave on a free boundary as in Section 5.7C. For what particular angles of incidence will the only reflected wave be transverse? (Take $\nu = \frac{1}{3}$.)

5.29. Consider a transverse elastic wave incident on a traction-free plane surface and polarized normal to the plane of incidence. Show that the boundary condition can be satisfied with only a reflected transverse wave that is similarly polarized. What is the relation of the amplitudes, wavelengths, and direction of propagation of the incident and reflected wave?

5.30. Consider the problem of Section 5.7C and determine the characteristics of the reflected waves if the boundary ($X_2 = 0$) is fixed (no motion). How are the results different from the case of a free boundary?

5.31. A longitudinal elastic wave is incident on a fixed boundary.
(a) Show that in general there are two reflected waves, one longitudinal and the other transverse (polarized in plane normal to incident plane).
(b) Find, as in Section 5.7C, the amplitude ratio of reflected to incident elastic waves.

5.32. Do the previous problem for a free boundary.

5.33. Verify that the thickness stretch vibration given by Eq. 5.28 does satisfy the longitudinal wave equation.

5.34. Do Example 5.11 if the right face ($X_1 = l$) is free.

5.35. (a) Find the thickness stretch vibration if the $X_1 = 0$ face is being forced by a traction $\mathbf{t} = (\beta \cos \omega t)\mathbf{e}_1$ and the right-hand face ($X_1 = l$) is fixed.
(b) Find the resonant frequencies.

5.36. (a) Find the thickness-shear vibration if the left-hand face ($X_1 = 0$) has a forced displacement $\mathbf{u} = (\alpha \cos \omega t)\mathbf{e}_3$ and the right-hand face ($X_1 = l$) is fixed.
(b) Find the resonant frequencies.

5.37. Do the previous problem if the forced displacement is given by $\mathbf{u} = \alpha(\cos \omega t \mathbf{e}_2 + \sin \omega t \mathbf{e}_3)$. Describe the particle motion throughout the plate.

5.38. Determine the total elongation of a steel bar 76 cm long if the tensile stress is 0.1 GPa and $E_Y = 207$ GPa.

5.39. A cast iron bar, 4ft (122 cm) long and $1\frac{1}{2}$ in. (3.81 cm) in diameter is pulled by equal and opposite axial forces P at its ends.
(a) Find the maximum normal and shearing stresses if $P = 20{,}000$ lb (89000 N)
(b) Find the total elongation and lateral contraction ($E_Y = 15 \times 10^6$ psi (103 GPa), $\nu = 0.25$).

5.40. A steel bar ($E_Y = 207$ GPa) of 6 cm² cross-section and 6 m length is acted on by the indicated (Fig. P5.1) axially applied forces. Find the total elongation of the bar.

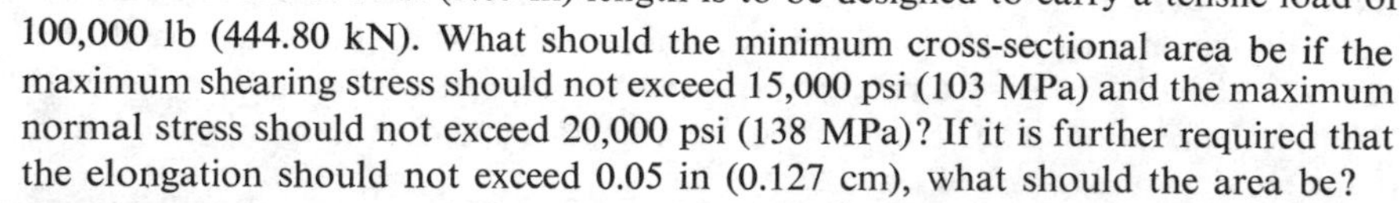

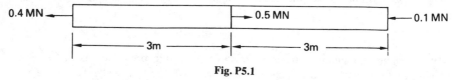

Fig. P5.1

5.41 A steel bar of 10-ft (3.05 m) length is to be designed to carry a tensile load of 100,000 lb (444.80 kN). What should the minimum cross-sectional area be if the maximum shearing stress should not exceed 15,000 psi (103 MPa) and the maximum normal stress should not exceed 20,000 psi (138 MPa)? If it is further required that the elongation should not exceed 0.05 in (0.127 cm), what should the area be?

5.42. Consider a bar of cross-sectional area A that is stretched by a tensile force P at each end.
(a) Determine the normal and shearing stresses on a plane with a normal vector

that makes an angle α with the cylindrical axis. For what values of α are the normal and shearing stresses equal?
(b) The load carrying capacity of the bar is based on the shearing stress on the plane defined by $\alpha = \alpha_0$ remaining less than τ_0. Sketch how the maximum load will depend on the angle α_0.

5.43. Consider a cylindrical bar that is acted upon by an axial stress $T_{11} = \sigma$.
(a) What will the state of stress in the bar be if the lateral surface is constrained so that there is no contraction or expansion?
(b) Show that the effective Young's modulus $E'_Y = T_{11}/E_{11}$ is given by

$$E'_Y = \frac{(1-\nu)}{(1-2\nu)(1+\nu)} E_Y.$$

(c) Evaluate the effective modulus for Poisson's ratio equal to $\frac{1}{3}$ and $\frac{1}{2}$.

5.44. Let the state of stress in a tension specimen be given by $T_{11} = \sigma$, all other $T_{ij} = 0$.
(a) Find the components of the deviatoric stress $\mathbf{T}^0 = \mathbf{T} - \frac{1}{3} T_{kk} \mathbf{I}$.
(b) Find the scalar invariants of $\mathbf{T}^0$.

5.45. Three identical steel rods support the load P, as shown in Fig. P5.2. How much load does each rod carry? Neglect the weights of the rod and the rigid bar.

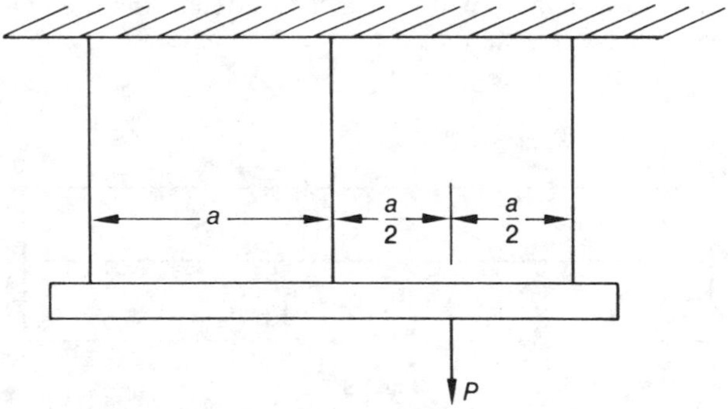

Fig. P5.2

5.46. Solve the previous problem if the cross-sectional area of the middle bar is twice that of the left- and right-hand bars.

5.47. Let the axis of a cylindrical bar be vertical and initially coincide with the x_1-axis. If $X_1 = 0$ corresponds to the lower face, then the body force is given by $\rho\mathbf{B} = -\rho g \mathbf{e}_1$. Assume that the stress distribution induced by the body force alone is of the form

$$T_{11} = \rho g X_1$$

and all other $T_{ij} = 0$.

(a) Show that the stress tensor is a possible state of stress in the presence of the body force mentioned above.
(b) If this possible state of stress is the actual distribution of stress in the cylindrical bar, what surface tractions should act on the lateral face and the pair of end faces in order to produce this state of stress.

5.48. A circular steel shaft is subjected to twisting couples of 2700 N·m. The allowable tensile stress is 0.124 GPa. If the allowable shearing stress is 0.6 times the allowable tensile stress, what is the minimum allowable diameter?

5.49. A circular steel shaft is subjected to twisting couples of 5000 ft-lb (6780 N.m). Determine the shaft diameter if the maximum shear stress is not to exceed 10,000 psi (69 MPa) and the angle of twist is not to exceed 1.5° in 20 diameters of length $\mu = 12\times10^6$ psi (82.7 GPa).

5.50. Demonstrate that the elastic solution for the solid circular bar in torsion is also valid for a circular cylindrical tube in torsion. If a is the outside radius and b is the inside radius, how must Eq. (5.42) for the twist per unit length be altered?

5.51. In Example 5.16, if the radius of the left portion is a_1 and the radius of the right portion is a_2, what is the twisting moment produced in each portion of the shaft? Both shafts are of the same material.

5.52 Solve the previous problem if $a_1 = 3.0$ cm, $a_2 = 2.5$ cm, $l_1 = l_2 = 75$cm, and $M_t = 700$ N·m.

5.53. For the circular shaft shown in Fig. P5.3, determine the twisting moment produced in each part of the shaft.

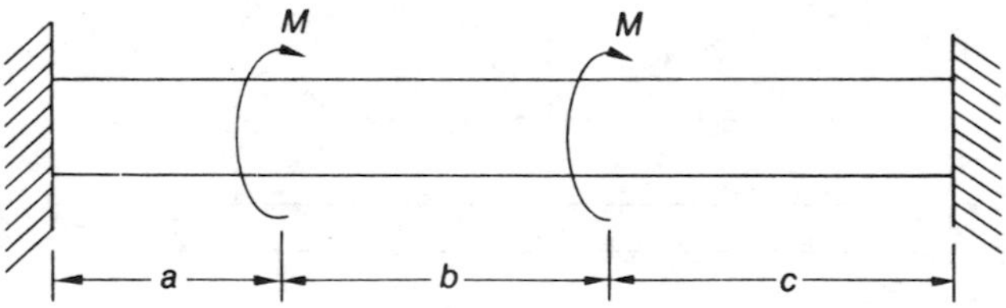

Fig. P5.3

5.54. A circular bar of one-inch (2.54 cm) radius is under the action of an axial tensile load of 30,000 lb (133 kN) and a twisting couple of 25,000 in.-lb (2830 N·m).
(a) Determine the stress throughout the bar.
(b) Find the maximum normal and shearing stress that occurs over all locations and all cross-sectional planes throughout the bar.

5.55. Show that for any cylindrical bar of non-circular cross-section in torsion that the stress vector at all points along the lateral boundary acting on any of the normal cross-sectional planes must be tangent to the boundary. (Hint: Use $\mathbf{T} = \mathbf{T}^T$.)

5.56. Demonstrate that the displacement and stress for the elliptic bar in torsion may also be used for an elliptic tube, if the inside boundary is defined by

$$\frac{X_2^2}{a^2}+\frac{X_3^2}{b^2}=k^2,$$

where $k < 1$.

5.57. Compare the twisting torque which can be transmitted by a shaft with an elliptical cross-section having a major axis equal to twice the minor axis with a shaft of circular cross-section having a diameter equal to the major axis. Both shafts are of the same material. Also compare the unit twist under the same twisting moment.

5.58. Repeat the previous problem, except that the circular shaft has a diameter equal to the minor axis of the elliptical shaft.

5.59. (a) For an elliptic bar in torsion, show that the magnitude of the maximum shearing stress varies linearly along radical lines ($X_2 = kX_3$) and reaches a maximum on the outer boundary.

(b) Show that on the boundary the maximum shearing stress is given by

$$(T_s)_{\max}=\frac{2M_t}{\pi a^2 b^3}\sqrt{b^4+X_3^2(a^2-b^2)}$$

so that the greatest shearing stress does occur at the end of the minor axis.

5.60. Consider the torsion of a cylindrical bar with an equilateral triangular cross-section as in Fig. P5.4.

(a) Show that a warping function $\phi = \alpha(3X_2^2X_3 - X_3^3)$ generates an equilibrium stress field.

(b) Determine the constant α in order to satisfy the traction-free lateral boundary condition. Demonstrate that the entire lateral surface is traction-free.

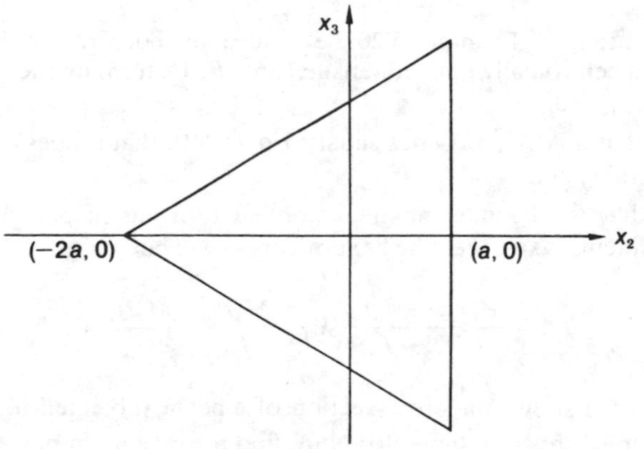

Fig. P5.4

(c) Write out explicitly the stress distribution generated by this warping function. Evaluate the maximum shearing stress at the triangular corners and along the line $X_3 = 0$ in a cross-section. Along the line $X_3 = 0$, where does the greatest shearing stress occur?

5.61. An alternate manner of formulating the problem of the torsion of a cylinder of noncircular cross-section employs a stress function $\psi(X_2, X_3)$ such that the stresses are given by

$$T_{12} = \frac{\partial\psi}{\partial X_3}, \quad T_{13} = -\frac{\partial\psi}{\partial X_2},$$

and all other $T_{ij} = 0$.

(a) Demonstrate that the equilibrium equations are identically satisfied for any choice of ψ.

(b) Show that if ψ satisfies the equation

$$\frac{\partial^2\psi}{\partial X_2{}^2} + \frac{\partial^2\psi}{\partial X_3{}^2} = \text{constant},$$

then the stress will correspond to a compatible strain field for simply-connected cross-sectional areas.

(c) Show that the lateral boundary condition requires that $\nabla\psi$ be in the same direction as the outward normal. In other words, the values of ψ on the outer boundary is a constant.

5.62. A beam of circular cross-section is subjected to pure bending. The magnitude of each end couple is 14,000 N·m. If the maximum normal stress is not to exceed 0.124 GPa, what should be the diameter?

5.63 The rectangular beam of Example 5.20 has a width b and a height $1.2b$. If the right-hand couple is given by $\mathbf{M}=24{,}000\mathbf{e}_2$ ft-lb (32,500 N·m), determine the dimension b in order that the maximum shearing stress does not exceed 600 psi (4.14 MPa).

5.64 Let the beam of Example 5.20 be loaded by both the indicated bending moment and a centroidally applied tensile force P. Determine the magnitude at P in order that $T_{11} \geqslant 0$.

5.65. Verify that if $\phi(X_1, X_2)$ *does* satisfy Eq. (5.57), that it does correspond to a compatible strain field.

5.66. Show that the bending moment applied to a bar in pure bending is not referred to principal axes, then the flexural stress will be

$$T_{11} = \frac{M_2 I_{zz} + M_3 I_{zy}}{I_{zz} I_{yy} - I_{zy}^2} X_3 - \frac{M_3 I_{yy} + M_2 I_{zy}}{I_{zz} I_{yy} - I_{zy}^2} X_2.$$

5.67. Figure P5.5 shows the cross-section of a beam subjected to pure bending. If the end couples are given by $\pm 10^4$ N·m, find the maximum normal stress.

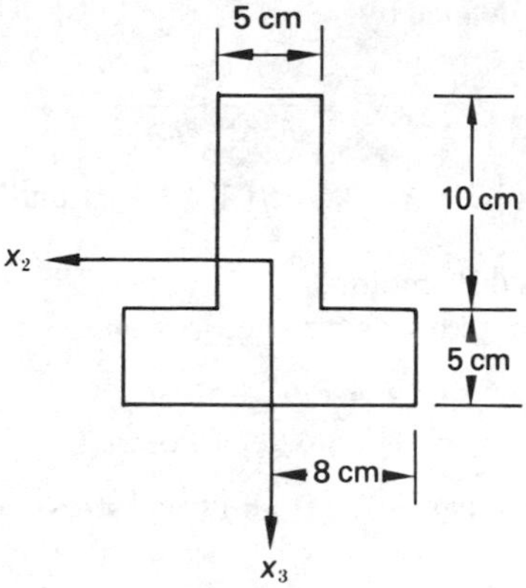

Fig. P5.5

5.68. Consider the stress function

$$\phi = \alpha_1 X_1{}^2 + \alpha_2 X_1 X_2 + \alpha_3 X_2{}^2.$$

(a) Verify that this stress function is a possible one for plane strain.
(b) Determine the stresses and sketch the boundary tractions on the rectangular boundary, $X_1 = 0$, $X_1 = a$, $X_2 = 0$, $X_2 = b$.

5.69. Consider the stress function $\phi = \alpha X_1{}^2 X_2$.
(a) Is this a possible stress function for plane strain?
(b) Determine the stresses.
(c) Determine and sketch the boundary traction on the boundary defined by

$$X_1 = 0,\ X_1 = a,\ X_2 = 0,\ X_2 = b.$$

5.70. Consider the stress function $\phi = \alpha X_1{}^4 + \beta X_2{}^4$.
(a) Is this a possible stress function for plane strain?
(b) Determine and sketch the boundary tractions on the rectangular boundary of the previous problem.

5.71. Consider the stress function $\phi = \alpha X_1 X_2{}^2 + \beta X_1 X_2{}^3$.
(a) Is this a possible stress function for plane strain?
(b) Determine the stresses.
(c) Find the condition necessary for the traction on $X_2 = b$ to vanish and sketch the stress traction on the remaining boundaries $X_2 = 0$, $X_1 = 0$, $X_1 = a$.

5.72. (a) Show that the equation (sometimes called the bi-harmonic equation) that ϕ must satisfy to generate a possible elastic solution may be written as

$$\nabla^2(\nabla^2\phi) = 0,$$

where the operator ∇^2 is defined by

$$\nabla^2 \equiv \frac{\partial^2}{\partial X_1{}^2} + \frac{\partial^2}{\partial X_2{}^2}.$$

(b) Consider $\phi = \alpha \log (X_1{}^2 + X_2{}^2)^{1/2} + \beta(X_1{}^2 + X_2{}^2)$ and show that it satisfies the equation of part (a).

(c) Determine the stress distribution.

(d) Let the cylindrical coordinate base vectors be given by

$$\mathbf{e}_r = \cos\theta \mathbf{e}_1 + \sin\theta \mathbf{e}_2,$$
$$\mathbf{e}_\phi = -\sin\theta \mathbf{e}_1 + \cos\theta \mathbf{e}_2.$$

Determine for the stress of part (c) the cylindrical stress components

$$T_{rr} = \mathbf{e}_r \cdot \mathbf{T}\mathbf{e}_r,$$

$$T_{r\phi} = \mathbf{e}_r \cdot \mathbf{T}\mathbf{e}_\phi,$$

$$T_{\phi\phi} = \mathbf{e}_\phi \cdot \mathbf{T}\mathbf{e}_\phi.$$

(e) Find the boundary tractions on the cylindrical boundaries $X_1{}^2 + X_2{}^2 = a^2$ and $X_1{}^2 + X_2{}^2 = b^2$.

6

Newtonian Viscous Fluid

Substances such as water and air are examples of a fluid. Mechanically speaking they are different from a piece of steel or concrete in that they are unable to sustain shearing stresses without continuously deforming. For example, if water or air is placed between two parallel plates with say, one of the plates fixed and the other plate applying a shearing stress, it will deform indefinitely with time if the shearing stress is not removed. Also, in the presence of gravity, the fact that water at rest always conforms to the shape of its container is a demonstration of its inability to sustain shearing stress at rest. Based on this notion of fluidity, we define a fluid to be a class of idealized materials which, when in rigid body motion (including the state of rest), cannot sustain any shearing stress. Water is also an example of the so-called liquid which undergoes negligible density changes under a wide range of loads, whereas air is an example of a gas which does otherwise. This aspect of behavior is generalized into the concept of incompressible and compressible fluids. However, as it will be seen that under certain conditions (low Mach number flow) air can be treated as incompressible and under other conditions (acoustic) water has to be treated as compressible.

In this chapter, we study a special model of fluid, which has the property that the stress associated with the motion depends linearly on the instantaneous value of the rate of deformation. This model of fluid is known as a Newtonian fluid or linearly viscous fluid which has been found to describe adequately the mechanical behavior of many real fluids under a wide range of situations. However, some fluids, such as polymeric solutions, require a more general model (Non-Newtonian Fluid) for a more adequate description. A very general model of non-Newtonian fluid will be discussed in a later chapter.

6.1 FLUIDS

Based on the notion of fluidity discussed in the previous paragraphs, we define a **fluid** to be a class of idealized materials which, when in rigid body motions (including the state of rest), cannot sustain any shearing stresses. In other words, when a fluid is in a rigid body motion, the stress vector on any plane is normal to the plane. Thus every plane is a principal plane, or equivalently, every direction is an eigenvector of the stress tensor. It then follows that the eigenvalues are all the same.† As a consequence, on all planes passing through a point, not only are there no shearing stresses but also the normal stresses are all the same. In other words, for a fluid in rigid body motion

$$\mathbf{T} = -p\mathbf{I}. \tag{6.1a}$$

Or in component form

$$T_{ij} = -p\delta_{ij}. \tag{6.1b}$$

The scalar p is the magnitude of the compressive normal stress and is known as the "**hydrostatic pressure**."

6.2 COMPRESSIBLE AND INCOMPRESSIBLE FLUIDS

What one generally calls "liquids" such as water and mercury have the property that their density remains essentially unchanged under a wide range of pressures. Idealizing this property, we define an "**incompressible fluid**" to be one for which the density of every particle remains the same at all times regardless of the state of stress. That is, for an incompressible fluid

$$\frac{D\rho}{Dt} = 0. \tag{6.2}$$

It then follows from the equation of mass conservation

$$\frac{D\rho}{Dt} + \rho\frac{\partial v_k}{\partial x_k} = 0$$

that

$$\frac{\partial v_k}{\partial x_k} = 0, \tag{6.3}$$

†From $\mathbf{Tn}_1 = \lambda_1\mathbf{n}_1$, $\mathbf{Tn}_2 = \lambda_2\mathbf{n}_2$, we get $(\lambda_1 - \lambda_2)\mathbf{n}_1 \cdot \mathbf{n}_2 = 0$ [Section 2B13, Eq. (v)]. Since $\mathbf{n}_1$ and $\mathbf{n}_2$ are arbitrary, therefore $\lambda_1 = \lambda_2$.

or,

$$\text{div}\,\mathbf{v} = 0. \tag{6.3a}$$

An incompressible fluid need not have a spatially uniform density (e.g., salt water with nonuniform concentrations with depth). If the density is also uniform, it is referred to as "homogeneous fluid," for which $\rho =$ constant everywhere.

Substances such as air and vapors which change their density appreciably with pressure are often treated as compressible fluids. Of course, it is not hard to see that there are situations where water has to be regarded as compressible and air may be regarded as incompressible. However, for theoretical studies, it is convenient to regard the incompressible and compressible fluid as two distinct kinds of fluids.

6.3 EQUATIONS OF HYDROSTATICS

With $T_{ij} = -p\delta_{ij}$, the equations of equilibrium

$$\frac{\partial T_{ij}}{\partial x_j} + \rho B_i = 0, \tag{6.4}$$

where B_i are components of body forces per unit mass become

$$\frac{\partial p}{\partial x_i} = \rho B_i \tag{6.5a}$$

or

$$\nabla p = \rho \mathbf{B}. \tag{6.5b}$$

In the case where B_i are the components of the weight per unit mass, if we let the positive x_3-axis be pointing vertically downward, we have, $B_1 = B_2 = 0$ and $B_3 = g$, so that Eqs. (6.5) read

$$\frac{\partial p}{\partial x_1} = 0, \tag{6.6a}$$

$$\frac{\partial p}{\partial x_2} = 0, \tag{6.6b}$$

$$\frac{\partial p}{\partial x_3} = \rho g. \tag{6.6c}$$

If the fluid is in a state of rigid body motion (rate of deformation = 0) Eq.(6.5) has to be modified to include the acceleration term, see example 6.2.

EXAMPLE 6.1

Let $x_3 = 0$ be the surface of a homogeneous fluid and let the positive x_3-axis be pointing vertically downward. Find the pressure distribution in the liquid.

Solution. We have

$$\frac{\partial p}{\partial x_1} = 0.$$

$$\frac{\partial p}{\partial x_2} = 0,$$

$$\frac{\partial p}{\partial x_3} = \rho g = \text{constant.}$$

Thus p depends only on x_3 and in fact

$$p = \rho g x_3 + p_0.$$

where p_0 is the atmospheric pressure.

EXAMPLE 6.2

A tank containing a homogeneous fluid moves horizontally to the right with a constant acceleration a, find the angle θ of the inclination of the free surface.

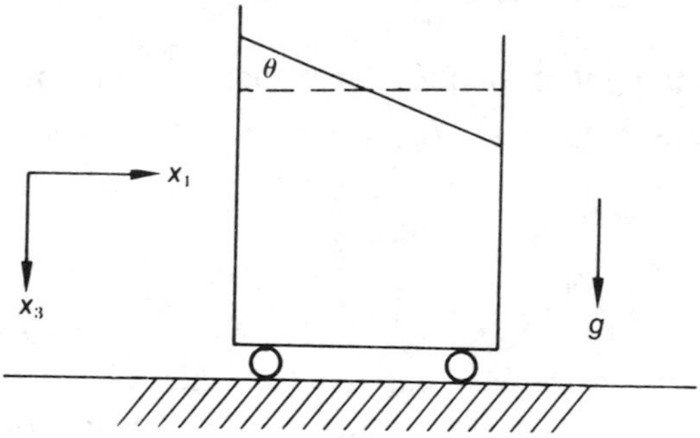

Fig. 6.1

Solution. The equations of motion are

$$\rho a = -\frac{\partial p}{\partial x_1}, \tag{i}$$

$$0 = -\frac{\partial p}{\partial x_2}, \tag{ii}$$

$$0 = -\frac{\partial p}{\partial x_3} + \rho g. \tag{iii}$$

From (ii), p is independent of x_2, from Eq. (i)

$$p = -\rho a x_1 + f(x_3) \tag{iv}$$

and from Eqs. (iii) and (iv)

$$\frac{\partial p}{\partial x_3} = \frac{df}{\partial x_3} = \rho g.$$

Thus

$$f(x_3) = \rho g x_3 + \text{constant},$$

i.e.,

$$p = -\rho a x_1 + \rho g x_3 + c.$$

If at the instant of interest the origin of the axes is located at a point on the free surface, then, $c = p_0$, the ambient pressure. On the surface $p = p_0$, thus the surface is a plane given by

$$\rho g x_3 = \rho a x_1,$$

i.e.,

$$x_3 = \frac{a}{g} x_1$$

and

$$\tan\theta = \frac{dx_3}{dx_1} = \frac{a}{g}.$$

EXAMPLE 6.3

For minor altitude differences, the atmosphere can be assumed to have constant temperature. Find the pressure and density distribution for this case.

Solution. Let the positive x_3-axis be pointing vertically upward. Then

$$\frac{\partial p}{\partial x_1} = 0, \tag{i}$$

$$\frac{\partial p}{\partial x_2} = 0, \tag{ii}$$

$$\frac{\partial p}{\partial x_3} = -\rho g. \tag{iii}$$

From Eqs. (i) and (ii), we see p is a function of x_3 only, thus Eq. (iii) becomes

$$\frac{dp}{dx_3} = -\rho g. \tag{iv}$$

Assuming that p, ρ, and θ (absolute temperature) are related by the equation of state for ideal gas, we have

$$p = \rho R\theta, \tag{v}$$

where R is the gas constant, Eq. (iv) becomes

$$\frac{dp}{p} = -\frac{g}{R\theta} dx_3.$$

Integrating, we get

$$\ln p = -\frac{g}{R\theta} x_3 + \ln p_0,$$

where p_0 is the pressure at the ground $(x_3 = 0)$, thus

$$p = p_0 e^{-(g/R\theta)x_3}, \tag{vi}$$

and from Eq. (v), if ρ_0 is the density at $x_3 = 0$, we have

$$\rho = \rho_0 e^{-(g/R\theta)x_3}. \tag{vii}$$

6.4 NEWTONIAN FLUID

Since the state of stress for a fluid under rigid body motion (including rest) is given by an isotropic tensor, therefore in dealing with a fluid in general motion, it is natural to decompose the stress tensor into two parts

$$T_{ij} = -p\delta_{ij} + T'_{ij}, \tag{6.7}$$

where the values of T'_{ij} depend on the rate and/or higher rates of deformation in such a way that they are zero when the fluid is under rigid body motion (i.e., zero rates of deformation) and p is a scalar whose value is not to depend explicitly on these rates.

We now define a class of idealized materials called "Newtonian Fluids" as follows:

I. For a material point, the values of T'_{ij} at any time t depend linearly on the components of the rate of deformation tensor

$$D_{ij} = \frac{1}{2}\left(\frac{\partial v_i}{\partial x_j} + \frac{\partial v_j}{\partial x_i}\right)$$

at that time and not on any other kinematic quantities.

II. The fluid is isotropic.

Following the same arguments made in connection with isotropic linear elastic materials, we obtain that for a Newtonian fluid, (also known as linearly viscous fluid) the most general form of T'_{ij} is, with $\Delta \equiv D_{11} + D_{22} + D_{33} = D_{kk}$,

$$T'_{ij} = \lambda\Delta\delta_{ij} + 2\mu D_{ij}, \tag{6.8}$$

where λ and μ are material constants (different from those of an elastic body) having the dimension of (Force)(Time)/(Length)^2. The stress tensor T'_{ij} is known as the "**viscous stress tensor**." Thus, the total stress tensor is

$$T_{ij} = -p\delta_{ij} + \lambda\Delta\delta_{ij} + 2\mu D_{ij}, \tag{6.9}$$

i.e.,

$$T_{11} = -p + \lambda\Delta + 2\mu D_{11}, \tag{6.9a}$$

$$T_{22} = -p + \lambda\Delta + 2\mu D_{22}, \tag{6.9b}$$

$$T_{33} = -p + \lambda\Delta + 2\mu D_{33}. \tag{6.9c}$$

$$T_{12} = 2\mu D_{12}, \tag{6.9d}$$

$$T_{13} = 2\mu D_{13}, \tag{6.9e}$$

$$T_{23} = 2\mu D_{23}. \tag{6.9f}$$

The scalar p in the above equations is called the "**pressure**." It is a somewhat ambiguous terminology. As is seen from the above equations, when D_{ij} are nonzero, p is neither the total compressive normal stress on any plane unless the viscous components happen to be zero, nor is it, in general, the mean normal compressive stress, $T_{kk}/3$, (*see* next section). Of course, if $D_{ij} = 0$ (e.g., at rest), p is indeed the total normal stress on any plane passing through a point. For fluid theory, it is only necessary to remember that the isotropic tensor $-p\delta_{ij}$ is that part of T_{ij} which does not depend explicitly on any rates of deformation.

6.5 INTERPRETATION OF λ AND μ

Consider the shear flow given by the velocity field

$$v_1 = f(x_2), \quad v_2 = 0, \quad v_3 = 0.$$

For this flow

$$D_{11} = D_{22} = D_{33} = D_{13} = D_{23} = 0$$

and

$$D_{12} = \frac{1}{2}\frac{df}{dx_2},$$

so that

$$T_{11} = T_{22} = T_{33} = -p, \quad T_{13} = T_{23} = 0$$

and

$$T_{12} = \mu \frac{df}{dx_2}. \tag{6.10}$$

Thus, μ is the proportionality constant relating the shearing stress to the velocity gradient. It is known as the (first) **coefficient of viscosity**.

From Eq. (6.8), we have, for a general velocity field

$$\tfrac{1}{3} T'_{ii} = (\lambda + \tfrac{2}{3}\mu)\Delta. \tag{6.11}$$

Thus, $(\lambda + \frac{2}{3}\mu)$ is the proportionality constant relating the viscous mean normal stress to the rate of change of volume. It is known as the **coefficient of bulk viscosity**. The total mean normal stress is given by

$$\tfrac{1}{3} T_{ii} = -p + (\lambda + \tfrac{2}{3}\mu)\Delta \tag{6.12}$$

and it is clear that the so-called pressure is in general not the mean normal stress.

EXAMPLE 6.4

Given the following velocity field

$$v_1 = -x_1 - x_2, \quad v_2 = x_2 - x_1, \quad v_3 = 0,$$

for a Newtonian liquid with viscosity $\mu = 0.982$ mPa·s (2.05×10^{-5} lb sec/ft^2). For a plane whose normal is in the $\mathbf{e}_1$-direction, (a) find the excess of the total normal compressive stress over the pressure p, and (b) find the magnitude of the shearing stress.

Solution. (a) From

$$T_{11} = -p + 2\mu D_{11}, \qquad (\Delta = 0).$$

we have

$$(-T_{11}) - p = -2\mu D_{11}.$$

Now

$$D_{11} = \frac{\partial v_1}{\partial x_1} = -1\ (\text{sec})^{-1}.$$

Therefore

$$(-T_{11}) - p = -2(0.982)(-1) = 1.96 \text{ mPa}.$$

(b)

$$T_{12} = 2\mu D_{12} = \mu\left(\frac{\partial v_1}{\partial x_2} + \frac{\partial v_2}{\partial x_1}\right) = -2\mu$$

$$= -1.96 \text{ mPa}$$

$$T_{13} = 2\mu D_{13} = \mu\left(\frac{\partial v_1}{\partial x_3} + \frac{\partial v_3}{\partial x_1}\right) = 0.$$

Thus, the magnitude of shearing stress $= 1.96$ mPa.

6.6 INCOMPRESSIBLE NEWTONIAN FLUID

For an incompressible fluid, $\Delta = 0$ at all times. Thus the constitutive equation for such a fluid becomes

$$T_{ij} = -p\delta_{ij} + 2\mu D_{ij} \tag{6.13}$$

and the pressure $p = -T_{ii}/3$ which has the meaning of mean normal compressive stress. It is not to depend explicitly either on kinematic quantities or on thermodynamic variables (note the density of a particle does not change with time) and is to be regarded as one of the fundamental dynamic variables.

Since

$$D_{ij} = \frac{1}{2}\left(\frac{\partial v_i}{\partial x_j} + \frac{\partial v_j}{\partial x_i}\right), \tag{6.14}$$

where v_i are velocity components, the constitutive equations can be written

$$T_{ij} = -p\delta_{ij} + \mu\left(\frac{\partial v_i}{\partial x_j} + \frac{\partial v_j}{\partial x_i}\right), \tag{6.15}$$

i.e.,

$$T_{11} = -p + 2\mu\frac{\partial v_1}{\partial x_1}, \tag{6.15a}$$

$$T_{22} = -p + 2\mu\frac{\partial v_2}{\partial x_2}, \tag{6.15b}$$

$$T_{33} = -p + 2\mu\frac{\partial v_3}{\partial x_3}, \tag{6.15c}$$

$$T_{12} = \mu\left(\frac{\partial v_1}{\partial x_2} + \frac{\partial v_2}{\partial x_1}\right), \tag{6.15d}$$

$$T_{13} = \mu\left(\frac{\partial v_1}{\partial x_3} + \frac{\partial v_3}{\partial x_1}\right), \tag{6.15e}$$

$$T_{23} = \mu\left(\frac{\partial v_2}{\partial x_3} + \frac{\partial v_3}{\partial x_2}\right). \tag{6.15f}$$

Substituting the constitutive equation [Eq. (6.15)] into the equation of motion

$$\rho\left(\frac{\partial v_i}{\partial t} + v_j\frac{\partial v_i}{\partial x_j}\right) = \rho B_i + \frac{\partial T_{ij}}{\partial x_j}$$

and noting that

$$\frac{\partial T_{ij}}{\partial x_j} = -\frac{\partial p}{\partial x_j}\delta_{ij} + \mu\frac{\partial^2 v_i}{\partial x_j \partial x_j} + \mu\frac{\partial^2 v_j}{\partial x_j \partial x_i}$$

$$= -\frac{\partial p}{\partial x_i} + \mu\frac{\partial^2 v_i}{\partial x_j \partial x_j},$$

$$\left(\text{note}\ \frac{\partial^2 v_j}{\partial x_j \partial x_i} = \frac{\partial}{\partial x_i}\left(\frac{\partial v_j}{\partial x_j}\right) = \frac{\partial}{\partial x_i}\Delta = 0\right),$$

we obtain the following equations of motion in terms of velocity components

$$\rho\left(\frac{\partial v_i}{\partial t} + v_j\frac{\partial v_i}{\partial x_j}\right) = \rho B_i - \frac{\partial p}{\partial x_i} + \mu\frac{\partial^2 v_i}{\partial x_j \partial x_j}, \tag{6.16}$$

i.e.,

$$\rho\left(\frac{\partial v_1}{\partial t}+v_1\frac{\partial v_1}{\partial x_1}+v_2\frac{\partial v_1}{\partial x_2}+v_3\frac{\partial v_1}{\partial x_3}\right)=\rho B_1-\frac{\partial p}{\partial x_1}+\mu\left(\frac{\partial^2}{\partial x_1{}^2}+\frac{\partial^2}{\partial x_2{}^2}+\frac{\partial^2}{\partial x_3{}^2}\right)v_1, \tag{6.16a}$$

$$\rho\left(\frac{\partial v_2}{\partial t}+v_1\frac{\partial v_2}{\partial x_1}+v_2\frac{\partial v_2}{\partial x_2}+v_3\frac{\partial v_2}{\partial x_3}\right)=\rho B_2-\frac{\partial p}{\partial x_2}+\mu\left(\frac{\partial^2}{\partial x_1{}^2}+\frac{\partial^2}{\partial x_2{}^2}+\frac{\partial^2}{\partial x_3{}^2}\right)v_2, \tag{6.16b}$$

$$\rho\left(\frac{\partial v_3}{\partial t}+v_1\frac{\partial v_3}{\partial x_1}+v_2\frac{\partial v_3}{\partial x_2}+v_3\frac{\partial v_3}{\partial x_3}\right)=\rho B_3-\frac{\partial p}{\partial x_3}+\mu\left(\frac{\partial^2}{\partial x_1{}^2}+\frac{\partial^2}{\partial x_2{}^2}+\frac{\partial^2}{\partial x_3{}^2}\right)v_3, \tag{6.16c}$$

or, in invariant form,

$$\rho\left[\frac{\partial \mathbf{v}}{\partial t}+(\nabla\mathbf{v})\,\mathbf{v}\right]=\rho\mathbf{B}-\nabla p+\mu\,\mathbf{div}\,(\nabla\mathbf{v}). \tag{6.16d}$$

These are known as the Navier–Stokes equations of motion for incompressible Newtonian fluid. There are four unknown functions v_1, v_2, v_3, and p in the three equations. The fourth equation is supplied by the continuity equation $\Delta=0$, i.e.,

$$\frac{\partial v_1}{\partial x_1}+\frac{\partial v_2}{\partial x_2}+\frac{\partial v_3}{\partial x_3}=0, \tag{6.17a}$$

or, in invariant form,

$$\text{div}\,\mathbf{v}=0. \tag{6.17b}$$

EXAMPLE 6.5

If all particles have their velocity vectors parallel to a fixed direction, the flow is said to be a parallel flow (or unidirectional flow). Show that for parallel flows of an incompressible linearly viscous fluid, the total normal compressive stress on any plane parallel to and perpendicular to the direction of flow is the pressure p.

Solution. Let the direction of the flow be the x_1-axis, then

$$v_2=0,\quad v_3=0$$

and from the equation of continuity $\partial v_1/\partial x_1=0$. Thus the velocity field for a parallel flow is

$$v_1=v_1(x_2,x_3),\quad v_2=0,\quad v_3=0.$$

For this flow, $D_{11}=D_{22}=D_{33}=0$, thus

$$T_{11}=-p,$$

$$T_{22}=-p,$$

$$T_{33}=-p.$$

In cylindrical coordinates, with v_r, v_ϕ, v_z denoting velocity components in r-, ϕ-, z-directions, Navier-Stokes equations for incompressible Newtonian fluid take the following form

$$\frac{\partial v_r}{\partial t}+v_r\frac{\partial v_r}{\partial r}+\frac{v_\phi}{r}\frac{\partial v_r}{\partial \phi}+v_z\frac{\partial v_r}{\partial z}-\frac{v_\phi{}^2}{r}=-\frac{1}{\rho}\frac{\partial p}{\partial r}+B_r+\frac{\mu}{\rho}\left\{\frac{\partial^2 v_r}{\partial r^2}+\frac{1}{r^2}\frac{\partial^2 v_r}{\partial \phi^2}+\frac{\partial^2 v_r}{\partial z^2}+\frac{1}{r}\frac{\partial v_r}{\partial r}-\frac{2}{r^2}\frac{\partial v_\phi}{\partial \phi}-\frac{v_r}{r^2}\right\}, \tag{6.18a}$$

$$\frac{\partial v_\phi}{\partial t}+v_r\frac{\partial v_\phi}{\partial r}+\frac{v_\phi}{r}\frac{\partial v_\phi}{\partial \phi}+v_z\frac{\partial v_\phi}{\partial z}+\frac{v_r v_\phi}{r}=-\frac{1}{\rho r}\frac{\partial p}{\partial \phi}+B_\phi+\frac{\mu}{\rho}\left\{\frac{\partial^2 v_\phi}{\partial r^2}+\frac{1}{r^2}\frac{\partial^2 v_\phi}{\partial \phi^2}+\frac{\partial^2 v_\phi}{\partial z^2}+\frac{1}{r}\frac{\partial v_\phi}{\partial_r}+\frac{2}{r^2}\frac{\partial v_r}{\partial \phi}-\frac{v_\phi}{r^2}\right\}, \tag{6.18b}$$

$$\frac{\partial v_z}{\partial t}+v_r\frac{\partial v_z}{\partial r}+\frac{v_\phi}{r}\frac{\partial v_z}{\partial \phi}+v_z\frac{\partial v_z}{\partial z}=-\frac{1}{\rho}\frac{\partial p}{\partial z}+B_z+\frac{\mu}{\rho}\left\{\frac{\partial^2 v_z}{\partial r^2}+\frac{1}{r^2}\frac{\partial^2 v_z}{\partial \phi^2}+\frac{\partial^2 v_z}{\partial z^2}+\frac{1}{r}\frac{\partial v_z}{\partial r}\right\}. \tag{6.18c}$$

The equation of continuity takes the form

$$\frac{1}{r}\frac{\partial}{\partial r}(rv_r)+\frac{1}{r}\frac{\partial v_\phi}{\partial \phi}+\frac{\partial v_z}{\partial z}=0. \tag{6.19}$$

6.7 BOUNDARY CONDITIONS

On a rigid boundary, we shall impose the "nonslip" condition (also known as adherence condition), i.e., the fluid layer next to a rigid surface moves with that surface, in particular if the surface is at rest the velocity of the fluid at the surface is zero. The nonslip condition is well supported by experiments for many fluids. Even when the fluid does not wet the bounding surface (as is the case of mercury on glass), the condition is found to be appropriate and even for some non-Newtonian fluid (for which the constitutive equations are more involved than what we have considered) the nonslip condition has been confirmed by experiment.

6.8 STREAMLINE, PATHLINE, STEADY, UNSTEADY, LAMINAR, AND TURBULENT FLOW

(i) *Streamline*

A **streamline** (at time t) is a curve, having the property that at each of its points, the tangent line has the direction of the velocity vector of the particle instantaneously at the point. Experimentally, streamlines on the surface of a fluid are often obtained by making a short-time exposure photograph of the surface, sprinkled with reflecting particles. Each reflecting particle produces a short line on the photograph approximating the tangent line to a streamline. Mathematically, streamlines can be obtained from the velocity field $\mathbf{v}(\mathbf{x}, t)$ as follows: Let $\mathbf{x} = \mathbf{x}(s)$ be the parametric equation for the streamline at time t_0, which passes through a given point $\mathbf{x}_0$. Then an s can always be chosen such that†

$$\begin{cases} \dfrac{d\mathbf{x}}{ds} = \mathbf{v}(\mathbf{x}, t_0) \\ \mathbf{x}(0) = \mathbf{x}_0. \end{cases} \tag{6.20}$$

EXAMPLE 6.6

Given the velocity field

$$v_1 = \frac{x_1}{1+t}, \quad v_2 = x_2, \quad v_3 = 0.$$

find the streamline at time t_0 and passing through the point $(\alpha_1, \alpha_2, \alpha_3)$.

Solution. From $dx_1/ds = x_1/(1+t_0)$, we have

$$\int_{\alpha_1}^{x_1} \frac{dx_1}{x_1} = \frac{1}{1+t_0}\int_0^s ds.$$

Thus

$$\ln x_1 - \ln \alpha_1 = \frac{s}{1+t_0},$$

i.e.

$$x_1 = \alpha_1 e^{s/(1+t_0)}.$$

Similarly, from $dx_2/ds = x_2$, we have

$$\int_{\alpha_2}^{x_2} \frac{dx_2}{x_2} = \int_0^s ds.$$

Thus $x_2 = \alpha_2 e^s$. Obviously, $x_3 = \alpha_3$.

(ii) *Pathline*

A **pathline** is the path traversed by a fluid particle. To photograph a pathline, it is necessary to use long-time exposure of a reflecting particle. Mathematically, the pathline of a particle which is at $\mathbf{X}$ at time t_0, can be obtained from the velocity field $\mathbf{v}(\mathbf{x}, t)$ as follows: Let $\mathbf{x} = \mathbf{x}(t)$ be the pathline, then

$$\begin{cases} \dfrac{d\mathbf{x}}{dt} = \mathbf{v}(\mathbf{x}, t) \\ \mathbf{x}(t_0) = \mathbf{X}. \end{cases} \tag{6.21}$$

†Note, for arbitrary choice of the parameter s, the equation should read $d\mathbf{x}/ds = \beta\mathbf{v}(\mathbf{x}, t_0)$, e.g., if s is arc length, then $d\mathbf{x}/ds = \mathbf{v}/|\mathbf{v}|$, but one can always choose an s such that $\beta = 1$.

EXAMPLE 6.7

For the velocity field of the previous example, find the pathline for a particle which is at (X_1, X_2, X_3) at time t_0.

Solution. From $dx_1/dt = x_1/(1+t)$, we have

$$\int_{X_1}^{x_1} \frac{dx_1}{x_1} = \int_{t_0}^{t} \frac{dt}{1+t}.$$

Thus, $\ln x_1 - \ln X_1 = \ln(1+t) - \ln(1+t_0)$, i.e.,

$$x_1 = X_1 \frac{1+t}{1+t_0}.$$

Similarly, from $dx_2/dt = x_2$, we have

$$\int_{X_2}^{x_2} \frac{dx_2}{x_2} = \int_{t_0}^{t} dt.$$

Thus

$$x_2 = X_2 e^{t-t_0}$$

and obviously $x_3 = X_3$.

(iii) *Steady and Unsteady Flow*

A flow is called steady if at a *fixed* location nothing changes with time. Otherwise the flow is called unsteady. It is important to note, however, that in a steady flow, the velocity, acceleration, temperature, etc. of a given fluid particle in general changes with time. In other words, let ψ be any dependent variable, then in a steady flow $(\partial\psi/\partial t)_{\text{x-fixed}} = 0$, but $D\psi/Dt$ is in general not zero. For example, the steady flow given by the velocity field

$$v_1 = x_1, \quad v_2 = -x_2, \quad v_3 = 0$$

has an acceleration field given by

$$a_1 = \frac{\partial v_1}{\partial t} + v_1\frac{\partial v_1}{\partial x_1} + v_2\frac{\partial v_1}{\partial x_2} + v_3\frac{\partial v_1}{\partial x_3} = 0 + x_1(1) + 0 + 0 = x_1,$$

$$a_2 = \frac{\partial v_2}{\partial t} + v_1\frac{\partial v_2}{\partial x_1} + v_2\frac{\partial v_2}{\partial x_2} + v_3\frac{\partial v_2}{\partial x_3} = 0 + 0 + (-x_2)(-1) + 0 = x_2,$$

$$a_3 = 0.$$

We also note that for steady flow the pathlines coincide with the streamlines.

(iv) *Laminar and Turbulent Flow*

A laminar flow is a very orderly flow in which the fluid particles move in smooth layers, or laminae, sliding over particles in adjacent laminae without mixing with them. Such flow are generally realized at slow speed (when other things are fixed). For the case of water flowing through a tube of circular sections, it was found by Reynolds who observed the thin filaments of dye in the tube, that when the dimensionless parameter N_R (now known as Reynolds number) defined by $N_R = \bar{v}\rho d/\mu$, (where $\bar{v}$ is the average velocity in the pipe, d, the diameter of the pipe, and ρ and μ the density and viscosity of the fluid) is less than a certain value (approximately 2100) the thin filament of dye was maintained intact throughout the tube, forming straight lines parallel to the axis of the tube. Any accidental disturbances were rapidly obliterated. As the Reynolds number is increased the flow becomes increasingly sensitive to small perturbations until a stage is reached wherein the dye filament broke and diffused through the flowing water. This phenomenon of irregular intermingling of fluid particle in the flow is termed turbulent. In the case of pipe flow, the upper limit of the Reynolds number, beyond which the flow is turbulent, is indeterminate. Depending on the experimental setup and the initial quietness of the fluid, this upper limit can be as high as 100,000.

In the following sections, we restrict ourselves to the study of only laminar flows. It is therefore to be understood that the solutions to be presented are valid only within certain limits of some parameter (such as Reynolds number) governing the stability of the flow.

6.9 EXAMPLES OF LAMINAR FLOWS OF INCOMPRESSIBLE NEWTONIAN FLUID

A. Plane Couette Flow

The steady unidirectional flow, under zero pressure gradient in the flow direction, of an incompressible viscous fluid between two horizontal plates of infinite extent, one fixed and the other moving in its own plane with a constant velocity v_0, is known as the **plane Couette flow**.

Let x_1 be the direction of the flow. The velocity field for the plane Couette flow is then of the form

$$v_1 = v(x_2), \quad v_2 = 0, \quad v_3 = 0.$$

From the Navier–Stokes equations and the boundary conditions $v(0) = 0$ and $v(d) = v_0$, it can be shown (we leave it as an exercise) that

$$v(x_2) = \frac{v_0 x_2}{d}. \tag{6.22}$$

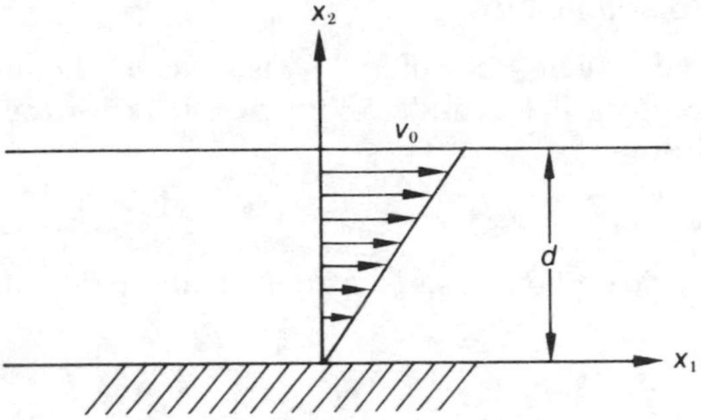

Fig. 6.2

B. Plane Poiseuille Flow

The steady unidirectional flow of an incompressible viscous fluid in a channel with two fixed parallel flat walls of infinite extent is known as the **plane Poiseuille flow**.

Let x_1 be the direction of the flow. The velocity field for the plane Poiseuille flow is then of the form

$$v_1 = v(x_2), \quad v_2 = 0, \quad v_3 = 0.$$

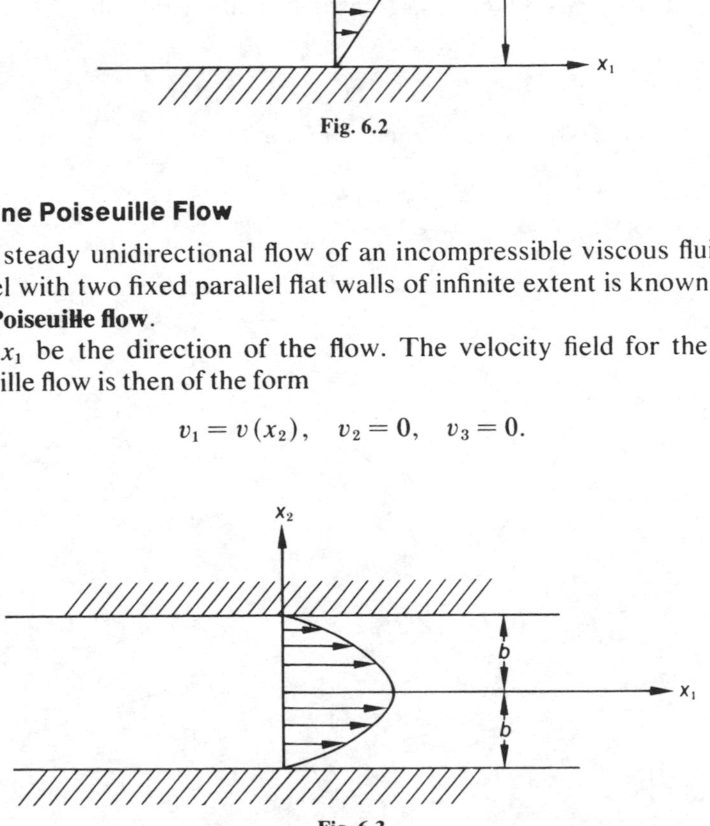

Fig. 6.3

From the Navier–Stokes equations and the boundary conditions $v(-b) = v(+b) = 0$, it can be shown (we leave it as an exercise) that $\partial p/\partial x_1 =$ a constant and

$$v(x_2) = -\frac{\partial p}{\partial x_1}\left(\frac{1}{2\mu}\right)(b^2 - x_2^2). \tag{6.23}$$

C. Hagen–Poiseuille Flow

The so-called Hagen–Poiseuille flow is a steady unidirectional axisymmetric flow in a circular cylinder. Thus, we look for the velocity field in the following form

$$v_1 = v(r), \quad r^2 = x_2^2 + x_3^2, \quad v_2 = v_3 = 0.$$

This velocity field obviously satisfies the equation of continuity

$$\frac{\partial v_1}{\partial x_1} + \frac{\partial v_2}{\partial x_2} + \frac{\partial v_3}{\partial x_3} = 0$$

for any $v_1(r)$. To satisfy the equation of motion, we have (*see* Fig. 6.4)

$$0 = -\frac{\partial p}{\partial x_1} + \rho g \sin\theta + \mu\left(\frac{\partial^2 v}{\partial x_2^2} + \frac{\partial^2 v}{\partial x_3^2}\right), \tag{i}$$

$$0 = -\frac{\partial p}{\partial x_2} - \rho g \cos\theta, \tag{ii}$$

$$0 = -\frac{\partial p}{\partial x_3}. \tag{iii}$$

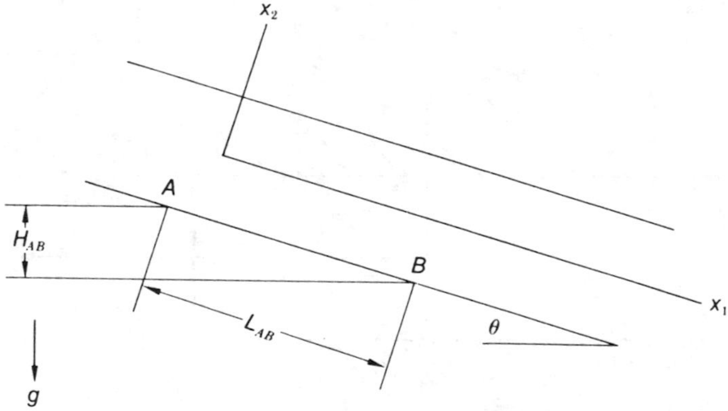

Fig. 6.4

From Eq. (iii) we have that p is independent of x_3. From Eq. (ii) we have, since $\rho g \cos\theta$ is a constant,

$$\frac{\partial}{\partial x_1}\left(\frac{\partial p}{\partial x_2}\right) = 0.$$

Interchanging the order of differentation, we have

$$\frac{\partial}{\partial x_2}\left(\frac{\partial p}{\partial x_1}\right)=0.$$

Furthermore, from Eq. (i), we have

$$\frac{\partial}{\partial x_1}\left(\frac{\partial p}{\partial x_1}\right)=\frac{\partial}{\partial x_1}\left[\rho g \sin\theta+\mu\left(\frac{\partial^2 v}{\partial x_2{}^2}+\frac{\partial^2 v}{\partial x_3{}^2}\right)\right]=0.$$

(Note: v is not a function of x_1.) Thus $\partial p/\partial x_1$ is independent of x_1, x_2, and x_3. In other words, $\partial p/\partial x_1=$ a constant so that $(p_B-p_A)/L_{AB}=\partial p/\partial x_1$ for any two points A and B as shown. Then Eq. (i) becomes

$$\left(\frac{\partial^2 v}{\partial x_2{}^2}+\frac{\partial^2 v}{\partial x_3{}^2}\right)=\frac{1}{\mu}\left(\frac{\partial p}{\partial x_1}-\rho g\sin\theta\right)$$

denoting the constant

$$\frac{1}{\mu}\left(\frac{\partial p}{\partial x_1}-\rho g\sin\theta\right)$$

by β we have

$$\frac{\partial^2 v}{\partial x_2{}^2}+\frac{\partial^2 v}{\partial x_3{}^2}=\beta. \tag{iv}$$

Now since $v_1=v(r)$, $r^2=x_2{}^2+x_3{}^2$,

$$\frac{\partial v}{\partial x_2}=\frac{dv}{dr}\frac{\partial r}{\partial x_2}=\frac{x_2}{r}\frac{dv}{dr},$$

$$\frac{\partial^2 v}{\partial x_2{}^2}=\frac{1}{r}\frac{dv}{dr}+x_2\left[\frac{d}{dr}\left(\frac{1}{r}\frac{dv}{dr}\right)\right]\frac{\partial r}{\partial x_2}=\frac{1}{r}\frac{dv}{dr}+\frac{x_2{}^2}{r}\frac{d}{dr}\left(\frac{1}{r}\frac{dv}{dr}\right).$$

Similarly,

$$\frac{\partial^2 v}{\partial x_3{}^2}=\frac{1}{r}\frac{dv}{dr}+\frac{x_3{}^2}{r}\frac{d}{dr}\left(\frac{1}{r}\frac{dv}{dr}\right).$$

Thus

$$\frac{\partial^2 v}{\partial x_2{}^2}+\frac{\partial^2 v}{\partial x_3{}^2}=\frac{2}{r}\frac{dv}{dr}+r\frac{d}{dr}\left(\frac{1}{r}\frac{dv}{dr}\right)=\frac{d^2 v}{dr^2}+\frac{1}{r}\frac{dv}{dr}=\frac{1}{r}\frac{d}{dr}\left(r\frac{dv}{dr}\right),$$

and Eq. (iv) becomes

$$\frac{1}{r}\frac{d}{dr}\left(r\frac{dv}{dr}\right)=\beta.$$

Thus

$$\frac{dv}{dr}=\beta\frac{r}{2}+\frac{b}{r}.$$

Integrating once more, we obtain

$$v=\frac{\beta r^2}{4}+b\ln r+c.$$

Since v must be bounded in the flow region, therefore $b=0$. Now, the nonslip condition in the cylindrical wall demands that

$$v=0 \qquad \text{at } r=\frac{d}{2}.$$

Thus

$$c=-\beta\left(\frac{d^2}{16}\right)$$

and

$$v=-\frac{\beta}{4}\left(\frac{d^2}{4}-r^2\right), \tag{6.24}$$

where

$$\beta=\frac{1}{\mu}\left(\frac{\partial p}{\partial x_1}-\rho g\sin\theta\right).$$

The above equation states that the velocity over the cross-section is distributed in the form of a paraboloid of revolution. The maximum velocity is (at $r=0$)

$$v_{\max}=-\frac{\beta d^2}{16}. \tag{6.25}$$

The mean velocity $\bar{v}$ is

$$\bar{v}=\frac{1}{(\pi d^2/4)}\int_A v\,dA=-\frac{\beta d^2}{32}=\frac{1}{2}v_{\max}, \tag{6.26}$$

and the volume rate of flow Q is

$$Q=\bar{v}\left(\frac{\pi d^2}{4}\right)=-\frac{\beta\pi d^4}{128}, \tag{6.27}$$

when $\theta=0$, (i.e., a horizontal pipe)

$$Q=-\frac{1}{\mu}\frac{\partial p}{\partial x_1}\left(\frac{\pi d^4}{128}\right). \tag{6.27a}$$

D. Plane Couette Flow of Two Layers of Incompressible Fluids

Let the viscosity and the density of the top layer be μ_1 and ρ_1 and those of the bottom layer be μ_2 and ρ_2. Let x_1 be the direction of flow and let $x_2=0$ be the interface. We look for steady unidirectional flows of the two

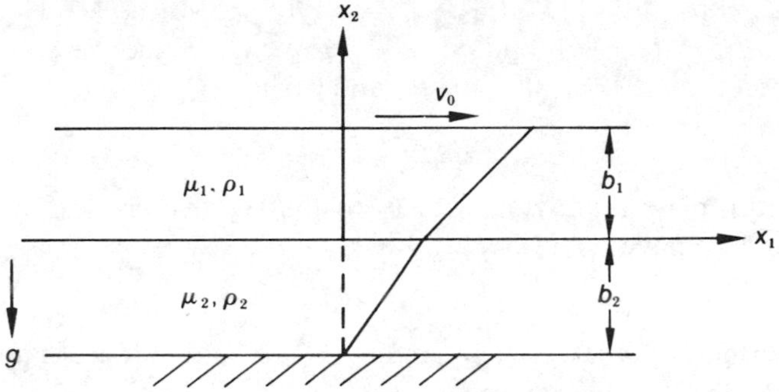

Fig. 6.5

layers between the infinite plates $x_2 = +b_1$ and $x_2 = -b_2$. The plate $x_2 = -b_2$ is fixed and the plate $x_2 = +b_1$ is moving on its own plane with velocity v_0. The pressure gradient in the flow direction is assumed to be zero.

Let the velocity distribution in the top layer be

$$v_1^{(1)} = v^{(1)}(x_2), \quad v_2^{(1)} = v_3^{(1)} = 0.$$

and that in the bottom layer be

$$v_1^{(2)} = v^{(2)}(x_2), \quad v_2^{(2)} = v_3^{(2)} = 0.$$

The equations of continuity are clearly satisfied. The Navier–Stokes equations give:

For layer 1,

$$0 = \mu_1 \frac{d^2 v^{(1)}}{dx_2^2}, \tag{i}$$

$$0 = -\frac{\partial p^{(1)}}{\partial x_2} - \rho_1 g, \tag{ii}$$

$$0 = -\frac{\partial p^{(1)}}{\partial x_3}. \tag{iii}$$

For layer 2,

$$0 = \mu_2 \frac{d^2 v^{(2)}}{dx_2^2}, \tag{iv}$$

$$0 = -\frac{\partial p^{(2)}}{\partial x_2} - \rho_2 g, \tag{v}$$

$$0 = -\frac{\partial p^{(2)}}{\partial x_3} \tag{vi}$$

From Eqs. (i) to (iii), $v^{(1)} = A_1x_2 + B_1$, and $p^{(1)} = -\rho_1gx_2 + C_1$. From Eqs. (iv) to (vi), $v^{(2)} = A_2x_2 + B_2$ and $p^{(2)} = -\rho_2gx_2 + C_2$. Since the bottom plate is fixed, therefore, $v^{(2)} = 0$ at $x = -b_2$, and we have

$$B_2 = A_2b_2. \tag{a}$$

Since the top plate is moving with v_0 to the right, therefore, $v^{(1)} = v_0$ at $x = +b_1$ and we have

$$B_1 = v_0 - A_1b_1. \tag{b}$$

At the interface $x_2 = 0$, we must have $v^{(1)} = v^{(2)}$ so that there is no slipping at the fluid interface. Therefore,

$$B_1 = B_2 \tag{c}$$

Furthermore, from Newton's third law, we have, on $x_2 = 0$, the stress vectors on the two laters are related by

$$\mathbf{t}^{(1)}_{-\mathbf{e}_2} = -\mathbf{t}^{(2)}_{+\mathbf{e}_2}.$$

In terms of stress tensors, we have $\mathbf{T}^{(1)}\mathbf{e}_2 = \mathbf{T}^{(2)}\mathbf{e}_2$. That is

$$T^{(1)}_{12} = T^{(2)}_{12}, \quad T^{(1)}_{22} = T^{(2)}_{22}, \text{ and } T^{(1)}_{32} = T^{(2)}_{32}.$$

In other words, these stress components must be continuous across the fluid interface.

Since

$$T^{(1)}_{12} = 2\mu_1 D^{(1)}_{12} = \mu_1 \frac{dv^{(1)}}{dx_2} = \mu_1 A_1,$$

$$T^{(2)}_{12} = 2\mu_2 D^{(2)}_{12} = \mu_1 \frac{dv^{(2)}}{dx_2} = \mu_2 A_2,$$

$T^{(1)}_{12} = T^{(2)}_{12}$ gives

$$\mu_1 A_1 = \mu_2 A_2 \tag{d}$$

Note that this condition means that the slope of the velocity profile is not continuous at $x_2 = 0$. Also

$$T^{(1)}_{22} = -p^{(1)} + 2\mu_1 D^{(1)}_{22} = -p^{(1)}$$

and

$$T^{(2)}_{22} = -p^{(2)},$$

so that $T^{(1)}_{22} = T^{(2)}_{22}$ at $x_2 = 0$ gives $C_1 = C_2 \equiv p_0$. Since $T^{(1)}_{32} = 0$ and $T^{(2)}_{32} = 0$, the condition $T^{(1)}_{32} = T^{(2)}_{32}$ is clearly satisfied.

From Eqs. (a, b, c, d), we obtain

$$A_1 = \frac{\mu_2 v_0}{(\mu_2 b_1 + \mu_1 b_2)},$$

$$A_2 = \frac{\mu_1}{\mu_2} A_1 = \frac{\mu_1 v_0}{(\mu_1 b_2 + \mu_2 b_1)},$$

and

$$B_2 = B_1 = \frac{\mu_1 b_2 v_0}{(\mu_2 b_1 + \mu_1 b_2)}.$$

Thus the velocity distributions are

$$v_1^{(1)} = \frac{1}{(\mu_2 b_1 + \mu_1 b_2)} (\mu_2 v_0 x_2 + \mu_1 v_0 b_2), \quad v_2^{(1)} = v_3^{(1)} = 0, \tag{6.28a}$$

and

$$v_1^{(2)} = \frac{1}{(\mu_2 b_1 + \mu_1 b_2)} (\mu_1 v_0 x_2 + \mu_1 v_0 b_2), \quad v_2^{(2)} = v_3^{(2)} = 0. \tag{6.28b}$$

Note that in the case of $b_2 = 0$, $v_1^{(1)} = (v_0/b_1)x_2$, which is the case of plane Couette flow of a single fluid.

E. Couette Flow

The laminar steady two-dimensional flow of an incompressible Newtonian fluid between two coaxial infinitely long cylinders caused by the rotation of either one or both cylinders with constant angular velocities is known as **Couette flow.**

For this flow, we look for the velocity field in the following form in cylindrical coordinates

$$v_r = 0, \quad v_\phi = v(r), \quad v_z = 0.$$

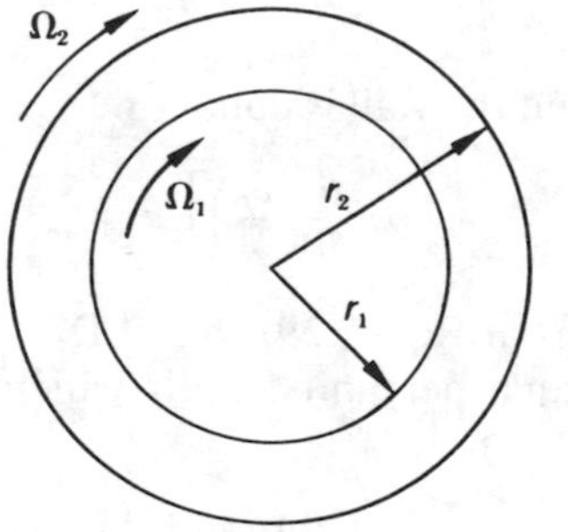

Fig. 6.6

This velocity field obviously satisfies the equation of continuity [Eq. (6.19)] regardless of what is $v(r)$.

In the absence of body forces and taking into account the rotational symmetry of the flow (i.e., nothing depends on ϕ), we have from the second equation of motion [Eq. (6.18b)]

$$\frac{d^2v}{dr^2}+\frac{1}{r}\frac{dv}{dr}-\frac{v}{r^2}=0.$$

It is easily verified that $v=r$ and $v=1/r$ satisfy the above equation. Thus the general solution is

$$v=Ar+\frac{B}{r},$$

where A and B are constants.

Let r_1 and r_2 denote the radii of the inner and outer cylinders, respectively, Ω_1 and Ω_2 their respective angular velocities. Then

$$r_1\Omega_1=Ar_1+\frac{B}{r_1}$$

and

$$r_2\Omega_2=Ar_2+\frac{B}{r_2},$$

from which the constants A and B are obtained to be

$$A=\frac{\Omega_2 r_2{}^2-\Omega_1 r_1{}^2}{r_2{}^2-r_1{}^2}, \quad B=\frac{r_1{}^2 r_2{}^2(\Omega_1-\Omega_2)}{r_2{}^2-r_1{}^2},$$

so that

$$v_\phi=v=\frac{1}{r_2{}^2-r_1{}^2}\left[r(\Omega_2 r_2{}^2-\Omega_1 r_1{}^2)-\frac{r_1{}^2 r_2{}^2}{r}(\Omega_2-\Omega_1)\right] \tag{6.29}$$

and $v_r=v_z=0$.

The shearing stress at the wall is equal to

$$\left[\mu r\frac{d}{dr}\left(\frac{v_\phi}{r}\right)\right]_{r=r_1,\,r_2}$$

and it is easily seen from Eq. (6.29) that if $\Omega_1=0$ and $\Omega_2=\Omega$ (or $\Omega_2=0$ and $\Omega_1=\Omega$), the torque per unit length which must be applied to the cylinders to maintain the flow is

$$M=\frac{4\pi\mu r_1{}^2 r_2{}^2\Omega}{r_2{}^2-r_1{}^2}. \tag{6.30}$$

F. Flow Near an Oscillating Plate

Let us consider the following unsteady two-dimensional parallel flow:

$$v_1 = v(x_2, t), \quad v_2 = 0, \quad v_3 = 0. \tag{6.31}$$

Omitting body forces and assuming a constant pressure field, the only nontrivial equation of motion is

$$\rho \frac{\partial v}{\partial t} = \mu \frac{\partial^2 v}{\partial x_2{}^2}. \tag{6.32}$$

It can be easily verified that

$$v = ae^{-\beta x_2} \cos (\omega t - \beta x_2 + \epsilon) \tag{6.33a}$$

satisfies the above equation if

$$\beta = \sqrt{\frac{\rho\omega}{2\mu}}. \tag{6.33b}$$

From Eq. (6.33a), the fluid velocity at $x_2 = 0$ is

$$v = a \cos (\omega t + \epsilon).$$

Thus, the solution (6.33a, b) represents the velocity field† of an infinite extent of liquid lying in the region $x_2 \geqslant 0$ and bounded by a plate at $x_2 = 0$ which executes simple harmonic oscillations of amplitude a and circular frequency ω. It represents a transverse wave of wavelength $2\pi/\beta$, propagated inward from the boundary with a phase velocity ω/β, but with rapidly diminishing amplitude (the falling off within a wavelength being in the ratio $e^{-2\pi} = 1/535$). Thus, we see that the influence of viscosity extends only to a short distance from the plate performing rapid oscillation of small amplitude a.

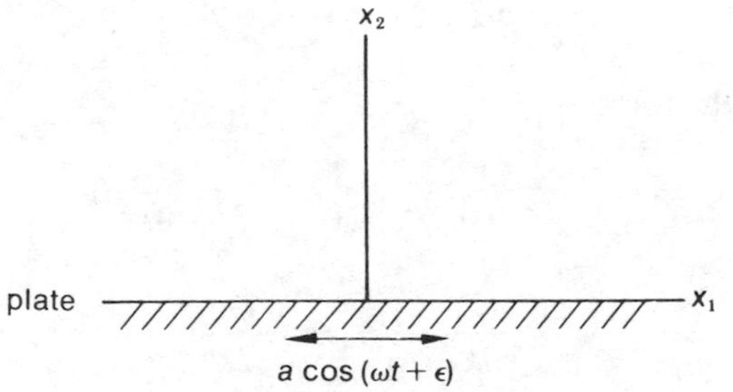

Fig. 6.7

† Steady-state oscillations.

6.10 RATE OF WORK DONE ON A PARTICLE

Referring to the infinitesimal rectangular parallepiped of Fig. 4.8, let us compute the rate at which work is done by the stress vectors and body force on the particle as it moves and deforms.

The rate at which work is done by the stress vectors $\mathbf{t}_{-\mathbf{e}_1}$ and $\mathbf{t}_{\mathbf{e}_1}$ on the pair of faces having $-\mathbf{e}_1$ and $\mathbf{e}_1$ as their respective normals is†

$$[(\mathbf{t}_{\mathbf{e}_1}\cdot\mathbf{v})_{x_1+dx_1,x_2,x_3}-(\mathbf{t}_{\mathbf{e}_1}\cdot\mathbf{v})_{x_1,x_2,x_3}]\,dx_2\,dx_3,$$

which is nothing but

$$\frac{\partial}{\partial x_1}(\mathbf{t}_{\mathbf{e}_1}\cdot\mathbf{v})\,dx_1dx_2dx_3. \tag{i}$$

Since $\mathbf{t}_{\mathbf{e}_1}\cdot\mathbf{v}=\mathbf{T}\mathbf{e}_1\cdot v_i\mathbf{e}_i=v_i\mathbf{e}_i\cdot\mathbf{T}\mathbf{e}_1=v_iT_{i1}$, the expression (i) becomes

$$\left[\frac{\partial}{\partial x_1}(v_iT_{i1})\right]dV, \tag{ii}$$

where $dV=dx_1dx_2dx_3$ denotes the differential volume. Similarly, the rate at which work is done by the stress vectors on the other two pairs of faces are

$$\left[\frac{\partial}{\partial x_2}(v_iT_{i2})\right]dV \quad\text{and}\quad \left[\frac{\partial}{\partial x_3}(v_iT_{i3})\right]dV.$$

Including the rate of work done by the body force ($\rho\mathbf{B}dV\cdot\mathbf{v}=\rho B_iv_idV$) the total rate of work done on the particle is

$$P=\frac{\partial}{\partial x_j}(v_iT_{ij})\,dV+\rho B_iv_idV. \tag{iii}$$

Using the relation

$$\frac{\partial}{\partial x_j}(v_iT_{ij})=v_i\frac{\partial T_{ij}}{\partial x_j}+T_{ij}\frac{\partial v_i}{\partial x_j},$$

Eq. (iii) takes the form

$$P=v_i\left[\frac{\partial T_{ij}}{\partial x_j}+\rho B_i\right]dV+T_{ij}\frac{\partial v_i}{\partial x_j}dV. \tag{6.34}$$

However,

$$\frac{\partial T_{ij}}{\partial x_j}+\rho B_i=\rho\frac{Dv_i}{Dt}$$

†The rate of work by $\mathbf{t}_{\mathbf{e}_1}$ is $(\mathbf{t}_{\mathbf{e}_1}dx_2dx_3)\cdot\mathbf{v}$ and that by $\mathbf{t}_{-\mathbf{e}_1}$ is $(\mathbf{t}_{-\mathbf{e}_1}dx_2dx_3)\cdot\mathbf{v}$, which is $(-\mathbf{t}_{\mathbf{e}_1}dx_2dx_3)\cdot\mathbf{v}$.

(equations of motion), therefore we have

$$P = v_i \frac{Dv_i}{Dt} \rho dV + T_{ij} \frac{\partial v_i}{\partial x_j} dV. \tag{6.35}$$

The first term in the right-hand side of Eq. (6.35) represents the rate of change of kinetic energy of the particle as is seen from the following:

$$\begin{aligned} \frac{D}{Dt}(KE) &\equiv \frac{D}{Dt}\left[\frac{1}{2}(\rho dV) v_i v_i\right] = \frac{1}{2}(\rho dV) v_i \frac{Dv_i}{Dt} + \frac{1}{2} v_i \frac{D}{Dt}[(\rho dV) v_i] \\ &= \frac{1}{2}(\rho dV) v_i \frac{Dv_i}{Dt} + \frac{1}{2}(\rho dV) v_i \frac{Dv_i}{Dt} + \frac{1}{2} v_i v_i \frac{D(\rho dV)}{Dt} \\ &= (\rho dV) v_i \frac{Dv_i}{Dt}. \end{aligned}$$

Thus,

$$P = \frac{D}{Dt}(KE) + T_{ij} \frac{\partial v_i}{\partial x_j} dV. \tag{6.36}$$

The second term in the right-hand side represents the rate at which work is done to change the volume and shape of the particle. With $T_{ij} = -p\delta_{ij} + T'_{ij}$, this term becomes

$$T_{ij} \frac{\partial v_i}{\partial x_j} dV = -p \frac{\partial v_i}{\partial x_i} dV + T'_{ij} \frac{\partial v_i}{\partial x_j} dV.$$

For an incompressible fluid, there is no volume change and

$$\begin{aligned} T_{ij} \frac{\partial v_i}{\partial x_j} &= T'_{ij} \frac{\partial v_i}{\partial x_j} = \mu\left(\frac{\partial v_i}{\partial x_j} + \frac{\partial v_j}{\partial x_i}\right)\frac{\partial v_i}{\partial x_j} \\ &= 2\mu D_{ij}(D_{ij} + W_{ij}) = 2\mu D_{ij} D_{ij} \\ &= 2\mu(D_{11}^2 + D_{22}^2 + D_{33}^2 + 2D_{12}^2 + 2D_{13}^2 + 2D_{23}^2). \end{aligned} \tag{6.37}†$$

This is the work per unit volume per unit time done to change the shape and this part of the work accumulates with time regardless of how the D_{ij}'s vary with time (it is zero only for rigid body motions). It represents the rate at which work is converted into heat. Thus $\mu \geq 0$ and the function

$$\Phi_{\text{inc}} = 2\mu(D_{11}^2 + D_{22}^2 + D_{33}^2 + 2D_{12}^2 + 2D_{13}^2 + 2D_{23}^2) \tag{6.38}$$

is known as the dissipation function for an incompressible Newtonian

†Since $D_{ij} = D_{ji}$ and $W_{ij} = -W_{ji}$ therefore $D_{ij}W_{ij} = 0$.

fluid. It can be readily seen that for a compressible fluid

$$T_{ij}\frac{\partial v_i}{\partial x_j} = -p\Delta + \lambda\Delta^2 + \Phi_{\text{inc}} \equiv -p\Delta + \Phi. \tag{6.39}$$

where $\Phi = \lambda(D_{11} + D_{22} + D_{33})^2 + \Phi_{\text{inc}}$, is the dissipation function for a compressible fluid.

EXAMPLE 6.8

For simple shearing flow with

$$v_1 = kx_2, \quad v_2 = 0, \quad v_3 = 0.$$

Find the rate at which work is converted into heat if the liquid inside of the plates is water with $\mu = 2 \times 10^{-5}$ lb sec/ft (0.958 mPa·s), and $k = 1$ reciprocal second.

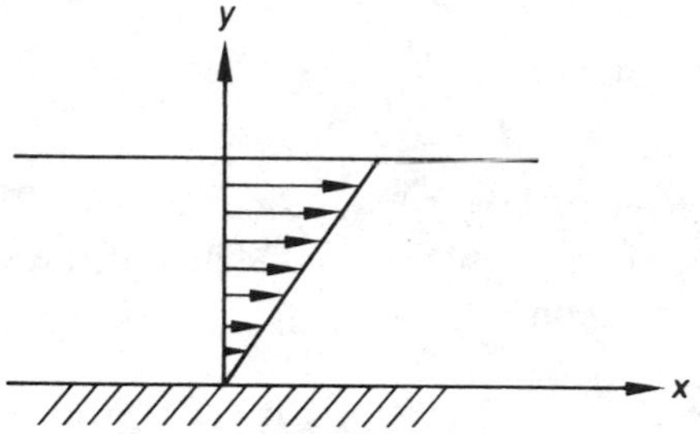

Fig. 6.8

Solution. Since the only nonzero component of the rate of deformation tensor is

$$D_{12} = \frac{k}{2}\,(\text{sec})^{-1}.$$

Thus from Eq. (6.38)

$$\Phi_{\text{inc}} = 4\mu D_{12}^2 = \mu k^2 = 2\times 10^{-5}\frac{\text{ft-lb}}{(\text{ft})^3}\Big/\text{sec.}\ (0.958 \times 10^{-3}\,\frac{\text{N·m}}{\text{m}^3}/\text{sec})$$

Thus, in one second per cubic feet of water the heat generated by viscosity is 2.5×10^{-8} B.T.U. (0.958×10^{-3} joule per cubic meter per second).

6.11 RATE OF HEAT FLOW INTO AN ELEMENT

Let **q** be a vector whose magnitude gives the rate of heat flow across a *unit area* and whose direction gives the direction of heat flow, then the net heat flow Q into a differential element can be computed as follows: Referring to the infinitesimal rectangular parallelepiped of Fig. 6.9, the net rate at which heat flows into the element across the face with $\mathbf{e}_1$ as its

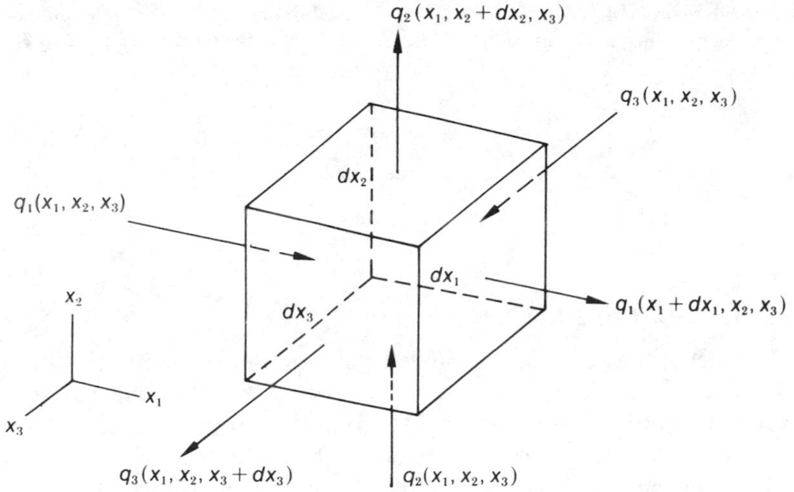

Fig. 6.9

outward normal is $\{(-\mathbf{q}\cdot\mathbf{e}_1)_{x_1+dx_1,x_2,x_3}\}dx_2\,dx_3$ and that across the face with $-\mathbf{e}_1$ as its outward normal is $\{(\mathbf{q}\cdot\mathbf{e}_1)_{x_1,x_2,x_3}\}dx_2dx_3$. Thus the net rate of heat inflow across the pair of faces is given by, with $q_1\equiv\mathbf{q}\cdot\mathbf{e}_1$,

$$-[q_1(x_1+dx_1,x_2,x_3)-q_1(x_1,x_2,x_3)]dx_2dx_3,$$

which is nothing but $-(\partial q_1/\partial x_1)dx_1dx_2dx_3$. Similarly, the net rate of heat inflow across the other two pairs of faces is

$$-\frac{\partial q_2}{\partial x_2}dx_1dx_2dx_3 \quad\text{and}\quad -\frac{\partial q_3}{\partial x_3}dx_1dx_2dx_3,$$

so that the total net rate of heat inflow is

$$Q=-\left(\frac{\partial q_1}{\partial x_1}+\frac{\partial q_2}{\partial x_2}+\frac{\partial q_3}{\partial x_3}\right)dV=-(\operatorname{div}\mathbf{q})\,dV. \qquad (6.40)$$

EXAMPLE 6.9

Using the Fourier heat conduction law $\mathbf{q}=-\kappa\nabla\theta$, where $\nabla\theta$ is the temperature gradient and κ is the coefficient of thermal conductivity, find the equation governing the steady-state distribution of temperature.

Solution. From Eq. (6.40) we have, per unit volume the net rate of inflow given by

$$-\left[\frac{\partial}{\partial x_1}\left(\kappa\frac{\partial\theta}{\partial x_1}\right)+\frac{\partial}{\partial x_2}\left(\kappa\frac{\partial\theta}{\partial x_2}\right)+\frac{\partial}{\partial x_3}\left(\kappa\frac{\partial\theta}{\partial x_3}\right)\right].$$

Now, if the boundaries of the body are kept at fixed temperatures, then when the steady-state is reached, the net rate of heat flow into any element in the body must be zero. Thus the desired equation is

$$\frac{\partial}{\partial x_1}\left(\kappa\frac{\partial\theta}{\partial x_1}\right)+\frac{\partial}{\partial x_2}\left(\kappa\frac{\partial\theta}{\partial x_2}\right)+\frac{\partial}{\partial x_3}\left(\kappa\frac{\partial\theta}{\partial x_3}\right)=0.$$

For constant κ, this reduces to the Laplace equation

$$\frac{\partial^2\theta}{\partial x_1^2}+\frac{\partial^2\theta}{\partial x_2^2}+\frac{\partial^2\theta}{\partial x_3^2}=0.$$

6.12 ENERGY EQUATION

Consider a particle with a differential volume dV at the position $\mathbf{x}$ at some time t. Let U denote its internal energy, KE its kinetic energy, Q the net rate of heat flow into the particle from its surroundings, and P the rate at which work is done on the particle by body forces and surface forces (i.e., P is the mechanical power input). Then, in the absence of other forms of energy input, the fundamental postulate of conservation of energy states that

$$\frac{D}{Dt}(U+KE)=P+Q. \tag{6.41}$$

From Eq. (6.36) of Section 6.10 and Eq. (6.40) of Section 6.11 we had

$$P=\frac{D}{Dt}(KE)+T_{ij}\frac{\partial v_i}{\partial x_j}dV \quad \text{and} \quad Q=-\frac{\partial q_i}{\partial x_i}dV,$$

and if we denote by u the internal energy per unit mass, so that $U=u(\rho dV)$ and

$$\frac{DU}{Dt}=\rho dV\frac{du}{dt} \qquad \left[\text{note}\,\frac{D}{Dt}(\rho dV)=0\right],$$

then the energy equation (6.41) becomes

$$\rho\frac{Du}{Dt}=T_{ij}\frac{\partial v_i}{\partial x_j}-\frac{\partial q_i}{\partial x_i}. \tag{6.42}$$

If the only heat flow taking place is that due to conduction governed by Fourier's law $\mathbf{q}=-\kappa\nabla\theta$, where θ is the temperature, then Eq. (6.42) becomes, assuming a constant coefficient of thermoconductivity κ,

$$\rho\frac{Du}{Dt}=T_{ij}\frac{\partial v_i}{\partial x_j}+\kappa\frac{\partial^2\theta}{\partial x_j\partial x_j}. \tag{6.43}$$

For an incompressible Newtonian fluid, if it is assumed that the internal energy per unit mass is given by $c\theta$, where c is the specific heat per unit mass, then Eq. (6.43) becomes

$$\rho c \frac{D\theta}{Dt} = \Phi_{\text{inc}} + \kappa \frac{\partial^2 \theta}{\partial x_j \partial x_j}, \tag{6.44}$$

where from Eq. (6.38), $\Phi_{\text{inc}} = 2\mu(D_{11}^2 + D_{22}^2 + D_{33}^2 + 2D_{12}^2 + 2D_{13}^2 + 2D_{23}^2)$, representing the heat generated through viscous forces.

There are many situations in which the heat generated through viscous action is very small compared with that arising from the heat conduction from the boundaries, in which case Eq. (6.44) simplifies to

$$\frac{D\theta}{Dt} = \alpha \frac{\partial^2 \theta}{\partial x_j \partial x_j}, \tag{6.45}$$

where $\alpha = \kappa/\rho c \equiv$ thermal diffusivity.

EXAMPLE 6.10

A fluid is at rest between two plates of infinite dimension. If the lower plate is kept at constant temperature θ_l and the upper plate at θ_u, find the steady-state temperature distribution.

Solution. The steady-state distribution is governed by the Laplace equation

$$\frac{\partial^2 \theta}{\partial x^2} + \frac{\partial^2 \theta}{\partial y^2} + \frac{\partial^2 \theta}{\partial z^2} = 0,$$

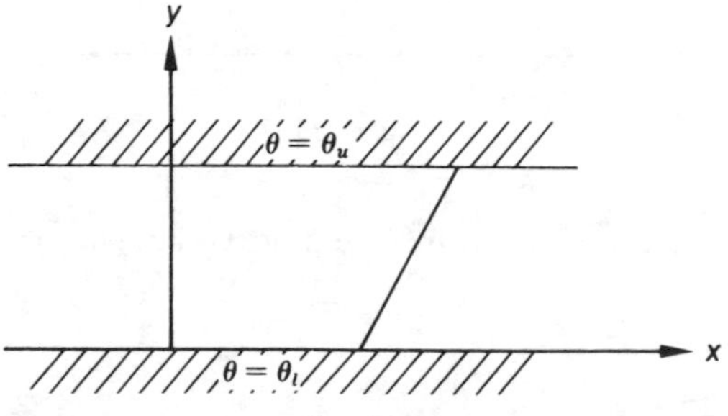

Fig. 6.10

which in this problem reduces to

$$\frac{d^2\theta}{dy^2} = 0.$$

Thus

$$\frac{d\theta}{dy} = C_1$$

and

$$\theta = C_1 y + C_2.$$

Using the boundary condition $\theta = \theta_l$ at $y = 0$ and $\theta = \theta_u$ at $y = d$, the constants of integration are determined to be

$$C_1 = \frac{\theta_u - \theta_l}{d},$$

$$C_2 = \theta_l.$$

It is noted here that when the values of θ are prescribed on the plates, the values of $d\theta/dy$ on the plates are completely determined. In fact, $d\theta/dy = (\theta_u - \theta_l)/d$. This serves to illustrate that, in steady-state heat conduction problem (governed by Laplace equation) it is in general not possible to prescribe both the values of θ and the normal derivatives of θ at the same points of the complete boundary unless they happen to be consistent with each other.

EXAMPLE 6.11

The plane Couette flow is given by the following velocity distribution:

$$v_1 = ky, \quad v_2 = 0, \quad v_3 = 0.$$

If the temperature at the lower plate is kept at θ_l and that at the upper plate at θ_u, find the steady-state temperature distribution.

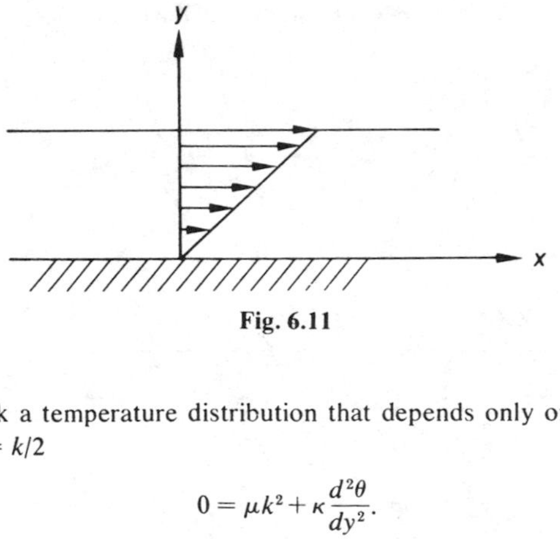

Fig. 6.11

Solution. We seek a temperature distribution that depends only on y. From Eq. (6.44) we have, since $D_{12} = k/2$

$$0 = \mu k^2 + \kappa \frac{d^2\theta}{dy^2}.$$

Thus

$$\frac{d^2\theta}{dy^2} = -\frac{\mu k^2}{\kappa},$$

which gives

$$\theta = -\frac{\mu k^2}{\kappa}\frac{y^2}{2} + C_1 y + C_2,$$

where C_1 and C_2 are constants of integration. Now at $y = 0$, $\theta = \theta_l$ and at $y = d$, $\theta = \theta_u$, therefore

$$\theta_l = C_2$$

and

$$\theta_u = -\frac{\mu k^2}{\kappa}\frac{d^2}{2} + C_1 d + \theta_l,$$

from which

$$C_1 = \frac{\theta_u}{d} + \frac{\mu k^2 d}{2\kappa} - \frac{\theta_l}{d}.$$

The temperature distribution is therefore given by

$$\theta = -\frac{\mu k^2}{2\kappa} y^2 + \left(\frac{\theta_u - \theta_l}{d} + \frac{\mu k^2 d}{2\kappa}\right) y + \theta_l.$$

6.13 VORTICITY VECTOR

We recall from Chapter 3 that the antisymmetric part of the velocity gradient ($\nabla \mathbf{v}$) is defined as the spin tensor $\mathbf{W}$. Being antisymmetric the tensor $\mathbf{W}$ is equivalent to a vector $\boldsymbol{\omega}$ in the sense that $\mathbf{W}\mathbf{x} = \boldsymbol{\omega} \times \mathbf{x}$ (*see* Section B11). In fact, $\boldsymbol{\omega} = -(W_{23}\mathbf{e}_1 + W_{31}\mathbf{e}_2 + W_{12}\mathbf{e}_3)$. Since $d\mathbf{v} = (\mathbf{D}+\mathbf{W})\,d\mathbf{x} = \mathbf{D}d\mathbf{x} + \mathbf{W}d\mathbf{x} = \mathbf{D}d\mathbf{x} + \boldsymbol{\omega} \times d\mathbf{x}$, the vector $\boldsymbol{\omega}$ is the angular velocity vector of that part of the motion, representing the rigid body rotation in the infinitesimal neighborhood of a material point. Another interesting interpretation of $\boldsymbol{\omega}$ is that it is the angular velocity vector of the principal axes of $\mathbf{D}$, which we show below:

Let $d\mathbf{x}$ be a material element in the direction of $\mathbf{n}$, at time t, i.e.,

$$\mathbf{n} = \frac{d\mathbf{x}}{ds},$$

where ds is the length of $d\mathbf{x}$. Now

$$\frac{D}{Dt}\mathbf{n} = \frac{D}{Dt}\left(\frac{d\mathbf{x}}{ds}\right) = \frac{1}{ds}\left(\frac{D}{Dt}d\mathbf{x}\right) - \frac{1}{ds^2}\left[\frac{D}{Dt}(ds)\right]d\mathbf{x}. \tag{6.46}$$

But from Eq. (3.20) and Eq. (3.22), we have

$$\frac{D}{Dt}(d\mathbf{x}) = (\nabla \mathbf{v})\,d\mathbf{x} = (\mathbf{D}+\mathbf{W})\,d\mathbf{x}$$

and from Eq. (3.27),

$$\frac{1}{ds}\frac{D}{Dt}(ds) = \mathbf{n} \cdot \mathbf{D}\mathbf{n}.$$

Therefore

$$\frac{D}{Dt}\mathbf{n} = (\mathbf{D}+\mathbf{W})\mathbf{n} - (\mathbf{n} \cdot \mathbf{D}\mathbf{n})\mathbf{n}. \tag{6.47}$$

Now if $\mathbf{n}$ is an eigenvector of $\mathbf{D}$, then

$$\mathbf{D}\mathbf{n} = \lambda\mathbf{n},$$

so that

$$\mathbf{n}\cdot\mathbf{D}\mathbf{n}=\lambda$$

and Eq. (6.47) becomes

$$\frac{D}{Dt}\mathbf{n}=\mathbf{W}\mathbf{n}=\boldsymbol{\omega}\times\mathbf{n}, \tag{6.48}$$

which is the desired result.

In terms of the velocity field,

$$\boldsymbol{\omega}=\frac{1}{2}\left(\frac{\partial v_3}{\partial x_2}-\frac{\partial v_2}{\partial x_3}\right)\mathbf{e}_1+\frac{1}{2}\left(\frac{\partial v_1}{\partial x_3}-\frac{\partial v_3}{\partial x_1}\right)\mathbf{e}_2+\frac{1}{2}\left(\frac{\partial v_2}{\partial x_1}-\frac{\partial v_1}{\partial x_2}\right)\mathbf{e}_3. \tag{6.49}$$

Dropping the slightly awkward factor of $\frac{1}{2}$ in the right-hand side of Eq. (6.49), one defines the so-called "**vorticity vector**" $\boldsymbol{\zeta}$ as

$$\zeta_i=2\boldsymbol{\omega}=\left(\frac{\partial v_3}{\partial x_2}-\frac{\partial v_2}{\partial x_3}\right)\mathbf{e}_1+\left(\frac{\partial v_1}{\partial x_3}-\frac{\partial v_3}{\partial x_1}\right)\mathbf{e}_2+\left(\frac{\partial v_2}{\partial x_1}-\frac{\partial v_1}{\partial x_2}\right)\mathbf{e}_3, \tag{6.50}$$

The tensor $2\mathbf{W}$ is known as the vorticity tensor.

In indicial notation, the Cartesian components of $\boldsymbol{\zeta}$ are

$$\zeta_i=\epsilon_{ijk}\frac{\partial v_k}{\partial x_j} \tag{6.51a}$$

and in invariant notation,

$$\boldsymbol{\zeta}\equiv\text{curl}\,\mathbf{v}. \tag{6.51b}$$

Note also that

$$2W_{ij}=\frac{\partial v_i}{\partial x_j}-\frac{\partial v_j}{\partial x_i}=-\epsilon_{kij}\,\zeta_k\,. \tag{6.52}$$

EXAMPLE 6.12

Find the vorticity vector for the simple shearing flow

$$v_1=kx_2,\quad v_2=v_3=0.$$

Solution. We have

$$\zeta_1=\frac{\partial v_3}{\partial x_2}-\frac{\partial v_2}{\partial x_3}=0,\quad \zeta_2=\frac{\partial v_1}{\partial x_3}-\frac{\partial v_3}{\partial x_1}=0$$

and

$$\zeta_3=\frac{\partial v_2}{\partial x_1}-\frac{\partial v_1}{\partial x_2}=-k.$$

That is,

$$\boldsymbol{\zeta}=-k\mathbf{e}_3.$$

We see that the angular velocity vector $(=\boldsymbol{\zeta}/2)$ is normal to the x_1x_2-plane and the minus sign simply means that the spinning is clockwise looking from the positive side of x_3.

EXAMPLE 6.13

Find the distribution of the vorticity vector in the Couette flow discussed in Section 6.9E.

Solution. With $v_r = v_z = 0$ and $v_\phi = Ar + (B/r)$, it is obvious that the only nonzero vorticity component is in the z-direction.

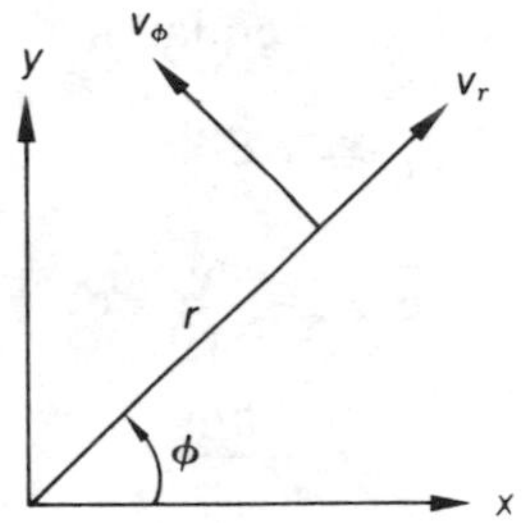

Fig. 6.12

Using

$$v_x = -v_\phi \sin\phi, \quad v_y = v_\phi \cos\phi, \quad v_z = 0$$

and $r^2 = x^2 + y^2$, $r\cos\phi = x$, $r\sin\theta = y$ and $\phi = \tan^{-1}(y/x)$; it can be demonstrated that

$$\zeta_z = \frac{\partial v_y}{\partial x} - \frac{\partial v_x}{\partial y} = \frac{dv_\phi}{dr} + \frac{v_\phi}{r} = \frac{1}{r}\frac{d}{dr}(rv_\phi).$$

Now

$$\frac{d}{dr}(rv_\phi) = \frac{d}{dr}(Ar^2 + B) = 2Ar.$$

Thus

$$\zeta_z = 2A = \frac{\Omega_2 r_2^2 - \Omega_1 r_1^2}{r_2^2 - r_1^2}.$$

EXAMPLE 6.14

Verify Eq. (6.52)

$$-\left(\frac{\partial v_i}{\partial x_j} - \frac{\partial v_j}{\partial x_i}\right) = \epsilon_{kij}\zeta_k.$$

Solution. For $i = j$, both sides are identically zero. For $i \neq j$, if $i = 1$ and $j = 2$, we have

$$-\left(\frac{\partial v_1}{\partial x_2} - \frac{\partial v_2}{\partial x_1}\right) = \epsilon_{112}\zeta_1 + \epsilon_{212}\zeta_2 + \epsilon_{312}\zeta_3 = \zeta_3,$$

if $i = 2, j = 3$, we have

$$-\left(\frac{\partial v_2}{\partial x_3} - \frac{\partial v_3}{\partial x_2}\right) = \epsilon_{123}\zeta_1 + \epsilon_{223}\zeta_2 + \epsilon_{323}\zeta_3 = \zeta_1$$

and if $i = 3, j = 1$, we have

$$-\left(\frac{\partial v_3}{\partial x_1} - \frac{\partial v_1}{\partial x_3}\right) = \epsilon_{131}\zeta_1 + \epsilon_{231}\zeta_2 + \epsilon_{331}\zeta_3 = \zeta_2.$$

The other three cases ($i = 2, j = 1$), ($i = 3, j = 2$) and ($i = 1, j = 3$) are also obviously true.

6.14 IRROTATIONAL FLOW

If the vorticity vector (or equivalently, vorticity tensor) corresponding to a velocity field is zero in some region and for some time interval, the flow is called **irrotational** in that region and that time interval.

Let $\Phi(x_1, x_2, x_3, t)$ be a scalar function and let the velocity components be derived from Φ by the following equation:

$$v_1 = -\frac{\partial \Phi}{\partial x_1}, \quad v_2 = -\frac{\partial \Phi}{\partial x_2}, \quad v_3 = -\frac{\partial \Phi}{\partial x_3}, \tag{6.53a}$$

i.e.,

$$v_i = -\frac{\partial \Phi}{\partial x_i}. \tag{6.53b}$$

Then

$$\zeta_1 = \frac{\partial v_3}{\partial x_2} - \frac{\partial v_2}{\partial x_3} = -\frac{\partial^2 \Phi}{\partial x_3 \partial x_2} + \frac{\partial^2 \Phi}{\partial x_2 \partial x_3} = 0$$

and similarly $\zeta_2 = \zeta_3 = 0$.

That is, a scalar function $\Phi(x_1, x_2, x_3, t)$ defines an irrotational flow field through the equations [Eq. (6.53)]. Obviously, not all arbitrary functions Φ will give rise to velocity fields that are physically possible. For one thing, the equation of continuity, expressing the principle of conservation of mass, must be satisfied. For an incompressible fluid, the equation of continuity reads

$$\frac{\partial v_i}{\partial x_i} = 0. \tag{6.54}$$

Thus, combining Eq. (6.53) with Eq. (6.54), we obtain the Laplacian equation for Φ:

$$\frac{\partial^2 \Phi}{\partial x_j \partial x_j} = 0, \tag{6.55a}$$

i.e.,

$$\frac{\partial^2 \Phi}{\partial x_1^2} + \frac{\partial^2 \Phi}{\partial x_2^2} + \frac{\partial^2 \Phi}{\partial x_3^3} = 0. \tag{6.55b}$$

In the next two sections, we shall discuss the conditions under which irrotational flows are dynamically possible for an inviscid and viscous fluid.

6.15 IRROTATIONAL FLOW OF AN INVISCID INCOMPRESSIBLE FLUID OF HOMOGENEOUS DENSITY

An inviscid fluid is defined by

$$T_{ij} = -p\delta_{ij}, \tag{6.56}$$

obtained by putting viscosity $\mu = 0$ in the constitutive equation for Newtonian viscous fluid.

The equations of motion for an inviscid fluid are

$$\rho\left(\frac{\partial v_i}{\partial t} + v_j\frac{\partial v_i}{\partial x_j}\right) = -\frac{\partial p}{\partial x_i} + \rho B_i \tag{6.57a}$$

or

$$\rho\frac{D\mathbf{v}}{Dt} = -\nabla p + \rho\mathbf{B}. \tag{6.57b}$$

Equations (6.57) are known as **Euler's equations of motion**. We now show that irrotational flows are always dynamically possible for an inviscid, incompressible fluid with homogeneous density provided that the body forces acting are derivable from a potential Ω by the formulas

$$B_i = -\frac{\partial \Omega}{\partial x_i}. \tag{6.58}$$

In the case of gravity force, with x_3-axis pointing vertically upward,

$$\Omega = gx_3, \tag{6.59a}$$

so that

$$B_1 = 0, \quad B_2 = 0, \quad \text{and } B_3 = -g. \tag{6.59b}$$

Using Eq. (6.58) and noting that $\rho =$ constant for homogeneous fluid, Eq. (6.57) can be written

$$\frac{\partial v_i}{\partial t} + v_j\frac{\partial v_i}{\partial x_j} = -\frac{\partial}{\partial x_i}\left(\frac{p}{\rho} + \Omega\right). \tag{6.60}$$

For an irrotational flow

$$\frac{\partial v_i}{\partial x_j} = \frac{\partial v_j}{\partial x_i},$$

so that

$$v_j\frac{\partial v_i}{\partial x_j} = v_j\frac{\partial v_j}{\partial x_i} = \frac{1}{2}\frac{\partial}{\partial x_i}v_jv_j.$$

That is,

$$v_j\frac{\partial v_i}{\partial x_j} = \frac{1}{2}\frac{\partial}{\partial x_i}v^2, \tag{6.61}$$

where $v^2 = v_j v_j = v_1{}^2 + v_2{}^2 + v_3{}^2$ is the square of the speed. Therefore Eq. (6.60) becomes

$$\frac{\partial}{\partial x_i}\left(-\frac{\partial \Phi}{\partial t} + \frac{v^2}{2} + \frac{p}{\rho} + \Omega\right) = 0. \tag{6.62}$$

Equation (6.62) states that

$$\left(-\frac{\partial \Phi}{\partial t} + \frac{v^2}{2} + \frac{p}{\rho} + \Omega\right)$$

is independent of x_1, x_2, and x_3. Thus

$$-\frac{\partial \Phi}{\partial t} + \frac{1}{2}v^2 + \frac{p}{\rho} + \Omega = f(t), \tag{6.63†}$$

where $f(t)$ is an arbitrary function of t.

If the flow is also steady then we have

$$\frac{1}{2}v^2 + \frac{p}{\rho} + \Omega = \text{constant}. \tag{6.64}$$

Equation (6.63) as well as the special case (6.64), is known as **Bernoulli's equation**. In addition to being a very useful formula in problems where the effect of viscosity can be neglected, the above derivation of the formula shows that irrotational flows are always dynamically possible under the conditions stated earlier. This is so because for whatever function Φ, so long as $v_i = -\partial\Phi/\partial x_i$, the dynamic equations of motion can always be integrated to give Bernoulli's equation from which the pressure distribution is obtained corresponding to which the equations of motion are all satisfied.

EXAMPLE 6.15

Given $\Phi = x^3 - 3xy^2$,

(a) show that Φ satisfies the Laplace equation;

(b) find the irrotational velocity field;

(c) find the pressure distribution for an incompressible homogeneous fluid, if at (0, 0, 0) $p = p_0$ and $\Omega = gz$.

(d) If the plane $y = 0$ is a solid boundary, find the tangential component of velocity on the plane.

Solution. (a) We have

$$\frac{\partial^2 \Phi}{\partial x^2} = 6x, \quad \frac{\partial^2 \Phi}{\partial y^2} = -6x, \quad \frac{\partial^2 \Phi}{\partial z^2} = 0,$$

†The integration function $f(t)$ can be absorbed into $\Phi(t)$ without changing the velocity field.

therefore

$$\frac{\partial^2 \Phi}{\partial x^2} + \frac{\partial^2 \Phi}{\partial y^2} + \frac{\partial^2 \Phi}{\partial z^2} = 6x - 6x = 0.$$

(b) From $v_i = -(\partial \Phi / \partial x_i)$, we have

$$v_1 = -\frac{\partial \Phi}{\partial x} = -3x^2 + 3y^2, \quad v_2 = -\frac{\partial \Phi}{\partial y} = 6xy, \quad v_3 = 0.$$

(c) From Bernoulli's equation [Eq. (6.64)]

$$\frac{1}{2}v^2 + \frac{p}{\rho} + \Omega = C,$$

we have at (0, 0, 0), $v_1 = 0$, $v_2 = 0$, $p = p_0$, and $\Omega = 0$. Thus

$$C = \frac{p_0}{\rho}$$

and

$$p = p_0 - \frac{\rho}{2}(v_1^2 + v_2^2) - \rho g z$$

or

$$p = p_0 - \frac{\rho}{2}[9(y^2 - x^2)^2 + 36x^2y^2] - \rho g z.$$

(d) On the plane $y = 0$, $v_1 = -3x^2$ and $v_2 = 0$. Now $v_2 = 0$ means that the normal components of velocity are zero on the plane, which is what it should be if $y = 0$ is a solid boundary. Since $v_1 = -3x^2$, the tangential components of velocity are not zero on the plane. That is, the fluid slips on the boundary. In inviscid theory, consistent with the assumption of zero viscosity, the slipping of fluid on a solid boundary is allowed. More discussion on this point will be given in the next section.

EXAMPLE 6.16

A liquid is being drained through a small opening as shown. Neglect viscosity and assume that the falling of the free surface is so slow that the flow can be treated as a steady one. Find the exit speed of the liquid jet as a function of h.

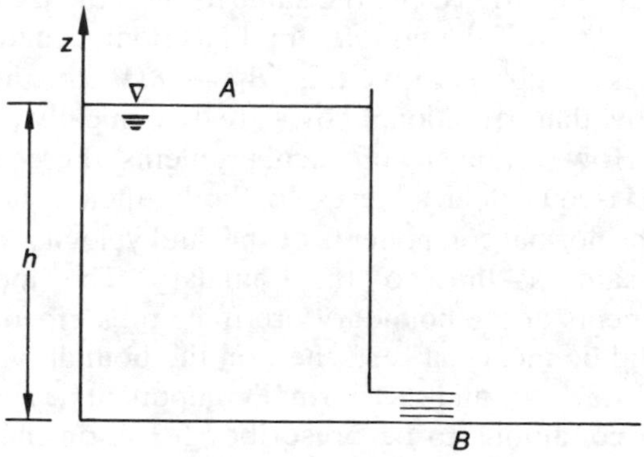

Fig. 6.13

Solution. For a point on the free surface such as the point A, $p = p_a$, $v \approx 0$, and $z = h$. Therefore, from Eq. (6.64),

$$\frac{1}{2}v^2 + \frac{p}{\rho} + gz = \frac{p_a}{\rho} + gh.$$

At a point on the exit jet, such as the point B, $z = 0$ and $p = p_a$. Thus

$$\frac{1}{2}v^2 + \frac{p_a}{\rho} = \frac{p_a}{\rho} + gh,$$

from which

$$v = \sqrt{2gh}.$$

This is the well-known Torricelli's formula.

6.16 IRROTATIONAL FLOWS AS SOLUTIONS OF NAVIER–STOKES EQUATION

For an incompressible Newtonian fluid, the equations of motion are the Navier–Stokes equations:

$$\frac{\partial v_i}{\partial t} + v_j \frac{\partial v_i}{\partial x_j} = -\frac{1}{\rho}\frac{\partial p}{\partial x_i} + \frac{\mu}{\rho}\frac{\partial^2 v_i}{\partial x_j \partial x_j} + B_i.$$

For irrotational flows

$$v_i = -\frac{\partial \Phi}{\partial x_i},$$

so that

$$\frac{\partial^2 v_i}{\partial x_j \partial x_j} = -\frac{\partial^2}{\partial x_j \partial x_j}\left(\frac{\partial \Phi}{\partial x_i}\right) = -\frac{\partial}{\partial x_i}\frac{\partial^2 \Phi}{\partial x_j \partial x_j}.$$

But from Eq. (6.55), $\partial^2\Phi/\partial x_j \partial x_j = 0$. Therefore, the terms involving viscosity in the Navier–Stokes equation drop out in the case of irrotational flows so that the equations take the same form as the Euler equation for an inviscid fluid. Thus, if the *viscous* fluid has homogeneous density and if the body forces are conservative (i.e., $B_i = -\partial\Omega/\partial x_i$), the results of the last sections show that irrotational flows are dynamically possible also for a viscous fluid. However, in any physical problems, there are always solid boundaries. A viscous fluid adheres to the boundary so that both the tangential and the normal components of the fluid velocity at the boundary should be the same as those of the boundary. This means that both velocity components at the boundary are to be prescribed. For example, if $y = 0$ is a solid boundary at rest, then on the boundary, the tangential components $v_x = v_z = 0$, and the normal components $v_y = 0$. For irrotational flow, the conditions to be prescribed for Φ on the boundary are $\Phi =$ constant at $y = 0$ (so that $\partial\Phi/\partial x = \partial\Phi/\partial z = 0$) and $\partial\Phi/\partial y = 0$ at $y =$

0. But it is known (*see* Example 6.10, or from potential theory) that in general there does not exist solutions of Laplace equations satisfying both the conditions $\Phi =$ constant and $\nabla\Phi \cdot \mathbf{n} = \partial\Phi/\partial n = 0$ on the complete boundaries. Therefore, unless the motion of solid boundaries happens to be consistent with the requirements of irrotationality (*see* Example 6.17), vorticity will be generated on the boundary and diffuse into the flow field according to vorticity equations to be derived in the next section. However, in certain problems under suitable conditions, the vorticity generated by the solid boundaries is confined to a thin layer of fluid in the vicinity of the boundary so that outside of the layer the flows are irrotational if it originated from a state of irrotationality. We shall have more to say about this in the next two sections.

EXAMPLE 6.17

For Couette flow between two coaxial infinitely long cylinders, how should the ratio of the angular velocities of the two cylinders be so that the *viscous* fluid will be having irrotational flow.

Solution. From Example 6.13 of Section 6.13, the only nonzero vorticity component in the Couette flow is

$$\zeta_z = \frac{\Omega_2 r_2^2 - \Omega_1 r_1^2}{r_2^2 - r_1^2}.$$

If $\Omega_2 r_2^2 - \Omega_1 r_1^2 = 0$, the flow is irrotational. Thus

$$\frac{\Omega_2}{\Omega_1} = \frac{r_1^2}{r_2^2}.$$

It should be noted that even though the viscous terms drop out from the Navier–Stokes equations in the case of irrotational flows, it does not mean that there is no viscous dissipation in an irrotational flow of a viscous fluid. In fact, so long as there is one nonzero rate of deformation component, there is viscous dissipation [given by Eq. (6.38)] and the rate of work done to maintain the irrotational flow exactly compensates the viscous dissipations.

6.17 VORTICITY TRANSPORT EQUATION FOR INCOMPRESSIBLE VISCOUS FLUID WITH A CONSTANT DENSITY ρ

In this section, we derive the equation governing the vorticity vector for an incompressible homogeneous viscous fluid. First we assume that the body force is derivable from a potential, i.e., $B_i = -\partial\Omega/\partial x_i$. Now with $\rho =$

constant and $B_i = -\partial\Omega/\partial x_i$ the Navier–Stokes equation can be written

$$\frac{\partial v_i}{\partial t} + v_j \frac{\partial v_i}{\partial x_j} = -\frac{\partial}{\partial x_i}\left(\frac{p}{\rho} + \Omega\right) + \nu \frac{\partial^2 v_i}{\partial x_j \partial x_j}, \tag{6.65}$$

where $\nu = \mu/\rho$ is usually termed the kinematic viscosity. If we operate on Eq. (6.65) by the differential operator $\epsilon_{mni}(\partial/\partial x_n)$ [i.e., taking the curl of both sides of Eq. (6.65)], we have, since

$$\epsilon_{mni} \frac{\partial}{\partial x_n}\left(\frac{\partial v_i}{\partial t}\right) = \frac{\partial}{\partial t}\left(\epsilon_{mni} \frac{\partial v_i}{\partial x_n}\right) = \frac{\partial \zeta_m}{\partial t},$$

$$\epsilon_{mni} \frac{\partial}{\partial x_n}\left(v_j \frac{\partial v_i}{\partial x_j}\right) = \epsilon_{mni} \frac{\partial v_j}{\partial x_n}\frac{\partial v_i}{\partial x_j} + v_j \frac{\partial}{\partial x_j}\left(\epsilon_{mni} \frac{\partial v_i}{\partial x_n}\right)$$

$$= \epsilon_{mni} \frac{\partial v_j}{\partial x_n}\frac{\partial v_i}{\partial x_j} + v_j \frac{\partial \zeta_m}{\partial x_j},$$

$$\epsilon_{mni} \frac{\partial^2}{\partial x_n \partial x_i}\left(\frac{p}{\rho} + \Omega\right) = 0 \qquad \left[\text{i.e., curl grad}\left(\frac{p}{\rho} + \Omega\right) = 0\right], \tag{6.66†}$$

and

$$\epsilon_{mni} \frac{\partial}{\partial x_n}\left(\frac{\partial^2 v_i}{\partial x_j \partial x_j}\right) = \frac{\partial^2}{\partial x_j \partial x_j}\left(\epsilon_{mni} \frac{\partial v_i}{\partial x_n}\right) = \frac{\partial^2 \zeta_m}{\partial x_j \partial x_j}.$$

The Navier–Stokes equation therefore, takes the form

$$\frac{\partial \zeta_m}{\partial t} + v_j \frac{\partial \zeta_m}{\partial x_j} + \epsilon_{mni} \frac{\partial v_j}{\partial x_n}\frac{\partial v_i}{\partial x_j} = \nu \frac{\partial^2 \zeta_m}{\partial x_j \partial x_j}. \tag{6.67}$$

A more useful form can be obtained as follows. From Eq. (6.52), we have

$$\frac{\partial v_i}{\partial x_j} = \frac{\partial v_j}{\partial x_i} + \epsilon_{pji}\zeta_p.$$

Thus

$$\epsilon_{mni} \frac{\partial v_j}{\partial x_n}\frac{\partial v_i}{\partial x_j} = \epsilon_{mni} \frac{\partial v_j}{\partial x_n}\frac{\partial v_j}{\partial x_i} + \epsilon_{mni}\epsilon_{pji} \frac{\partial v_j}{\partial x_n}\zeta_p.$$

But

$$\epsilon_{mni} \frac{\partial v_j}{\partial x_n}\frac{\partial v_j}{\partial x_i} = 0 \tag{6.68‡}$$

†For example,

$$\epsilon_{1ni} \frac{\partial^2}{\partial x_n \partial x_i}\left(\frac{p}{\rho} + \Omega\right) = \epsilon_{123} \frac{\partial^2}{\partial x_2 \partial x_3}\left(\frac{p}{\rho} + \Omega\right) + \epsilon_{132} \frac{\partial^2}{\partial x_3 \partial x_2}\left(\frac{p}{\rho} + \Omega\right) = 0.$$

‡$\epsilon_{1ni} \dfrac{\partial v_j}{\partial x_n}\dfrac{\partial v_j}{\partial x_i} = \epsilon_{123} \dfrac{\partial v_j}{\partial x_2}\dfrac{\partial v_j}{\partial x_3} + \epsilon_{132} \dfrac{\partial v_j}{\partial x_3}\dfrac{\partial v_j}{\partial x_2} = 0$, etc.

and

$$\epsilon_{mni}\,\epsilon_{pji}\frac{\partial v_j}{\partial x_n}\zeta_p = (\delta_{mp}\delta_{nj} - \delta_{mj}\delta_{np})\frac{\partial v_j}{\partial x_n}\zeta_p$$

$$= \frac{\partial v_n}{\partial x_n}\zeta_m - \frac{\partial v_m}{\partial x_n}\zeta_n = -\frac{\partial v_m}{\partial x_n}\zeta_n. \qquad (6.69)\dagger$$

Therefore, we have

$$\frac{\partial \zeta_m}{\partial t} + v_j\frac{\partial \zeta_m}{\partial x_j} = \frac{\partial v_m}{\partial x_n}\zeta_n + \nu\frac{\partial^2 \zeta_m}{\partial x_j \partial x_j} \qquad (6.70)$$

or

$$\frac{D\zeta_m}{Dt} = \frac{\partial v_m}{\partial x_n}\zeta_n + \nu\frac{\partial^2 \zeta_m}{\partial x_j \partial x_j}, \qquad (6.71\text{a})$$

which can be written in the following invariant form;

$$\frac{D\boldsymbol{\zeta}}{Dt} = (\boldsymbol{\nabla}\mathbf{v})\boldsymbol{\zeta} + \nu\nabla^2\boldsymbol{\zeta}, \qquad (6.71\text{b})$$

where

$$\nabla^2 \equiv \frac{\partial^2}{\partial x_j \partial x_j}.$$

EXAMPLE 6.18

Reduce, from Eq. (6.71), the vorticity transport equation for two-dimensional flows.

Solution. Let the velocity field be

$$v_1 = v_1(x_1, x_2, t), \quad v_2 = v_2(x_1, x_2, t), \quad v_3 = 0.$$

Then

$$\boldsymbol{\zeta} = \left(\frac{\partial v_3}{\partial x_2} - \frac{\partial v_2}{\partial x_3}\right)\mathbf{e}_1 + \left(\frac{\partial v_1}{\partial x_3} - \frac{\partial v_3}{\partial x_1}\right)\mathbf{e}_2 + \left(\frac{\partial v_2}{\partial x_1} - \frac{\partial v_1}{\partial x_2}\right)\mathbf{e}_3$$

becomes

$$\boldsymbol{\zeta} = \left(\frac{\partial v_2}{\partial x_1} - \frac{\partial v_1}{\partial x_2}\right)\mathbf{e}_3 \equiv \zeta\mathbf{e}_3.$$

That is, the angular velocity vector ($= \zeta/2$) is perpendicular to the plane of flow as expected. Now

$$[(\boldsymbol{\nabla}\mathbf{v})\boldsymbol{\zeta}] = \begin{bmatrix} \frac{\partial v_1}{\partial x_1} & \frac{\partial v_1}{\partial x_2} & 0 \\ \frac{\partial v_2}{\partial x_1} & \frac{\partial v_2}{\partial x_2} & 0 \\ 0 & 0 & 0 \end{bmatrix} \begin{bmatrix} 0 \\ 0 \\ \zeta \end{bmatrix} = \begin{bmatrix} 0 \\ 0 \\ 0 \end{bmatrix}$$

Thus Eq. (6.71) reduces to the scalar equation

$$\frac{D\zeta}{Dt} = \nu\nabla^2\zeta, \qquad (6.72)$$

†Note $\partial v_n/\partial x_n = 0$ for incompressible fluid.

where

$$\frac{D}{Dt} \equiv \frac{\partial}{\partial t} + v_1 \frac{\partial}{\partial x_1} + v_2 \frac{\partial}{\partial x_2}$$

and

$$\nabla^2 \equiv \frac{\partial^2}{\partial x_1^2} + \frac{\partial^2}{\partial x_2^2}.$$

EXAMPLE 6.19

The velocity field for the plane Poiseuille flow is given by

$$v_1 = C\left(\frac{h^2}{4} - x_2^2\right), \quad v_2 = 0, \quad v_3 = 0.$$

(a) Find the vorticity components.
(b) Verify that Eq. (6.72) is satisfied.

Solution. (a) The only nonzero vorticity component is

$$\zeta_3 = \frac{\partial v_2}{\partial x_1} - \frac{\partial v_1}{\partial x_2} = 2Cx_2 \equiv \zeta.$$

(b) We have

$$\frac{D\zeta}{Dt} = \frac{\partial \zeta}{\partial t} + v_1 \frac{\partial \zeta}{\partial x_1} + v_2 \frac{\partial \zeta}{\partial x_2} = 0$$

and

$$\nabla^2 \zeta = \left(\frac{\partial^2}{\partial x_1^2} + \frac{\partial^2}{\partial x_2^2}\right)(2Cx_2) = 0,$$

so that Eq. (6.72) is satisfied.

6.18 CONCEPT OF A BOUNDARY LAYER

In this section we shall describe, qualitatively, the concept of viscous boundary layer by means of an analogy. In Example 6.18, we derived the vorticity equation for two-dimensional flow of an incompressible viscous fluid to be the following:

$$\frac{D\zeta}{Dt} = \nu \nabla^2 \zeta,$$

where ζ is the only nonzero vorticity component for the two-dimensional flow and ν is kinematic viscosity $(= \mu/\rho)$.

In Section 6.12, we saw that, if heat generated through viscous dissipation is neglected, the equation governing the temperature distribution in the flow field due to heat conduction through the boundaries of some hot body is given by

$$\frac{D\theta}{Dt} = \alpha \nabla^2 \theta, \quad \nabla^2 = \frac{\partial^2}{\partial x^2} + \frac{\partial^2}{\partial y^2}, \tag{6.45}$$

where θ is temperature and α the thermal diffusivity is related to conductivity κ, density ρ, and specific heat per unit mass c by the formulas $\alpha = \kappa/\rho c$.

Suppose now we have the problem of a uniform stream flowing past a hot body whose temperature in general varies along the boundary. Let the temperature at large distance from the body be θ_∞, then defining $\theta' = \theta - \theta_\infty$

$$\frac{D\theta'}{Dt} = \alpha \nabla^2 \theta',$$

with $\theta' = 0$ at $x^2 + y^2 \to \infty$. On the other hand, the distribution of vorticity around the body is governed by

$$\frac{D\zeta}{Dt} = \nu \nabla^2 \zeta,$$

$$\zeta = 0 \quad \text{at } x^2 + y^2 \to \infty,$$

where the variation of ζ being due to vorticity generated on the solid boundary and diffusing into the field is much the same as the variation of temperature, being due to heat diffusing from the hot body into the field.

Now, it is intuitively clear that in the case of the temperature distribution, the influence of the hot temperature of the body in the field depends on the speed of the stream. At very low speed, conduction dominates the convection of heat so that the influence will extend deep into the fluid in all directions as shown by the curve C_1 in Fig. 6.14, whereas at high speed, the heat is convected away by the fluid so rapidly that the region affected by the hot body will be confined to a thin layer in the immediate

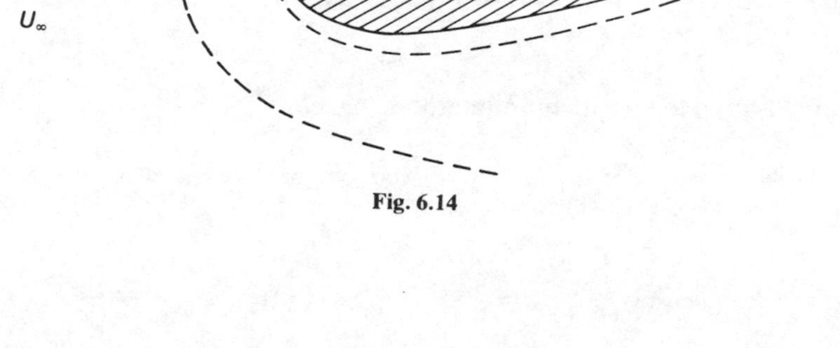

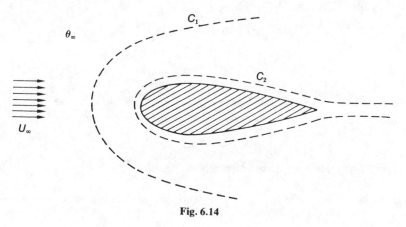

Fig. 6.14

neighborhood of the body and a tail of heated fluid behind it, as is shown by the curve C_2 in Fig. 6.14.

Analogously,† the influence of viscosity, which is responsible for the generation of vorticity on the boundary, depends on the speed at far upstream. At low speed, the influence will be deep into the field in all directions so that essentially the whole flow field is having vorticity. On the other hand, at high speed, the effect of viscosity is confined in a thin layer (known as a boundary layer) near the body and behind it. Outside of the layer, the flow is essentially irrotational. This concept enables one to solve a fluid flow problem by dividing the flow region into an irrotational external flow region and a viscous boundary layer and has proved to be very useful as it simplifies considerably the complexity of the mathematical problem involving the full Navier–Stokes equations. We shall not go into the methods of solution and of matching of the regions as they belong to the theory of boundary layer.

6.19 COMPRESSIBLE NEWTONIAN FLUID

For a compressible fluid, to be consistent with the state of stress corresponding to the state of rest and also to be consistent with the definition that p is not to depend explicitly on any kinematic quantities when in motion, we shall regard p as having the same value as the thermodynamic equilibrium pressure. Therefore, for a particular density ρ and temperature θ, the pressure is determined by the equilibrium equation of state

$$p = p(\rho, \theta), \tag{6.73}$$

(e.g., for ideal gas $p = R\rho\theta$). Thus,

$$T_{ij} = -p(\rho, \theta)\delta_{ij} + \lambda\Delta\delta_{ij} + 2\mu D_{ij}. \tag{6.74}$$

Since

$$\frac{T_{ii}}{3} = -p + (\lambda + \tfrac{2}{3}\mu)\,\Delta, \tag{6.75}$$

it is clear that the "pressure" p in this case does not have the meaning of mean normal compressive stress. It does have the meaning if

$$k = \lambda + \tfrac{2}{3}\mu = 0, \tag{6.76}$$

which is known to be true for monatomic gases.

†It should be noted that the analogy is incomplete because it is difficult to have the boundary condition of θ and ζ the same over the body.

Written in terms of μ and $k = \lambda + \frac{2}{3}\mu$ the constitutive equation reads

$$T_{ij} = -p\delta_{ij} - \tfrac{2}{3}\mu\Delta\delta_{ij} + 2\mu D_{ij} + k\Delta\delta_{ij}. \tag{6.77}$$

With T_{ij} given by the above equations, the equations of motion become (assuming constant μ and k)

$$\rho\frac{Dv_i}{Dt} = \rho B_i - \frac{\partial p}{\partial x_i} + \frac{1}{3}\mu\frac{\partial}{\partial x_i}\left(\frac{\partial v_j}{\partial x_j}\right) + \mu\frac{\partial^2 v_i}{\partial x_j \partial x_j} + k\frac{\partial}{\partial x_i}\left(\frac{\partial v_j}{\partial x_j}\right). \tag{6.78}$$

Equations (6.78) and (6.73) are four equations for six unknowns $v_1, v_2, v_3, p, \rho, \theta$, the fifth equation is given by the equation of continuity

$$\frac{D\rho}{Dt} + \rho\frac{\partial v_j}{\partial x_j} = 0. \tag{6.79}$$

The sixth equation is supplied by the energy equation

$$\rho\frac{Du}{Dt} = T_{ij}\frac{\partial v_i}{\partial x_j} + \kappa\frac{\partial^2\theta}{\partial x_j \partial x_j}. \tag{6.80}$$

For a given fluid the dependence of the internal energy u on ρ and θ is assumed to be the same as when the fluid is in the equilibrium, for example, for ideal gas

$$u = c_v\theta, \tag{6.81}$$

where c_v is the specific heat at constant volume.

In general, we have

$$u = u(\rho, \theta). \tag{6.82}$$

Equations (6.73), (6.78), (6.79), (6.80), and (6.82) form a system of seven equations for the seven unknowns $v_1, v_2, v_3, p, \rho, \theta$, and u.

6.20 ENERGY EQUATION IN TERMS OF ENTHALPY

Enthalpy per unit mass is defined as

$$h = u + \frac{p}{\rho}, \tag{6.83}$$

where u is the internal energy per unit mass, p the pressure, ρ the density.

Let $h_0 \equiv h + (v^2/2)$ (h_0 is known as stagnation enthalpy). We shall show that in terms of h_0, the energy equation becomes (neglecting body force)

$$\rho\frac{Dh_0}{Dt} = \frac{\partial p}{\partial t} + \frac{\partial}{\partial x_j}(T'_{ij}v_i - q_j), \tag{6.84}$$

where T'_{ij} is the viscous stress tensor, q_i the heat flux vector. First, by definition,

$$\frac{Dh_0}{Dt}=\frac{D}{Dt}\left(u+\frac{p}{\rho}+\frac{\mathbf{v}\cdot\mathbf{v}}{2}\right)=\frac{Du}{Dt}+\frac{D}{Dt}\left(\frac{p}{\rho}\right)+\mathbf{v}\cdot\frac{D\mathbf{v}}{Dt}.$$

From the energy equation [Eq. (6.42)], we have

$$\rho\frac{Du}{Dt}=T_{ij}\frac{\partial v_i}{\partial x_j}-\frac{\partial q_i}{\partial x_i}=(-p\delta_{ij}+T'_{ij})\frac{\partial v_i}{\partial x_j}-\frac{\partial q_i}{\partial x_i}.$$

That is,

$$\rho\frac{Du}{Dt}=-p\frac{\partial v_i}{\partial x_i}+T'_{ij}\frac{\partial v_i}{\partial x_j}-\frac{\partial q_i}{\partial x_i}.$$

Also, we have

$$\frac{D}{Dt}\left(\frac{p}{\rho}\right)=\frac{1}{\rho}\frac{Dp}{Dt}-\frac{p}{\rho^2}\frac{D\rho}{Dt}$$

and the equation of motion (in the absence of body force)

$$\rho\frac{Dv_i}{Dt}=\frac{\partial T_{ij}}{\partial x_j}=-\frac{\partial p}{\partial x_i}+\frac{\partial T'_{ij}}{\partial x_j}.$$

Thus

$$\rho\frac{Dh_0}{Dt}=-p\frac{\partial v_i}{\partial x_i}+T'_{ij}\frac{\partial v_i}{\partial x_j}-\frac{\partial q_i}{\partial x_i}+\frac{Dp}{Dt}-\frac{p}{\rho}\frac{D\rho}{Dt}-v_i\frac{\partial p}{\partial x_i}+\frac{\partial T'_{ij}}{\partial x_j}v_i.$$

Noting that

$$\frac{Dp}{Dt}-v_i\frac{\partial p}{\partial x_i}=\frac{\partial p}{\partial t}$$

and

$$-p\frac{\partial v_i}{\partial x_i}-\frac{p}{\rho}\frac{D\rho}{Dt}=-\frac{p}{\rho}\left(\frac{D\rho}{Dt}+\rho\frac{\partial v_i}{\partial x_i}\right)=0,$$

we have,

$$\rho\frac{Dh_0}{Dt}=T'_{ij}\frac{\partial v_i}{\partial x_j}+\frac{\partial T'_{ij}}{\partial x_j}v_i-\frac{\partial q_i}{\partial x_i}+\frac{\partial p}{\partial t}$$

or

$$\rho\frac{Dh_0}{Dt}=\frac{\partial}{\partial x_j}(T'_{ij}v_i)-\frac{\partial q_i}{\partial x_i}+\frac{\partial p}{\partial t}.$$

EXAMPLE 6.20

Show that for steady flow of an inviscid nonheat conducting fluid, if the flow originates from a homogeneous state, then (a) $h+(v^2/2)=$ constant, and (b) if the fluid is an ideal gas then

$$\frac{\gamma}{\gamma-1}\frac{p}{\rho}+\frac{v^2}{2}=\text{constant},$$

where $\gamma=C_p/C_v$, the ratio of specific heat under constant pressure and constant volume.

Solution. (a) Since the flow is steady, therefore $\partial p/\partial t = 0$. Since the fluid is inviscid and nonheat conducting, therefore $T'_{ij} = 0$ and $q_i = 0$. Thus the energy equation (6.84) reduces to

$$\frac{Dh_0}{Dt} = 0.$$

In other words, h_0 is a constant for each particle. But since the flow originates from a homogeneous state, therefore

$$h_0 \equiv h + \frac{v^2}{2} \equiv \frac{p}{\rho} + u + \frac{v^2}{2} = \text{constant} \tag{6.85}$$

in the whole flow field.
(b) For an ideal gas $p = \rho R\theta$, $u = c_v\theta$ and $R = c_p - c_v$, therefore

$$u = \frac{p}{\rho}\left(\frac{1}{\gamma - 1}\right),$$

where

$$\gamma = \frac{c_p}{c_v} \tag{6.86}$$

and

$$h_0 = \frac{p}{\rho}\left(\frac{\gamma}{\gamma - 1}\right) + \frac{v^2}{2} = \text{constant}. \tag{6.87}$$

6.21 ACOUSTIC WAVE

The propagation of sound can be approximated by considering the propagation of infinitesimal disturbances in a compressible inviscid fluid. For an inviscid fluid, neglecting body forces, the equations of motion are

$$\frac{\partial v_i}{\partial t} + v_j \frac{\partial v_i}{\partial x_j} = -\frac{1}{\rho}\frac{\partial p}{\partial x_i}. \tag{i}$$

Let us suppose that the fluid is initially at rest with

$$v_i = 0, \quad \rho = \rho_0, \quad p = p_0. \tag{ii}$$

Now suppose that the fluid is perturbed from rest, such that

$$v_i = v'_i(\mathbf{x}, t), \quad \rho = \rho_0 + \rho'(\mathbf{x}, t), \quad p = p_0 + p'(\mathbf{x}, t). \tag{iii}$$

Substituting into Eq. (i),

$$\frac{\partial v'_i}{\partial t} + v'_j \frac{\partial v'_i}{\partial x_j} = -\frac{1}{\rho_o(1 + \rho'/\rho_0)}\frac{\partial p'}{\partial x_i}. \tag{iv}$$

Since we assumed infinitesimal disturbances, the terms $v'_j(\partial v'_i/\partial x_j)$ and ρ'/ρ_0 are negligible and the equations of motion now take the linearized form

$$\frac{\partial v'_i}{\partial t} = -\frac{1}{\rho_0}\frac{\partial p'}{\partial x_i}. \tag{6.88}$$

In a similar manner, we consider the mass conservation equation

$$\frac{\partial \rho'}{\partial t} + v_i' \frac{\partial \rho'}{\partial x_i} + \rho_0(1+\rho'/\rho_0)\frac{\partial v_i'}{\partial x_i} = 0$$

and obtain the linearized equation

$$\frac{\partial v_i'}{\partial x_i} = -\frac{1}{\rho_0}\frac{\partial \rho'}{\partial t}. \tag{6.89}$$

Differentiating Eq. (6.88) by x_i and Eq. (6.89) by t, we eliminate the velocity,

$$\frac{\partial^2 p'}{\partial x_i \partial x_i} = \frac{\partial^2 \rho'}{\partial t^2}. \tag{6.90}$$

We further assume that the flow is barotropic, i.e., the pressure depends explicitly on density only, so that the pressure $p = p(\rho)$. Expanding $p(\rho)$ in a Taylor series about the rest value of pressure p_0, we have

$$p = p_0 + \left(\frac{dp}{d\rho}\right)_{\rho_0}(\rho - \rho_0) + \ldots. \tag{6.91}$$

Neglecting higher-order terms

$$p' = c_0{}^2 \rho', \tag{6.92}$$

where

$$c_0{}^2 = \left(\frac{dp}{d\rho}\right)_{\rho_0}. \tag{6.93}$$

Thus, for a barotropic fluid,

$$c_0{}^2 \frac{\partial^2 p'}{\partial x_i \partial x_i} = \frac{\partial^2 p'}{\partial t^2} \tag{6.94}$$

and

$$c_0{}^2 \frac{\partial^2 \rho'}{\partial x_i \partial x_i} = \frac{\partial^2 \rho'}{\partial t^2} \tag{6.95}$$

These equations are exactly analogous (for one-dimensional waves) to the elastic wave equations of Chapter 5. Thus, we conclude that the pressure and density disturbances will propagate with a speed $c_0 = \sqrt{(dp/d\rho)_{\rho_0}}$. We call c_0 the **speed of sound** at stagnation. For a fluid in general motion, the local speed of sound is defined to be

$$c = \sqrt{\frac{dp}{d\rho}}, \tag{6.96}$$

when **isentropic** relation of p and ρ is used, i.e.,

$$p = \beta \rho^\gamma, \tag{6.97}$$

where $\gamma = c_p/c_v$ (ratio of specific heats) and β is a constant

$$\frac{dp}{d\rho} = \beta\gamma\rho^{\gamma-1} = \gamma\frac{p}{\rho},$$

so that the speed of sound is

$$c = \sqrt{\gamma\frac{p}{\rho}}. \tag{6.98}$$

EXAMPLE 6.21

(a) Write an expression for a harmonic plane acoustic wave propagating in the $\mathbf{e}_1$-direction.
(b) Find the velocity disturbance.
(c) Compare $\partial v_i/\partial t$ to the neglected $v_j(\partial v_i/\partial x_j)$.

Solution. (a) Referring to the section on elastic waves, we simply write

$$p = \epsilon \sin\frac{2\pi}{l}(x_1 - c_0 t).$$

Note, we have dropped the awkward primes.
(b) Using Eq. (6.88) we have

$$\frac{\partial v_1}{\partial t} = -\frac{\epsilon}{\rho_0}\left(\frac{2\pi}{l}\right)\cos\left(\frac{2\pi}{l}\right)(x_1 - c_0 t).$$

Therefore, the velocity

$$v_1 = \frac{\epsilon}{\rho_0 c_0}\sin\left(\frac{2\pi}{l}\right)(x_1 - c_0 t)$$

is exactly the same form as the pressure wave.
(c) For the one-dimensional case, we wish to evaluate

$$\left|\frac{v_1\dfrac{\partial v_1}{\partial x_1}}{\partial v_1/\partial t}\right| = \frac{|v_1|\,(2\pi\epsilon/\rho_0 l c_0)}{\dfrac{\epsilon}{\rho_0}\left(\dfrac{2\pi}{l}\right)} = \frac{|v_1|}{c_0}.$$

Thus, the approximate is best when the disturbance has a velocity that is much smaller than the speed of sound.

EXAMPLE 6.22

Two fluids have a plane interface at $x_1 = 0$. Consider a plane acoustic wave that is normally incident on the interface and determine the amplitudes of the reflected and transmitted waves.

Solution. Let the fluid properties to the left of the interface ($x_1 < 0$) be denoted by ρ_1, c_1 and to the right ($x_1 > 0$) by ρ_2, c_2.

Now, let the incident pressure wave propagate to the right, as given by

$$p_I = \epsilon_I \sin\frac{2\pi}{l_I}(x_1 - c_1 t) \qquad (x_1 \leqslant 0).$$

This pressure wave results in a reflected wave

$$p_R = \epsilon_R \sin \frac{2\pi}{l_R}(x_1 + c_1 t) \qquad (x_1 \leq 0)$$

and a transmitted wave

$$p_T = \epsilon_T \sin \frac{2\pi}{l_T}(x_1 - c_2 t) \qquad (x_1 \geq 0).$$

We must now consider the conditions on the boundary $x_1 = 0$. First the total pressure defined must be the same, so that

$$(p_I + p_R)_{x_1=0} = (p_T)_{x_1=0}$$

or

$$\epsilon_I \sin \frac{2\pi c_1 t}{l_I} - \epsilon_R \sin \frac{2\pi c_1 t}{l_R} = \epsilon_T \sin \frac{2\pi c_2 t}{l_T}.$$

This equation will be satisfied for all time if

$$l_I = l_R = \frac{c_1}{c_2} l_T \tag{i}$$

and

$$\epsilon_I - \epsilon_R = \epsilon_T \tag{ii}$$

In addition, we require the normal velocities be continuous at all time on $x_1 = 0$, so that $(\partial v_1/\partial t)_{x_1=0}$ is also continuous. Thus by using Eq. (6.87),

$$-\left.\frac{\partial v_1}{\partial t}\right|_{x_1=0} = \frac{1}{\rho_1}\left(\frac{\partial p_I}{\partial x_1} + \frac{\partial p_R}{\partial x_1}\right)_{x_1=0} = \frac{1}{\rho_2}\left(\frac{\partial p_T}{\partial x_1}\right)_{x_1=0}.$$

Substituting for the pressure we obtain

$$\frac{1}{\rho_1}\left(\frac{\epsilon_I}{l_I} + \frac{\epsilon_R}{l_R}\right) = \frac{1}{\rho_2}\left(\frac{\epsilon_T}{l_T}\right). \tag{iii}$$

Combining Eqs. (i), (ii), and (iii) we obtain

$$\epsilon_T = \left(\frac{2}{1 + (\rho_1 c_1/\rho_2 c_2)}\right)\epsilon_I, \tag{iv}$$

$$\epsilon_R = \left(\frac{(\rho_1 c_1/\rho_2 c_2) - 1}{(\rho_1 c_1/\rho_2 c_2) + 1}\right)\epsilon_I. \tag{v}$$

Note, that for the special case $\rho_1 c_1 = \rho_2 c_2$,

$$\epsilon_T = \epsilon_I \quad \text{and} \quad \epsilon_R = 0.$$

6.22 IRROTATIONAL, BAROTROPIC FLOWS OF INVISCID COMPRESSIBLE FLUID

Consider an irrotational flow field given by

$$v_i = -\frac{\partial \Phi}{\partial x_i}. \tag{6.99}$$

In order that the mass conservation principle is obeyed, we must have

$$\frac{\partial \rho}{\partial t}+v_j\frac{\partial \rho}{\partial x_j}+\rho\frac{\partial v_j}{\partial x_j}=0. \tag{6.100}$$

In terms of Φ this equation becomes

$$\frac{\partial \rho}{\partial t}-\frac{\partial \Phi}{\partial x_j}\frac{\partial \rho}{\partial x_j}-\rho\frac{\partial^2 \Phi}{\partial x_j\,\partial x_j}=0. \tag{6.101}$$

The equations of motion for an inviscid fluid are the Euler's equation

$$\frac{\partial v_i}{\partial t}+v_j\frac{\partial v_i}{\partial x_j}=-\frac{1}{\rho}\frac{\partial p}{\partial x_i}+B_i. \tag{6.102}$$

We assume that the flow is **barotropic**, that is, the pressure is an explicit function of density (such as in isentropic or isothermal flow). Thus, in a barotropic flow,

$$p=p(\rho) \quad \text{and} \quad \rho=\rho(p). \tag{6.103}$$

Now

$$\frac{\partial}{\partial x_i}\left(\int\frac{1}{\rho}\,dp\right)=\left[\frac{d}{dp}\int\frac{1}{\rho}\,dp\right]\frac{\partial p}{\partial x_i}=\frac{1}{\rho}\frac{\partial p}{\partial x_i}. \tag{6.104}$$

Therefore, for barotropic flows of an inviscid fluid under conservative body forces (i.e., $B_i=-\partial\Omega/\partial x_1$), the equations of motion can be written

$$\frac{\partial v_i}{\partial t}+v_j\frac{\partial v_i}{\partial x_j}=-\frac{\partial}{\partial x_i}\left(\int\frac{dp}{\rho}+\Omega\right). \tag{6.105}$$

Comparing Eq. (6.105) with Eq. (6.60), we see immediately that under the conditions stated, irrotational flows are again always dynamically possible. In fact, the integration of Eq. (6.105) (in exactly the same way as was done in Section 6.15) gives the following Bernoulli equation

$$-\frac{\partial \Phi}{\partial t}+\int\frac{dp}{\rho}+\frac{v^2}{2}+\Omega=f(t), \tag{6.106}$$

which for steady flow, becomes

$$\int\frac{dp}{\rho}+\frac{v^2}{2}+\Omega=\text{constant}. \tag{6.107}$$

For most problems in gas dynamics, the body force is small compared with other forces and often neglected. We then have

$$\int\frac{dp}{\rho}+\frac{v^2}{2}=\text{constant}. \tag{6.108}$$

EXAMPLE 6.23

Show that for steady isentropic irrotational flows of an inviscid compressible fluid (body force neglected)

$$\frac{\gamma}{\gamma-1}+\frac{v^2}{2}=\text{constant}.$$

Solution. For an isentropic flow

$$p=\beta\rho^{\gamma}, \quad dp=\beta\gamma\rho^{\gamma-1}\,d\rho,$$

so that

$$\int\frac{dp}{\rho}=\beta\gamma\int\rho^{\gamma-2}\,d\rho=\beta\gamma\frac{\rho^{\gamma-1}}{\gamma-1}=\frac{\gamma}{\gamma-1}\frac{p}{\rho}.$$

Thus the Bernoulli equation [Eq. (6.108)] becomes

$$\frac{\gamma}{\gamma-1}\frac{p}{\rho}+\frac{v^2}{2}=\text{constant}.$$

We note that this is the same result as that obtained in Example 6.20, Eq. (6.87), by the use of the energy equation. In other words, under the conditions stated (inviscid, nonheat conducting, initial homogeneous state), the Bernoulli equation and the energy equation are the same.

EXAMPLE 6.24

Let p_0 denote the pressure at zero speed (called stagnation pressure). Show that for isentropic steady flow ($p/\rho^{\gamma}=\text{constant}$) of an ideal gas,

$$p_0=p\left[1+\frac{1}{2}(\gamma-1)\left(\frac{v}{c}\right)^2\right]^{\gamma/(\gamma-1)}, \tag{6.109}$$

where c is the local speed of sound.

Solution. Since (*see* previous example)

$$\frac{1}{2}v^2+\frac{\gamma}{\gamma-1}\frac{p}{\rho}=\text{constant}.$$

Thus

$$\frac{v^2}{2}=\frac{\gamma}{\gamma-1}\left(\frac{p_0}{\rho_0}-\frac{p}{\rho}\right).$$

Now $c^2=\gamma(p/\rho)$,

$$\frac{v^2}{2c^2}=\frac{1}{\gamma-1}\left(\frac{p_0}{p}\right)\left(\frac{\rho}{\rho_0}\right)-\frac{1}{\gamma-1}.$$

Since

$$\frac{\rho}{\rho_0}=\left(\frac{p}{p_0}\right)^{1/\gamma}=\left(\frac{p_0}{p}\right)^{-1/\gamma},$$

therefore

$$\left(\frac{\gamma-1}{2}\right)\left(\frac{v}{c}\right)^2=\left(\frac{p_0}{p}\right)^{(\gamma-1)/\gamma}-1.$$

Thus

$$\frac{p_0}{p}=\left[1+\left(\frac{\gamma-1}{2}\right)\left(\frac{v}{c}\right)^2\right]^{\gamma/(\gamma-1)}.$$

For small Mach number $M = (v/c)$, we can use binomial expansion to obtain from the above equation

$$\frac{p_0}{p} = 1 + \frac{\gamma}{2}\left(\frac{v}{c}\right)^2 + \frac{1}{8}\gamma\left(\frac{v}{c}\right)^4 + \dots$$

Noting that

$$\frac{p\gamma}{c^2} = \frac{p\gamma}{\gamma\left(\frac{p}{\rho}\right)} = \rho,$$

we have

$$p_0 = p + \frac{1}{2}\rho v^2 + \frac{1}{8}\rho v^2\left(\frac{v}{c}\right)^2 + \dots$$

or

$$p_0 = p + \frac{1}{2}\rho v^2\left[1 + \frac{1}{4}\left(\frac{v}{c}\right)^2 + \dots\right].$$

For small Mach number M, the above equation reads

$$p_0 = p + \tfrac{1}{2}\rho v^2,$$

which is the same as that for an incompressible fluid. In other words, for steady isentropic flow, the fluid may be considered as incompressible if the Mach number is small (say < 0.3).

EXAMPLE 6.25

For steady, barotropic irrotational flow, derive the equation for the velocity potential Φ. Neglect body force.

Solution. For steady flow, the equation of continuity is, with $v_i = -\partial\Phi/\partial x_i$,

$$\frac{\partial\Phi}{\partial x_i}\frac{\partial\rho}{\partial x_i} + \rho\frac{\partial^2\Phi}{\partial x_i\partial x_i} = 0, \tag{i}$$

and the equation of motion is

$$-\frac{\partial\Phi}{\partial x_j}\frac{\partial^2\Phi}{\partial x_j\partial x_i} = -\frac{1}{\rho}\frac{\partial p}{\partial x_i} = -\frac{1}{\rho}\frac{dp}{d\rho}\frac{\partial\rho}{\partial x_i}. \tag{ii}$$

Let $c^2 \equiv dp/d\rho$ (the local sound speed), then

$$-\frac{1}{\rho}\frac{\partial\rho}{\partial x_i} = \frac{1}{c^2}\left(\frac{\partial\Phi}{\partial x_j}\frac{\partial^2\Phi}{\partial x_j\partial x_i}\right). \tag{iii}$$

Substituting Eq. (iii) into Eq. (i), we obtain

$$-\frac{\partial\Phi}{\partial x_i}\frac{\partial\Phi}{\partial x_j}\frac{\partial^2\Phi}{\partial x_j\partial x_i} + c^2\frac{\partial^2\Phi}{\partial x_i\partial x_i} = 0 \tag{iv}$$

or

$$\left(c^2\delta_{ij} - \frac{\partial\Phi}{\partial x_i}\frac{\partial\Phi}{\partial x_j}\right)\frac{\partial^2\Phi}{\partial x_i\partial x_j} = 0. \tag{v}$$

In long form, Eq. (v) reads

$$\left[c^2 - \left(\frac{\partial\Phi}{\partial x_1}\right)^2\right]\frac{\partial^2\Phi}{\partial x_1^2} + \left[c^2 - \left(\frac{\partial\Phi}{\partial x_2}\right)^2\right]\frac{\partial^2\Phi}{\partial x_2^2} + \left[c^2 - \left(\frac{\partial\Phi}{\partial x_3}\right)^2\right]\frac{\partial^2\Phi}{\partial x_3^2}$$
$$-2\left(\frac{\partial\Phi}{\partial x_1}\frac{\partial\Phi}{\partial x_2}\frac{\partial^2\Phi}{\partial x_1\partial x_2} + \frac{\partial\Phi}{\partial x_2}\frac{\partial\Phi}{\partial x_3}\frac{\partial^2\Phi}{\partial x_2\partial x_3} + \frac{\partial\Phi}{\partial x_3}\frac{\partial\Phi}{\partial x_1}\frac{\partial^2\Phi}{\partial x_3\partial x_1}\right) = 0.$$

6.23 ONE-DIMENSIONAL FLOW OF A COMPRESSIBLE FLUID

In this section, we discuss some internal flow problems of a compressible fluid. The fluid will be assumed to be an ideal gas. The flow will be assumed to be one-dimensional in the sense that the pressure, temperature, density, velocity, etc. are uniform over any cross-section of the channel or duct in which the fluid is flowing. The flow will also be assumed to be steady and adiabatic.

In steady flow, the rate of mass flow is constant for all cross-sections. With A denoting the variable cross-sectional area, ρ the density and v the velocity, we have

$$\rho A v = C \text{ (a constant).} \tag{6.110}$$

To see the effect of area variation on the flow, we take the total derivative of Eq. (6.110), i.e.,

$$d\rho\,(Av) + \rho\,(dA)v + \rho A\,(dv) = 0.$$

Dividing the above equation by $\rho A v$, we obtain

$$\frac{d\rho}{\rho} + \frac{dA}{A} + \frac{dv}{v} = 0. \tag{6.111}$$

Thus

$$\frac{dA}{A} = -\frac{d\rho}{\rho} - \frac{dv}{v}. \tag{i}$$

Now the energy equation, for barotropic flow of an ideal gas is (*see* Eq. (6.108))

$$\frac{v^2}{2} + \int \frac{dp}{\rho} = \text{constant}.$$

Thus, we have through total differentiation of the above equation

$$v\,dv + \frac{dp}{\rho} = v\,dv + \frac{1}{\rho}\frac{dp}{d\rho}\,d\rho = 0.$$

But $\sqrt{(dp/d\rho)} = c$ (the speed of sound), thus

$$\frac{d\rho}{\rho} = -\frac{v\,dv}{c^2}. \tag{ii}$$

Combining Eqs. (i) and (ii), we get

$$\frac{dA}{A} = \frac{v\,dv}{c^2} - \frac{dv}{v} = \frac{dv}{v}\left(\frac{v^2}{c^2} - 1\right).$$

The ratio of flow speed to the local speed of sound is known as the **Mach**

number. Denoting the Mach number by M, the above equation can be written (known sometimes as **Hugoniot equation**)

$$\frac{dA}{A} = \frac{dv}{v}(M^2 - 1). \tag{6.112}$$

From the above equation, we see that for subsonic flow ($M < 1$), an increase in area produces a decrease in velocity, just as in the case of an incompressible fluid. On the other hand, for supersonic flow ($M > 1$), an increase in area produces an increase in velocity. Furthermore, the critical velocity ($M = 1$) can only be obtained at the smallest cross-sectional area where $dA = 0$.

We now study the flow in a converging nozzle and the flow in a converging-diverging nozzle, using one-dimensional assumptions.

(i) *Flow in a Converging Nozzle*

Let us consider the adiabatic flow of an ideal gas from a large tank (inside which the pressure p_1 and the density ρ_1 remain essentially unchanged) into a region of pressure p_R.

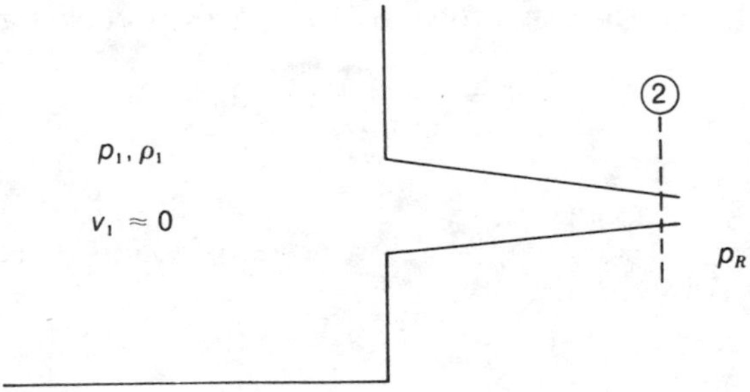

Fig. 6.15

Application of the energy equation, using the conditions inside the tank and at the section (2), we have

$$\frac{v_2^2}{2} + \frac{\gamma}{\gamma - 1}\frac{p_2}{\rho_2} = 0 + \frac{\gamma}{\gamma - 1}\frac{p_1}{\rho_1},$$

where p_2, ρ_2, and v_2 are pressure, density, and velocity at section (2). Thus

$$v_2^2 = \frac{2\gamma}{\gamma - 1}\left(\frac{p_1}{\rho_1} - \frac{p_2}{\rho_2}\right).$$

For adiabatic flow,

$$\left(\frac{p_2}{p_1}\right)^{1/\gamma} = \left(\frac{\rho_2}{\rho_1}\right),$$

therefore

$$v_2{}^2 = \frac{2\gamma}{\gamma-1}\frac{p_1}{\rho_1}\left[1-\left(\frac{p_2}{p_1}\right)^{(\gamma-1)/\gamma}\right] = \frac{2\gamma}{\gamma-1}\frac{p_2}{\rho_2}\left[\left(\frac{p_1}{p_2}\right)^{(\gamma-1)/\gamma}-1\right]. \tag{6.113}$$

Computing the rate of mass flow, we have

$$\text{rate of mass flow} \equiv \frac{dm}{dt} = A_2\rho_2 v_2 = A_2\left(\frac{p_2}{p_1}\right)^{1/\gamma}\rho_1 v_2.$$

Thus

$$\frac{dm}{dt} = A_2\left(\frac{p_2}{p_1}\right)^{1/\gamma}\rho_1\left\{\frac{2\gamma}{\gamma-1}\frac{p_1}{\rho_1}\left[1-\left(\frac{p_2}{p_1}\right)^{(\gamma-1)/\gamma}\right]\right\}^{1/2}$$

or

$$\frac{dm}{dt} = A_2\left\{\frac{2\gamma}{\gamma-1}p_1\rho_1\left[\left(\frac{p_2}{p_1}\right)^{2/\gamma}-\left(\frac{p_2}{p_1}\right)^{(\gamma+1)/\gamma}\right]\right\}^{1/2}. \tag{6.114}$$

For given p_1, ρ_1, and A_2, we see dm/dt depends only on p_2. When $p_2 = 0$, dm/dt is zero and when $p_2 = p_1$, dm/dt also is zero.

Figure 6.16 shows the curve of dm/dt versus p_2/p_1 according to Eq. (6.114). It can be easily established that $(dm/dt)_{max}$ occurs at

$$p_2 = \left(\frac{2}{\gamma+1}\right)^{\gamma/(\gamma-1)} p_1 \tag{6.115}$$

and at this pressure p_2,

$$v_2{}^2 = \frac{2\gamma}{\gamma-1}\frac{p_2}{\rho_2}\left(\frac{\gamma+1}{2}-1\right) = \gamma\frac{p_2}{\rho_2} = \text{speed of sound at section (2).}$$

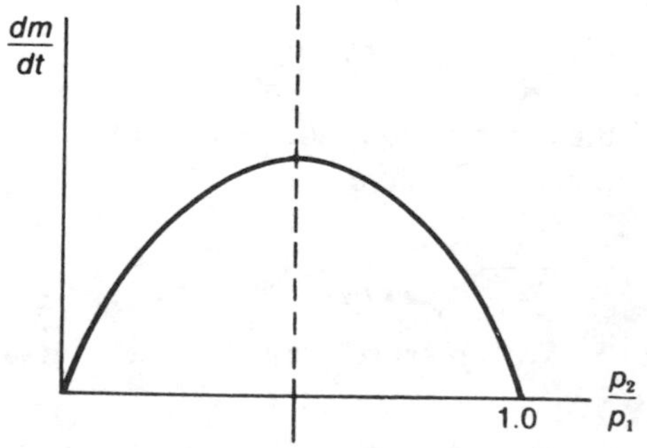

Fig. 6.16

The pressure p_2 given by Eq. (6.115) is known as the critical pressure p_c. The pressure p_2 at section (2) can never be less than p_c (which depends only on p_1) because otherwise the flow will become supersonic at section (2)† which is impossible in view of the conclusion reached earlier that to have $M = 1$, dA must be zero, and to have $M > 1$, dA must be increasing (divergent nozzle). Thus for the case of a convergent nozzle, p_2 can never be less than p_R, the pressure surrounding the exit jet. When $p_R \geqslant p_c$, $p_2 = p_R$ and when $p_R < p_c$, $p_2 = p_c$. In other words, the relation between dm/dt and p_R/p_1 is given as

$$\frac{dm}{dt} = A_2\left\{\frac{2\gamma}{\gamma-1}p_1\rho_1\left[\left(\frac{p_2}{p_1}\right)^{2/\gamma} - \left(\frac{p_2}{p_1}\right)^{(\gamma+1)/\gamma}\right]\right\}^{1/2} \quad \text{for } p_R \geqq p_c \tag{6.116a}$$

and

$$\frac{dm}{dt} = A_2\left\{\frac{2\gamma}{\gamma-1}p_1\rho_1\left[\left(\frac{2}{\gamma+1}\right)^{2/(\gamma-1)} - \left(\frac{2}{\gamma+1}\right)^{(\gamma+1)/(\gamma-1)}\right]\right\}^{1/2} \quad \text{for } p_R \leqq p_c$$
$$= \text{a constant.} \tag{6.116b}$$

Figure 6.17 shows this relationship.

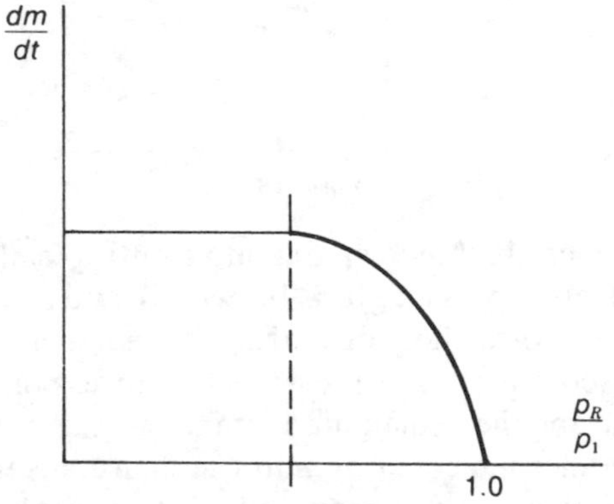

Fig. 6.17

(ii) *Flow in a Convergent-Diverging Nozzle*

For a compressible fluid coming out from a large supply tank, in order to increase the speed, a converging nozzle is needed. From (i) we have seen

†Note: $\gamma > 1$, v_2^2 is a monotonically decreasing function of p_2. (*See* Eq. 6.113.)

that the maximum attainable Mach number is unity in a converging passage. We have also concluded at the beginning of this section that in order to have the Mach number larger than unity, the cross-sectional area must increase in the direction of the flow. Thus, in order to make supersonic flow possible from a supply tank, the fluid must flow in a converging-diverging nozzle as shown in Fig. 6.18. The flow in the converging part of the nozzle is always subsonic regardless of the receiver pressure p_R ($< p_1$). The flow in the diverging passage is subsonic for certain range of p_R/p_1 (*see* curves *a* and *b* in Fig. 6.18). There is a value of p_R at which the flow at the throat is sonic, the flow corresponding to this case is known as choked flow (curve *c*). Further reduction of p_R cannot affect the condition at the throat and produces no change in flow rate. There is one receiver

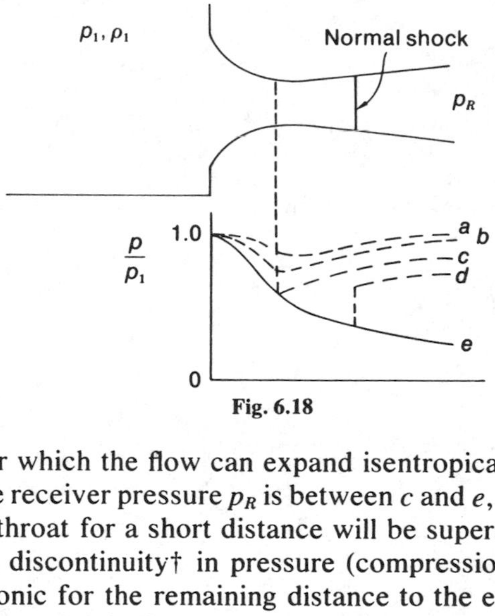

Fig. 6.18

pressure p_R for which the flow can expand isentropically to p_R (the solid curve *e*). If the receiver pressure p_R is between *c* and *e*, such as *d*, the flow following the throat for a short distance will be supersonic. This is then followed by a discontinuity† in pressure (compression shock) and flow becomes subsonic for the remaining distance to the exit. If the receiver pressure is below that indicated by *e* in the figure, a series of expansion waves and oblique shock waves occur outside the nozzle.

PROBLEMS

6.1 In Fig. P6.1, the gate *AB* is rectangular, 60 cm wide and 4 m long. The gate is hinged at the upper edge *A*. Neglect the weight of the gate. Find the reactional force at *B*. Take the specific weight of water to be 9800 N/m³ (62.4lb/ft³).

†That is the increase in pressure takes place in a very short distance.

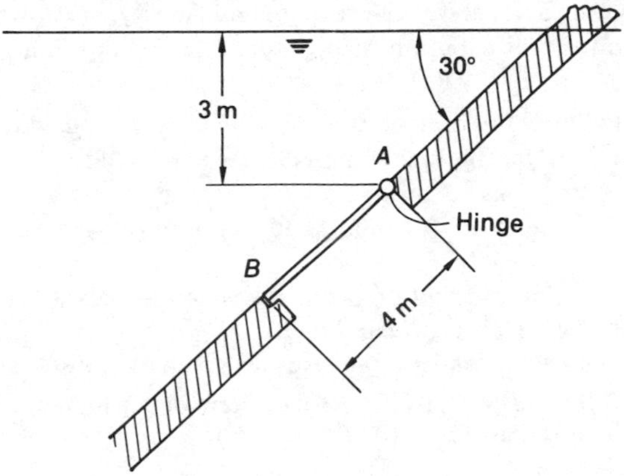

Fig. P6.1

6.2. A glass of water moves vertically upward with a constant acceleration **a**. Find the pressure at a point whose depth from the surface of the water is h.

6.3. A glass of water moves with a constant acceleration **a** in the direction shown in Fig. P6.2. Find the pressure at the point A. Take the atmospheric pressure to be p_a.

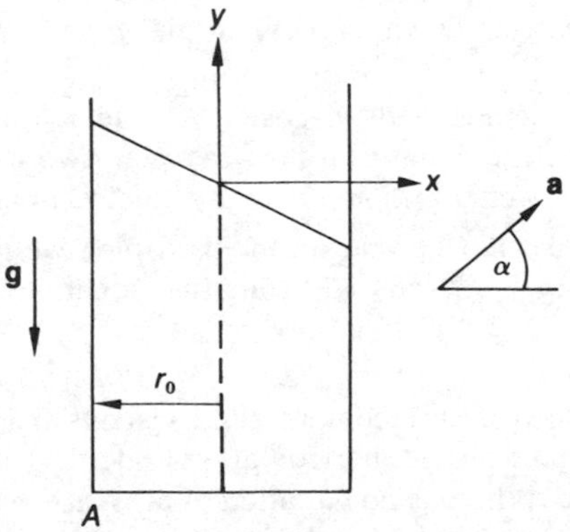

Fig. P6.2

6.4. A liquid in a container rotates with a constant angular velocity ω about a vertical axis. Find the shape of the liquid surface.

6.5. In astrophysical applications, an atmosphere having the relation between the density ρ and pressure p given by

$$\frac{p}{p_0} = \left(\frac{\rho}{\rho_0}\right)^n,$$

where p_0 and ρ_0 are some reference pressure and density, is known as a polytropic atmosphere. Find the distribution of pressure and density in a polytropic atmosphere.

6.6. For a parallel flow of an incompressible linearly viscous fluid, if we take the flow direction to be $\mathbf{e}_3$, (a) show that the velocity field is of the form

$$v_1 = 0, \quad v_2 = 0, \quad \text{and} \quad v_3 = v(x_1, x_2).$$

(b) If $v(x_1, x_2) = x_2$, find the total normal stress on the plane whose normal is in the direction of $\mathbf{e}_2 + \mathbf{e}_3$ in terms of μ and p.
(c) On what planes are the total normal stresses given by the so-called "pressure"?

6.7. Given the following velocity field for a Newtonian incompressible fluid with a viscosity $\mu = 0.96$ mPa·s (2×10^{-5} lb sec/ft²),

$$v_1 = x_1^2 - x_2^2, \quad v_2 = -2x_1x_2, \quad v_3 = 0.$$

At the point (1, 2, 1) and on the plane whose normal is in the direction of $\mathbf{e}_1$,
(a) find the excess of the total normal compressive stress over the pressure p,
(b) find the magnitude of the shearing stress.

6.8. Do Problem 6.7 except that the plane has a normal in the direction of $3\mathbf{e}_1 + 4\mathbf{e}_2$.

6.9. Obtain the steady unidirectional flow of an incompressible viscous fluid layer of uniform depth d flowing down an inclined plane which makes an angle θ with the horizontal.

6.10. Two layers of liquids with viscosity μ_1 and μ_2 and density ρ_1 and ρ_2, respectively, and with equal depths b flow between two fixed horizontal parallel plates. Find the velocity distribution for steady unidirectional flow.

6.11. For Hagen–Poiseuille flow in an inclined pipe we define the piezometric head $h = p/\rho g - x_1 \sin\theta + x_2 \cos\theta$. From the equations of motion show that
(a) $h = h(x_1)$
(b) $(dh/dx_1) =$ constant.

6.12. Consider the flow of an incompressible viscous fluid through the annular space between two concentric horizontal cylinders. The radii are a and b.
(a) Find the flow field if there is no variation of pressure in the axial direction and if the inner and outer cylinders have velocities v_a and v_b, respectively.
(b) Find the flow field if there is a pressure gradient in the axial direction and both cylinders are fixed.

6.13. Show that for the velocity field

$$v_1 = v(y, z), \quad v_2 = v_3 = 0$$

the Navier–Stokes equations, with $\rho\mathbf{B} = 0$, reduce to

$$\frac{\partial^2 v}{\partial y^2} + \frac{\partial^2 v}{\partial z^2} = \frac{1}{\mu}\frac{dp}{dx} = \beta = \text{constant}.$$

6.14. Referring to Problem 6.13, consider a pipe having an elliptic cross-section given by $y^2/a^2+z^2/b^2=1$. Assuming that,

$$v=A(y^2/a^2+z^2/b^2)+B,$$

find A and B.

6.15. Referring to Problem 6.13, consider an equilateral triangular cross-section defined by the planes $z+b/(2\sqrt{3})=0$, $z+\sqrt{3}y-b/\sqrt{3}=0$, $z-\sqrt{3}y-b/\sqrt{3}=0$. Assuming that,

$$v=A(z+b/(2\sqrt{3}))(z+\sqrt{3}y-b/\sqrt{3})(z-\sqrt{3}y-b/\sqrt{3})+B$$

find A and B.

6.16. Verify Eq. (6.30).

6.17. Determine the temperature distribution in plane Poiseuille flow where the bottom plate is kept at a constant temperature θ_1 and the top plate at θ_2. Include the heat generated by viscous dissipation.

6.18. Determine the temperature distribution in the laminar flow between coaxial cylinders (Couette flow) if the temperatures at the inner and outer cylinders are kept at the constant values of θ_1 and θ_2, respectively.

6.19. Express the constitutive equation for an incompressible linearly viscous fluid in cylindrical coordinates (use the result of Problem 3.48).

6.20. Obtain Navier–Stokes equations in cylindrical coordinates given in Section 6.6 by using the results of Problem 6.19, Problem 3.46, and Problem 4.26.

6.21. Given the velocity field of a linearly viscous fluid

$$v_1=kx_1,\quad v_2=-kx_2,\quad v_3=0.$$

(a) show that the velocity field is irrotational.
(b) Find the stress tensor.
(c) Find the acceleration field.
(d) Show that the velocity field satisfies the Navier–Stokes equations by finding the pressure distribution directly from the equations. Neglect body forces. Take $p=p_0$ at the origin.
(e) Use the Bernoulli equation to find the pressure distribution.
(f) Find the rate of dissipation of mechanical energy into heat.
(g) If $x_2=0$ is a fixed boundary, what condition is not satisfied by the velocity field?

6.22. Do Problem 6.21 for the following velocity field:

$$v_1=k(x_1{}^2-x_2{}^2),\quad v_2=-2kx_1x_2,\quad v_3=0.$$

6.23. Obtain the vorticity components for the plane Poiseuille flow.

6.24. Obtain the vorticity components for the Hagen–Poiseuille flow (use the result of Problem 3.49).

6.25. For two-dimensional flow of an incompressible fluid, we can express the velocity components in terms of a scalar function ψ (known as the Lagrange stream function) by the relation

$$v_1 = \frac{\partial \psi}{\partial y}, \quad v_2 = -\frac{\partial \psi}{\partial x}.$$

(a) Show that the equation of conservation of mass is automatically satisfied for any $\psi(x, y)$ which has continuous second partial derivatives.

(b) A streamline is defined to be a line whose tangent at every point coincides with the instantaneous directions of the fluid velocity at that point. Show that for two-dimensional flow of an incompressible fluid $\psi =$ constant are streamlines, where ψ is the Lagrange stream function.

(c) If the velocity field is irrotational, then $v_i = -\partial\phi/\partial x_i$. Show that the curves $\phi =$ constant and $\psi =$ constant are orthogonal to each other.

(d) Obtain the only nonzero vorticity component in terms of ψ.

6.26. Show that

$$\psi = V_0 y\left(1 - \frac{a^2}{x^2+y^2}\right)$$

represents a two-dimensional irrotational flow of an inviscid fluid. Sketch the stream lines in the region $x^2 + y^2 \geq a^2$.

6.27. Referring to Fig. P6.3, compute the maximum possible flow of water. Take the atmospheric pressure to be 93.1 kPa (13.5 lb/in²), the specific weight of water 9800 N/m³ (62.4 lb/ft³), and vapor pressure 17.2 kPa (2.5 lb/in²). Assume the fluid to be inviscid. Find l for this rate of discharge.

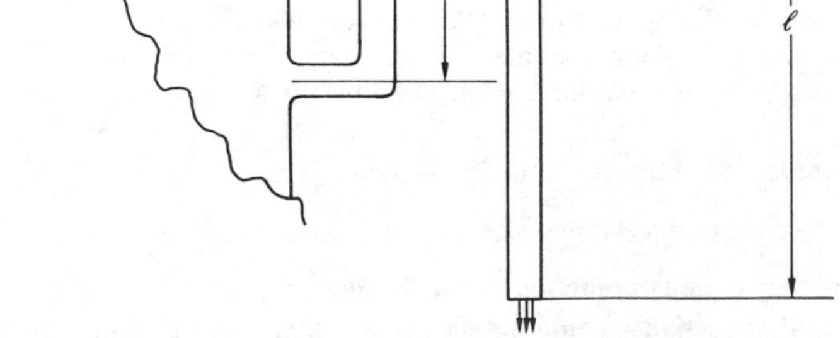

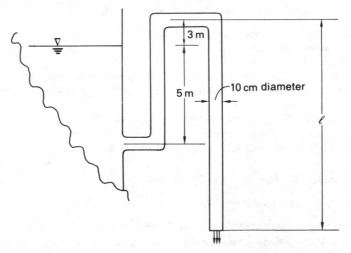

Fig. P6.3

6.28 Water flows upward through a vertical pipeline which tapers from ten-inch (25.4 cm) to six-inch (15.2 cm) diameter in a distance of 6 ft (1.83 m). If the pressures at the beginning and end of the constriction are 30 lb/in² (207 kPa) and 25 lb/in² (172 kPa), respectively, what is the flow rate? Assume the fluid to be inviscid.

6.29. Verify that the equation of conservation of mass is automatically satisfied if the velocity components are given by

$$v_1 = \frac{1}{\rho}\frac{\partial \psi}{\partial y}, \quad v_2 = -\frac{1}{\rho}\frac{\partial \psi}{\partial x}, \quad v_3 = 0,$$

where ψ is any function of x and y having continuous second partial derivatives.

6.30. Verify that the equation of conservation of mass is automatically satisfied if the velocity components in cylindrical coordinates are given by

$$v_r = -\frac{1}{\rho r}\frac{\partial \psi}{\partial z}, \quad v_z = \frac{1}{\rho r}\frac{\partial \psi}{\partial r}, \quad v_\phi = 0,$$

where $\psi(r, z)$ is any function of r and z having continuous second partial derivatives (*see* Problem 3.50).

6.31. Consider the following velocity field in cylindrical coordinates:

$$v_r = v(r), \quad v_\phi = v_z = 0.$$

(a) Show that $v(r) = A/(\rho r)$, where A is a constant (*see* Problem 3.50) so that the equation of conservation of mass is satisfied.
(b) If the rate of mass flow through a circular cylindrical surface of radius r and unit length is Q_m, determine the constant A in terms of Q_m.

6.32. Derive Eq. (6.78).

6.33. Show that for a compressible Newtonian fluid

$$T_{ij}\frac{\partial v_i}{\partial x_j} = -p\Delta + \Phi,$$

where $\Phi = \lambda\Delta^2 + \Phi_{\text{inc}}$.

6.34. Consider the following velocity and density field for a compressible Newtonian fluid:

$$v_1 = v(y), \quad v_2 = v_3 = 0, \quad \rho = \rho(y).$$

(a) Verify that the equation of conservation of mass is satisfied.
(b) If viscosity is assumed to be a constant, independent of temperature, find $v(y)$ for plane Couette motion. Neglect body forces.
(c) Use the energy equation to find the enthalpy distribution if the temperatures at the fixed and moving plates are θ_1 and θ_2, respectively. Take $H = c_p\theta$ and assume μ, c_p, and κ are all constants independent of temperature. Include viscous dissipation.

6.35. Show that for one-dimensional, steady, adiabatic flow of an ideal gas, the ratio of temperature θ_1/θ_2 at sections (1) and (2) is given by

$$\frac{\theta_2}{\theta_1}=\frac{1+\frac{1}{2}(\gamma-1)M_1^2}{1+\frac{1}{2}(\gamma-1)M_2^2},$$

where γ is the ratio of specific heat, M_1 and M_2 are local Mach numbers at 1 and 2, respectively.

6.36. Show that for a compressible fluid in isothermal flow with no external work,

$$\frac{dM^2}{M^2}=2\frac{dv}{v},$$

where M is the Mach number. (Assume perfect gas.)

6.37. Show that for a perfect gas flowing through a constant area duct at constant temperature conditions,

$$\frac{dp}{p}=-\frac{1}{2}\frac{dM^2}{M^2}.$$

6.38. For the flow of a compressible inviscid fluid around a thin body in a uniform stream of speed V_0 in the x_1-direction, we let the velocity potential be

$$\Phi=-V_0(x_1+\Phi_1),$$

where Φ_1 is assumed to be very small. Show that for steady flow the equation governing Φ_1 is, with $M_0\equiv V_0/c_0$,

$$(1-M_0^2)\frac{\partial^2\Phi_1}{\partial x_1^2}+\frac{\partial^2\Phi_1}{\partial x_2^2}+\frac{\partial^2\Phi_1}{\partial x_3^2}=0,$$

(*see* Example 6.25).

7

Integral Formulation of General Principles

In Sections 3.9, 4.7, 4.4, and 6.12, the field equations expressing the principle of conservation of mass, of linear momentum, of moment of momentum, and of energy were derived by the consideration of differential elements in the continuum [Eqs. (3.29), (4.16), (4.5), and (6.42)]. In the form of differential equations, the principles are sometimes referred to as "local principles." In this chapter, we shall formulate the principles in terms of an arbitrary fixed part of the continuum. The principles are then in integral form, which is sometimes referred to as the global principles. Under the assumption of smoothness of functions involved, the two forms are completely equivalent and in fact the requirement that the global theorem be valid for each and every part of the continuum results in the differential form of the balance equations.

The purpose of the present chapter is twofold: (1) To provide an alternate approach to the formulation of field equations expressing the general principles, and (2) to apply the global theorems to obtain approximate solutions of some engineering problems, using the concept of control volumes, moving or fixed.

We shall begin by proving Green's theorem, from which the divergence theorem will be introduced through a generalization (without proof), which we shall need later in the chapter.

7.1 GREEN'S THEOREM

Let $P(x, y)$, $\partial P/\partial x$ and $\partial P/\partial y$ be continuous functions of x and y in a closed region R bounded by C. Let $\mathbf{n} = n_x\mathbf{e}_1 + n_y\mathbf{e}_2$ be the unit outward normal of C. Then **Green's theorem**† states that

$$\int_C P n_x \, ds = \int_R \frac{\partial P}{\partial x} \, dA \tag{7.1}$$

and

$$\int_C P n_y ds = \int_R \frac{\partial P}{\partial y} \, dA. \tag{7.2}$$

For the proof, let us assume for simplicity that the region R is such that every straight line through an interior point and parallel to either axis cuts the boundary in exactly two points. Figure 7.1 shows one such region. Let

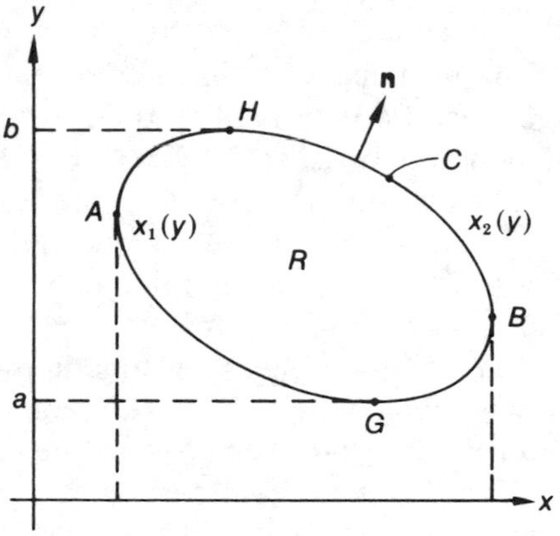

Fig. 7.1

a and b be the least and the greatest values of y on C (points G and H in the figure). Let $x = x_1(y)$ and $x = x_2(y)$ be equations for the boundaries HAG and GBH, respectively. Then

$$\int_R \frac{\partial P}{\partial x} \, dA = \int_a^b \left[\int_{x_1(y)}^{x_2(y)} \frac{\partial P}{\partial x} \, dx \right] dy.$$

Now

$$\int_{x_1(y)}^{x_2(y)} \frac{\partial P}{\partial x} \, dx = P(x, y) \Big]_{x_1(y)}^{x_2(y)} = P[x_2(y), y] - P[x_1(y), y].$$

†The theorem is valid under less restrictive conditions on the first partial derivative.

Thus

$$\int_R \frac{\partial P}{\partial x}\,dA = \int_a^b P[x_2(y), y]\,dy - \int_a^b P[x_1(y), y]\,dy$$

$$= \int_{GBH} P\,dy - \int_{GAH} P\,dy.$$

Since

$$\int_{GAH} P\,dy = -\int_{HAG} P\,dy.$$

Thus

$$\int_R \frac{\partial P}{\partial x}\,dA = \int_{GBH} P\,dy + \int_{HAG} P\,dy.$$

That is,

$$\int_R \frac{\partial P}{\partial x}\,dA = \oint P\,dy = -\oint P\,dy.$$

Let s be the arc length measured along the boundary C and let $x = x(s)$ and $y = y(s)$ be the parametric equations for the boundary curve. Then, $dy/ds = +n_x$ if s is measured along the curve in the counterclockwise direction and $dy/ds = -n_x$ if s is measured along the curve in the clockwise direction. Thus

$$\int_R \frac{\partial P}{\partial x}\,dA = \int_C P n_x\,ds.$$

Equation (7.2) can be proved in a similar manner.

EXAMPLE 7.1

For $P(x, y) = xy^2$, evaluate $\int_C P(x, y)\, n_x ds$ along the closed path $OABC$ (Fig. 7.2). Also evaluate the area integral $\int_R (\partial P/\partial x)\, dA$. Compare the results.

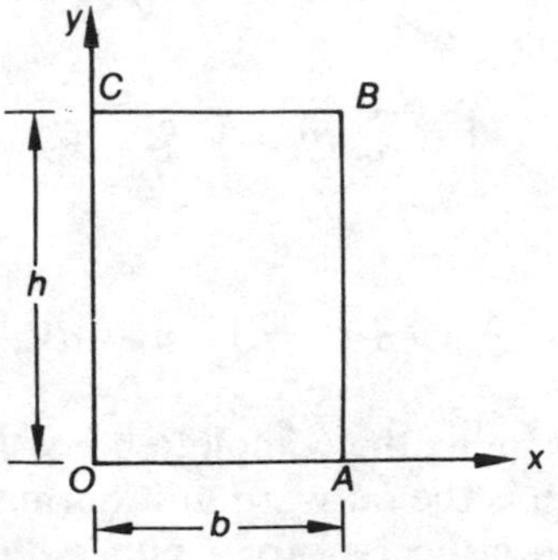

Fig. 7.2

Solution. We have,

$$\int_C P(x,y)n_x ds = \int_{OA} x(0)^2(0)\,ds + \int_{AB} by^2(1)\,dy + \int_{BC} xh^2(0)\,ds$$

$$+ \int_{CO} (0)y^2(-1)\,ds = \int_0^h by^2 dy = \frac{bh^3}{3}.$$

On the other hand,

$$\int_R \frac{\partial P}{\partial x} dA = \int_R y^2\, dA = \int_0^h y^2 b\, dy = \frac{bh^3}{3}.$$

Thus, we see

$$\int_C P n_x\, ds = \int_A \frac{\partial P}{\partial x} dA.$$

7.2 DIVERGENCE THEOREM

Let $\mathbf{v} = v_1(x,y)\mathbf{e}_1 + v_2(x,y)\mathbf{e}_2$ be a vector field. Applying Eqs. (7.1) and (7.2) to v_1 and v_2 and adding, we have

$$\int_C (v_1 n_1 + v_2 n_2)\, ds = \int_R \left(\frac{\partial v_1}{\partial x} + \frac{\partial v_2}{\partial y}\right) dA. \tag{7.3a}$$

In indicial notation, Eq. (7.3a) becomes

$$\int_C v_i n_i\, ds = \int_R \frac{\partial v_i}{\partial x_i} dA \tag{7.3b}$$

and in invariant notation,

$$\int_C \mathbf{v}\cdot\mathbf{n}\, ds = \int_R \operatorname{div}\mathbf{v}\, dA. \tag{7.3c}$$

The following generalization not only appears natural, but can indeed be proved (we omit proof)

$$\int_S v_i n_i\, dS = \int_R \frac{\partial v_i}{\partial x_i} dV, \tag{7.4a}$$

or, in invariant notation,

$$\int_S \mathbf{v}\cdot\mathbf{n}\, dS = \int_R \operatorname{div}\mathbf{v}\, dV, \tag{7.4b}$$

where S is a surface forming the complete boundary of a bounded closed region R in space and $\mathbf{n}$ is the outward unit normal of S. Equation (7.4) is known as the **divergence theorem** (or **Gauss's theorem**). The theorem is valid if the components of $\mathbf{v}$ are continuous and have continuous first partial derivatives in R. It is also valid under less restrictive conditions on the derivatives.

Next, if T_{ij} are components of a tensor $\mathbf{T}$, then the application of Eq. (7.4a) gives

$$\int_S T_{ij} n_j \, dS = \int_R \frac{\partial T_{ij}}{\partial x_j} dV \tag{7.5a}$$

or, in invariant notation,

$$\int_S \mathbf{T}\mathbf{n} \, dS = \int_R \operatorname{div} \mathbf{T} \, dV, \tag{7.5b}$$

where the Cartesian components of div $\mathbf{T}$ are, by definition, $\partial T_{ij}/\partial x_j$.

Equation (7.5) is the divergence theorem for a tensor field. It is obvious that for tensor fields of higher order, Eq. (7.5b) is also valid provided the Cartesian components of div $\mathbf{T}$ are defined to be $\partial T_{ij\ldots l}/\partial x_l$.

EXAMPLE 7.2

Let $\mathbf{T}$ be a stress tensor field and let S be a closed surface. Show that the resultant force of the distributive forces on S is given by $\int_V (\operatorname{div} \mathbf{T}) dV$.

Solution. Let $\mathbf{f}$ be the resultant force, then

$$\mathbf{f} = \int_S \mathbf{t} \, dS,$$

where $\mathbf{t}$ is the stress vector. But $\mathbf{t} = \mathbf{T}\mathbf{n}$, therefore from the divergence theorem, we have

$$\mathbf{f} = \int_S \mathbf{t} \, dS = \int_S \mathbf{T}\mathbf{n} \, dS = \int_V (\operatorname{div} \mathbf{T}) \, dV, \tag{7.6a}$$

i.e.,

$$f_i = \int_V \frac{\partial T_{ij}}{\partial x_j} dV. \tag{7.6b}$$

EXAMPLE 7.3

Referring to Example 7.2, also show that the resultant moment, about a fixed point O, of the distributive forces on S is given by

$$\int \{\mathbf{x} \times (\operatorname{div} \mathbf{T}) + 2\mathbf{t}^A\} dV,$$

where $\mathbf{x}$ is the position vector from the fixed point O and $\mathbf{t}^A$ is the dual vector of the antisymmetric part of $\mathbf{T}$ (*see* Section 2B11, Chapter 2 for definition of the vector of an antisymmetric tensor).

Solution. Let $\mathbf{m}$ denote the resultant moment about O. Then

$$\mathbf{m} = \int_S \mathbf{x} \times \mathbf{t} \, dS.$$

Let m_i be the components of $\mathbf{m}$, then

$$m_i = \int_S \epsilon_{ijk} x_j t_k \, dS = \int_S \epsilon_{ijk} x_j T_{kp} n_p \, dS.$$

Using the divergence theorem, Eq. (7.5a), we have

$$m_i = \int_V \frac{\partial}{\partial x_p}(\epsilon_{ijk}x_j T_{kp})\,dV.$$

Now

$$\frac{\partial}{\partial x_p}(\epsilon_{ijk}x_j T_{kp}) = \epsilon_{ijk}\left(\frac{\partial x_j}{\partial x_p}T_{kp} + x_j\frac{\partial T_{kp}}{\partial x_p}\right)$$

$$= \epsilon_{ijk}\left(\delta_{jp}T_{kp} + x_j\frac{\partial T_{kp}}{\partial x_p}\right) = \epsilon_{ipk}T_{kp} + \epsilon_{ijk}x_j\frac{\partial T_{kp}}{\partial x_p}.$$

Noting that $\epsilon_{ipk}T_{kp} = -\epsilon_{ikp}T_{kp}$ are components of twice the dual vector of the antisymmetric part of **T**, and $\epsilon_{ijk}x_j(\partial T_{kp}/\partial x_p)$ are components of $\mathbf{x}\times(\text{div}\,\mathbf{T})$, we have

$$\mathbf{m} = \int_S \mathbf{x}\times\mathbf{t}\,dS = \int \{\mathbf{x}\times(\text{div}\,\mathbf{T}) + 2\mathbf{t}^A\}\,dV. \tag{7.7}$$

EXAMPLE 7.4

Referring to Example 7.2, show that the total power (rate of work done) by the stress vector on S is given by, with $\text{tr}\,(\mathbf{A}) = A_{ii}$,

$$\int_V \{(\text{div}\,\mathbf{T})\cdot\mathbf{v} + \text{tr}\,(\mathbf{T}^T\nabla\mathbf{v})\}\,dV,$$

where **v** is the velocity field.

Solution. Let P be the total power, then

$$P = \int_S \mathbf{t}\cdot\mathbf{v}\,dS = \int_S \mathbf{Tn}\cdot\mathbf{v}\,dS.$$

But $\mathbf{Tn}\cdot\mathbf{v} = \mathbf{n}\cdot\mathbf{T}^T\mathbf{v}$ by definition of transpose. Thus,

$$P = \int_S \mathbf{n}\cdot(\mathbf{T}^T\mathbf{v})\,dS.$$

Application of the divergence theorem gives

$$P = \int_V \text{div}\,(\mathbf{T}^T\mathbf{v})\,dV.$$

Now

$$\text{div}\,(\mathbf{T}^T\mathbf{v}) = \frac{\partial(T_{ji}v_j)}{\partial x_i} = \frac{\partial T_{ji}}{\partial x_i}v_j + T_{ji}\frac{\partial v_j}{\partial x_i}$$

$$= (\text{div}\,\mathbf{T})\cdot\mathbf{v} + \text{tr}\,(\mathbf{T}^T\nabla\mathbf{v}). \tag{7.8}†$$

Thus

$$P = \int_S \mathbf{t}\cdot\mathbf{v}\,dS = \int_V \{(\text{div}\,\mathbf{T})\cdot\mathbf{v} + \text{tr}\,(\mathbf{T}^T\nabla\mathbf{v})\}\,dV.$$

7.3 INTEGRALS OVER A CONTROL VOLUME AND INTEGRALS OVER A MATERIAL VOLUME

Consider first a one-dimensional problem in which the motion is given by (with arbitrary reference time t_0)

$$\begin{aligned} x_1 &= x_1(X_1, t) \qquad \text{with } X_1 = x_1(X_1, t_0),\\ x_2 &= X_2,\\ x_3 &= X_3, \end{aligned} \tag{7.9}$$

†*See also* the definition of div **T** in Section 2B20, Eq. (B59).

and the density field is given by

$$\rho = \rho(x_1, t). \tag{7.10}$$

The integral

$$m(t, a, b) = \int_{x_1=a}^{x_1=b} \rho(x_1, t)(A\,dx_1) \tag{7.11}$$

gives the total mass at time t within the spatially fixed cylindrical volume of constant cross-sectional area A and bounded by the end faces $x_1 = a$ and $x_1 = b$.

On the other hand, if $X_1 = a$ and $X_1 = b$ are those particles which at time t_0 are at $x_1 = a$ and $x_1 = b$, respectively, i.e.,

$$x_1(a, t_0) = a,$$
$$x_1(b, t_0) = b, \tag{7.12}$$

then the integral

$$M(t, a, b) = \int_{x_1(a,t)}^{x_1(b,t)} \rho(x_1, t)A\,dx_1 \tag{7.13}$$

gives the total mass at time t of that part of material which at time t_0 is instantaneously coincidental with that inside the fixed boundary surface considered in m, i.e.,

$$m(t_0, a, b) = M(t_0, a, b). \tag{7.14}$$

But at other times $m \neq M$, and in particular,

$$m(t_0 + dt, a, b) \neq M(t_0 + dt, a, b).$$

In other words, at time $t = t_0$

$$\left(\frac{\partial m}{\partial t}\right) \neq \left(\frac{\partial M}{\partial t}\right),$$

as is obvious from the fact that the limits of the integral in Eq. (7.13) are time-dependent. Physically $\partial m/\partial t$ gives the rate at which mass is increasing inside the fixed volume (known as **control volume**) bounded by the cylindrical lateral surface and the end faces $x = a$ and $x = b$, whereas $\partial M/\partial t \equiv DM/Dt$ gives the mass increase rate of that part of material which at time $t = t_0$ is coincidental with that in the control volume. They should again obviously be different, since the principle of conservation of mass demands that $DM/Dt = 0$, whereas the mass within the control volume need not be so.

The above example serves to illustrate the two types of volume integrals which we shall employ in the following sections. We shall use V_c to indicate a **control volume** and V_m to indicate a **material volume**. That is, the integrals $\int_{V_c} \psi(x, t)\, dV$ and $\int_{V_m} \psi(x, t)\, dV$ are over control and material volumes, respectively.

7.4 PRINCIPLE OF CONSERVATION OF MASS

The **global principle of conservation of mass** can be formulated in the following two ways:

(i) The time rate at which mass is increasing inside a control volume = the **mass influx** (i.e., net rate of mass inflow) through the control volume, i.e.,

$$\frac{\partial}{\partial t} \int_{Vc} \rho(\mathbf{x}, t)\, dV = -\int_{Sc} (\rho \mathbf{v} \cdot \mathbf{n})\, dS, \tag{7.15}$$

where the minus sign in the right-hand side integral is due to the convention that $\mathbf{n}$ is an outward unit normal.

Now, since V_c is independent of time, therefore

$$\frac{\partial}{\partial t} \int_{Vc} \rho(\mathbf{x}, t)\, dV = \int_{Vc} \frac{\partial \rho}{\partial t}\, dV \tag{7.16}$$

and, by divergence theorem,

$$\int_{Sc} (\rho \mathbf{v} \cdot \mathbf{n})\, dS = \int_{Vc} \operatorname{div}(\rho \mathbf{v})\, dV, \tag{7.17}$$

thus, Eq. (7.15) can be written

$$\int_{Vc} \left[\frac{\partial \rho}{\partial t} + \operatorname{div}(\rho \mathbf{v})\right] dV = 0. \tag{7.18}$$

By demanding that the above equation be valid for all V_c we arrive at the equation of continuity derived before [*see* Section 3.9, Eq. (3.29)]:

$$\frac{\partial \rho}{\partial t} + \operatorname{div}(\rho \mathbf{v}) = 0. \tag{7.19}$$

(ii) The total mass of a fixed part of material should remain constant at all times. That is,

$$\frac{D}{Dt} \int_{Vm} \rho(x, t)\, dV = 0. \tag{7.20}$$

Now

$$\frac{D}{Dt} \int_{Vm} \rho(x, t)\, dV = \int_{Vm=Vc} \frac{D}{Dt} (\rho\, dV)$$

$$= \int_{Vc} \frac{D\rho}{Dt}\, dV + \int_{Vc} \rho \frac{D}{Dt} (dV).$$

Note that at a given instant t, $V_c = V_m$, and

$$\frac{1}{dV}\frac{D}{Dt}dV = \text{div } \mathbf{v}. \quad (\textit{see } \text{Eq. } 3.28)$$

therefore Eq. (7.20) can be written

$$\int_{V_c}\left(\frac{D\rho}{Dt} + \rho \text{ div } \mathbf{v}\right)dV = 0, \tag{7.21}$$

which is the same as Eq. (7.18). [Note $D\rho/Dt = (\partial\rho/\partial t) + (\nabla\rho)\cdot\mathbf{v}$.]

EXAMPLE 7.5

For the following velocity and density fields:

$$v_1 = \frac{x_1}{1+t}, \quad v_2 = v_3 = 0, \quad \rho = \frac{\rho_0}{1+t} \quad (\rho_0 = \text{constant}).$$

(a) Check that the equation of continuity is satisfied.
(b) Compute the total mass and the rate of increase of mass inside a cylindrical control volume of cross-sectional area A and having as its end faces the planes $x_1 = 1$ and $x_1 = 3$.
(c) Compute the net rate of inflow of mass into the control volume of part (b).

Solution. (a) We have

$$\frac{D\rho}{Dt} + \rho \text{ div } \mathbf{v} = \frac{\partial\rho}{\partial t} + v_1\frac{\partial\rho}{\partial x_1} + \rho\frac{\partial v_1}{\partial x_1} = -\frac{\rho_0}{(1+t)^2} + \frac{x_1}{(1+t)}(0) + \frac{\rho_0}{(1+t)^2} = 0.$$

Thus, the equation of continuity is satisfied.
(b) The total mass inside the control volume at time t is

$$m(t) = \int_{Vc}\rho(x,t)\,dV = \int_{x_1=1}^{x_1=3}\frac{\rho_0}{1+t}A\,dx_1 = \frac{A\rho_0}{1+t}(2) = \frac{2A\rho_0}{1+t}$$

and the rate at which the mass is increasing inside the control volume at time t is

$$\frac{\partial m}{\partial t} = -\frac{2A\rho_0}{(1+t)^2},$$

i.e., the mass is decreasing.
(c) There is neither inflow nor outflow through the lateral surface of the control volume. Through the end face $x_1 = 1$, the rate of inflow (mass influx) is

$$(\rho A v)_{x_1=1} = \frac{\rho_0 A}{(1+t)^2}.$$

On the other hand, the mass outflux through the end face $x_1 = 3$, is

$$(\rho A v)_{x_1=3} = \frac{3\rho_0 A}{(1+t)^2}.$$

Thus, the net mass influx is clearly

$$-\frac{2\rho_0 A}{(1+t)^2},$$

which is the same as the result of part (b).

EXAMPLE 7.6

(a) Check that the motion (note reference time t_0)

$$x_1 = \frac{1+t}{1+t_0} X_1, \quad x_2 = X_2, \quad x_3 = X_3$$

corresponds to the velocity field given in the previous example.
(b) If the density field is again given by $\rho = \rho_0/(1+t)$, find the total mass at time t of the material which at time t_0 is in the control volume of the previous example.
(c) Compute also, the total linear momentum for the fixed part of material considered in part (b).

Solution. (a) First we note that when $t = t_0$,

$$x_1 = X_1, \quad x_2 = X_2, \quad x_3 = X_3,$$

i.e., $t = t_0$ is the (arbitrary) reference time. Now

$$v_1 = \frac{Dx_1}{Dt} = \frac{X_1}{1+t_0}, \quad v_2 = \frac{Dx_2}{Dt} = 0, \quad v_3 = \frac{Dx_3}{Dt} = 0,$$

and

$$X_1 = \frac{1+t_0}{1+t} x_1,$$

therefore we have

$$v_1 = \frac{x_1}{1+t}, \quad v_2 = v_3 = 0.$$

(b) The particles which were at $x_1 = 1$ and $x_1 = 3$ when $t = t_0$ have the material coordinate $X_1 = 1$, and $X_1 = 3$, respectively. Thus, the total mass at time t is

$$M = \int_{x_1 = \frac{1+t}{1+t_0}}^{x_1 = 3\frac{1+t}{1+t_0}} \frac{\rho_0}{1+t} A\,dx_1 = \frac{\rho_0 A}{1+t}\left(\frac{3(1+t)}{1+t_0} - \frac{1+t}{1+t_0}\right) = \frac{2\rho_0 A}{1+t_0}.$$

We see that this time-dependent integral turns out to be independent of time. This is because the chosen density and velocity field satisfies the equation of continuity so that, the total mass of a fixed part of material is guaranteed to remain constant.
(c) Total linear momentum is, since $v_2 = v_3 = 0$,

$$\mathbf{P} = \int_{x_1 = \frac{1+t}{1+t_0}}^{\frac{3(1+t)}{1+t_0}} \rho v_1 A\,dx_1 \mathbf{e}_1 = \frac{A\rho_0}{(1+t)^2} \int_{\frac{1+t}{1+t_0}}^{\frac{3(1+t)}{1+t_0}} x_1\,dx_1 \mathbf{e}_1$$

$$= \frac{A\rho_0}{(1+t)^2}\left[\frac{9(1+t)^2}{2(1+t_0)^2} - \frac{(1+t)^2}{2(1+t_0)^2}\right] \mathbf{e}_1 = \frac{4A\rho_0}{(1+t_0)^2} \mathbf{e}_1.$$

The fact that $\mathbf{P}$ is also a constant is accidental. This happens because the given motion happens to be accelerationless, which corresponds to no net force acting on the material volume. In general, the linear momentum for a fixed part of material is a function of time.

7.5 PRINCIPLE OF LINEAR MOMENTUM

The **global principle of linear momentum** states that the total force (surface and body forces) acting on any fixed part of material is equal to the rate of change of linear momentum of the part. That is, with ρ denoting density, $\mathbf{v}$ velocity, $\mathbf{t}$ stress vector, and $\mathbf{B}$ body force per unit mass, the theorem states

$$\frac{D}{Dt}\int_{V_m} \rho \mathbf{v} dV = \int_{S_c} \mathbf{t} dS + \int_{V_c} \rho \mathbf{B} dV. \tag{7.22}$$

Note again that at a given instant t, V_c is the same as V_m. Now, since $(D/Dt)(\rho dV) = 0$, therefore

$$\frac{D}{Dt}\int_{V_m} \rho \mathbf{v} dV = \int_{V_m = V_c} \frac{D}{Dt}(\rho \mathbf{v} dV) = \int_{V_c} \frac{D\mathbf{v}}{Dt}\rho dV. \tag{7.23}$$

Making use of Eq. (7.23) and converting the surface integral in Eq. (7.22) into a volume integral (*see* Example 7.2), Eq. (7.22) becomes

$$\int_{V_c}\left(\rho\frac{D\mathbf{v}}{Dt} - \operatorname{div}\mathbf{T} - \rho\mathbf{B}\right) dV = 0, \tag{7.24}$$

from which, the following field equation of motion is again rediscovered:

$$\rho\frac{D\mathbf{v}}{Dt} = \operatorname{div}\mathbf{T} + \rho\mathbf{B}. \tag{7.25}$$

On the other hand, we also have

$$\begin{aligned}
\frac{D}{Dt}\int_{V_m} \rho v_i dV &= \int_{V_c}\frac{D(\rho v_i)}{Dt} dV + \int_{V_c} \rho v_i\left[\frac{D}{Dt}(dV)\right] \\
&= \int_{V_c}\left\{\frac{\partial(\rho v_i)}{\partial t} + v_j\frac{\partial(\rho v_i)}{\partial x_j} + \rho v_i\frac{\partial v_j}{\partial x_j}\right\} dV \\
&= \int_{V_c}\frac{\partial(\rho v_i)}{\partial t} dV + \int_{V_c}\frac{\partial}{\partial x_j}(\rho v_i v_j) dV.
\end{aligned} \tag{7.26}$$

But, by divergence theorem,

$$\int_{V_c}\frac{\partial}{\partial x_j}(\rho v_i v_j) dV = \int_{S_c}\rho v_i v_j n_j dS = \int_{S_c}\rho v_i(\mathbf{v}\cdot\mathbf{n}) dS. \tag{7.27}$$

Thus

$$\frac{D}{Dt}\int_{V_m}\rho\mathbf{v} dV = \int_{V_c}\frac{\partial(\rho\mathbf{v})}{\partial t} dV + \int_{S_c}\rho\mathbf{v}(\mathbf{v}\cdot\mathbf{n}) dS. \tag{7.28}$$

Equation (7.28) is sometimes referred to as **Reynolds transport theorem** for linear momentum. Using this result, the momentum principle takes the form of the following **balance equation**:

$$\int_{S_c} \mathbf{t}\,dS + \int_{V_c} \rho \mathbf{B}\,dV = \frac{\partial}{\partial t}\int_{V_c} \rho \mathbf{v}\,dV + \int_{S_c} \rho \mathbf{v}(\mathbf{v}\cdot\mathbf{n})\,dS. \qquad (7.29)$$

In words, it states that the **total force exerted on a fixed part of material instantaneously in a control volume V_c = time rate of change of total linear momentum inside the control volume + net outflux of linear momentum through the control surface S_c**. Equation (7.29) is very useful for obtaining approximate results in many engineering problems.

EXAMPLE 7.7

A homogeneous rope of total length l and total mass m slides down from the corner of a smooth table. Find the motion of the rope and the tension at the corner.

Solution. Let x denote the portion of rope that has slid down the corner at time t. Then the portion that remains on the table at time t is $l-x$. Consider the control volume shown as $(V_c)_1$ in Figure 7.3. Then the momentum in the horizontal direction inside the control volume at any time t is, with $\dot{x}$ denoting dx/dt,

$$\frac{m}{l}(l-x)\dot{x}$$

and the net momentum outflux is $((m/l)\dot{x})\dot{x}$. Thus if T denotes the tension at the cornerpoint of the rope at time t, we have

$$T = \frac{d}{dt}\left\{\frac{m}{l}(l-x)\dot{x}\right\} + \frac{m}{l}\dot{x}^2 = \frac{m}{l}(-\dot{x})\dot{x} + \frac{m}{l}(l-x)\ddot{x} + \frac{m}{l}\dot{x}^2,$$

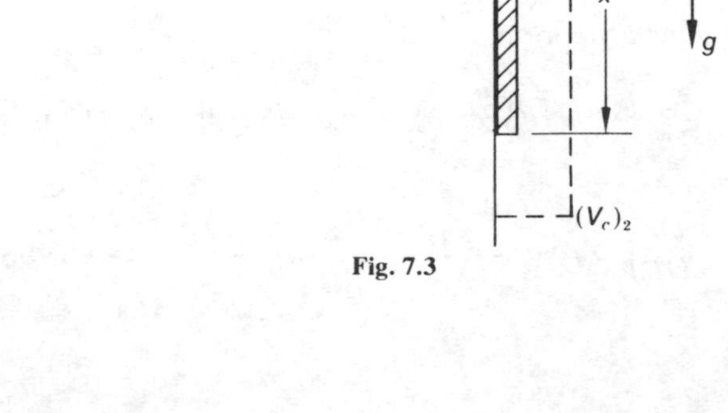

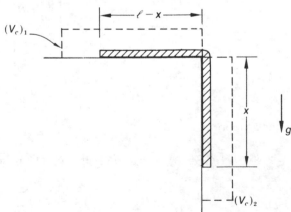

Fig. 7.3

i.e.,

$$T = \frac{m}{l}(l-x)\ddot{x}, \tag{i}$$

as expected.

On the other hand, by considering the control volume $(V_c)_2$ (*see* Fig. 7.3), we have, the momentum in the downward direction $((m/l)x)\dot{x}$ and the momentum influx in the same direction $((m/l)\dot{x})\dot{x}$. Thus

$$-T + \left(\frac{m}{l}x\right)g = \frac{d}{dt}\left(\frac{m}{l}x\dot{x}\right) - \frac{m}{l}\dot{x}^2.$$

i.e.,

$$-T + \frac{m}{l}xg = \frac{m}{l}x\ddot{x}. \tag{ii}$$

From Eqs. (i) and (ii), we have

$$\frac{m}{l}(l-x)\ddot{x} = \frac{m}{l}xg - \frac{m}{l}x\ddot{x},$$

i.e.,

$$\ddot{x} - \frac{g}{l}x = 0. \tag{iii}$$

The general solution of Eq. (iii) is

$$x = C_1 e^{\sqrt{(g/l)}t} + C_2 e^{-\sqrt{(g/l)}t}.$$

Thus, if the rope starts from rest with an initial overhang of x_0, we have

$$x_0 = C_1 + C_2,$$

$$0 = C_1 - C_2,$$

which gives $C_1 = C_2 = x_0/2$, and the solution is

$$x = \frac{x_0}{2}\left(e^{\sqrt{(g/l)}t} + e^{-\sqrt{(g/l)}t}\right).$$

The tension at the corner is given by

$$T = \frac{m}{l}(l-x)\ddot{x} = \frac{m}{l}(l-x)\left(\frac{g}{l}x\right).$$

We note that the motion can also be obtained by considering the whole rope as a system. In fact, the total linear momentum of the rope at any time t is

$$\frac{m}{l}(l-x)\dot{x}\mathbf{e}_1 + \frac{m}{l}x\dot{x}\mathbf{e}_2,$$

its rate of change is

$$\frac{m}{l}[(l-x)\ddot{x} - \dot{x}^2]\mathbf{e}_1 + \frac{m}{l}(x\ddot{x} + \dot{x}^2)\mathbf{e}_2,$$

and the total resultant force on the rope is

$$\frac{m}{l}xg\mathbf{e}_2.$$

Thus, equating the force to the rate of change of momentum for the whole rope, we obtain

$$(l-x)\ddot{x} - \dot{x}^2 = 0,$$

and

$$x\ddot{x}+\dot{x}^2 = gx.$$

Eliminating $\dot{x}^2$ from the above two equations, we arrive at Eq. (iii) again.

EXAMPLE 7.8

Figure 7.4 shows a steady jet of water impinging onto a curved vane in a tangential direction. Neglect the effect of weight and assume that the upstream and downstream regions such as A and B have uniform speed v_0. Find the resultant force (above that due to the atmospheric pressure) exerted on the vane by the jet. The volume flow rate is Q.

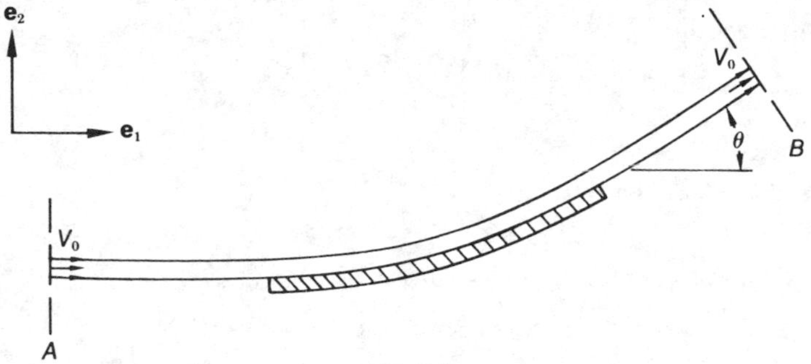

Fig. 7.4

Solution. Let us take as control volume that portion of the jet bounded by the planes at A and B. If we assume that the flow near A as well as B is uniform with speed v_0, then the stress vector on the plane A as well as B is normal to the plane with a magnitude equal to the atmospheric pressure which we take to be zero. Thus the only forces acting on the material in the control volume is that from the vane to the jet. Let **F** be the resultant of these forces. Since the flow is steady, the rate of increase of momentum inside the control volume is zero. The rate of outflow of linear momentum across B is $\rho Q(v_0 \cos\theta\, \mathbf{e}_1 + v_0 \sin\theta\, \mathbf{e}_2)$ and the rate of inflow of linear momentum across A is $\rho Q(v_0\mathbf{e}_1)$. Thus

$$\mathbf{F} = \rho Q[v_0(\cos\theta - 1)\mathbf{e}_1 + v_0 \sin\theta\, \mathbf{e}_2].$$

That is,

$$F_x = -\rho Q v_0(1-\cos\theta),$$

$$F_y = \rho Q v_0 \sin\theta,$$

and the force components on the vane by the jet are equal and opposite to F_x and F_y.

EXAMPLE 7.9

For boundary layer flow of water over a flat plate, if the velocity profile and that of the horizontal components at the leading and the trailing edges of the plate respectively assumed to be those shown in Fig. 7.5, find the shear force acting on the fluid by the plate. Assume that the flow is steady and that the pressure is uniform in the whole flow.

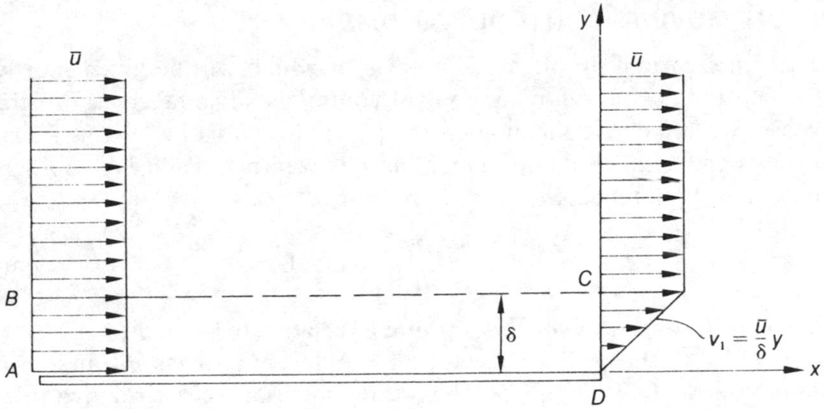

Fig. 7.5

Solution. Consider the control volume $ABCD$. Since the pressure is assumed to be uniform and since the flow outside of the boundary layer δ is essentially uniform in horizontal velocity components with very small vertical velocity components (so that the shearing stress on BC is negligible), therefore the net force on the control volume is the shearing force from the plate. Denoting this force (per unit width in z-direction) by $F\mathbf{e}_1$, we have from the momentum principle

$$F = \text{net outflux of } x\text{-momentum through } ABCD,$$

i.e.,

$$F = \int_{S_c} v_1(\rho\mathbf{v}\cdot\mathbf{n})dS = -\int_0^\delta \bar{u}(\rho\bar{u})dy + \int_{BC} \bar{u}(\rho\mathbf{v}\cdot\mathbf{n})dS + \int_0^\delta \rho\left(\frac{\bar{u}^2y^2}{\delta^2}\right)dy + \int_{AD}(0)dS.$$

Thus

$$F = -\rho\bar{u}^2\delta + \frac{\rho\bar{u}^2\delta}{3} + \bar{u}\int_{BC}(\rho\mathbf{v}\cdot\mathbf{n})dS.$$

From principle of conservation of mass, we have

$$\int_{BC}(\rho\mathbf{v}\cdot\mathbf{n})dS - \int_0^\delta \rho\bar{u}\,dy + \int_0^\delta \rho\frac{\bar{u}}{\delta}y\,dy = 0,$$

i.e.,

$$\int_{BC}(\rho\mathbf{v}\cdot\mathbf{n})dS = \rho\bar{u}\delta - \frac{\rho\bar{u}\delta}{2} = \frac{\rho\bar{u}\delta}{2}.$$

Thus

$$F = -\rho\bar{u}^2\delta + \frac{\rho\bar{u}^2\delta}{3} + \frac{\rho\bar{u}^2\delta}{2} = -\frac{\rho\bar{u}^2\delta}{6}.$$

That is, the force per unit width on the fluid by the plate is acting to the left with a magnitude of $\rho\bar{u}^2\delta/6$.

7.6 ON MOVING CONTROL VOLUME

There are certain problems, examples of which will be given shortly, for which the use of a moving control volume is advantageous. For this purpose, we first derive the momentum principle valid for a moving frame.

From a course in rigid body mechanics, we learned that if F_1 and F_2 are two frames of reference, then for any vector $\mathbf{b}$,

$$\left(\frac{d\mathbf{b}}{dt}\right)_{F_1} = \left(\frac{d\mathbf{b}}{dt}\right)_{F_2} + \boldsymbol{\omega} \times \mathbf{b}, \tag{7.30}$$

where $\boldsymbol{\omega}$ is the angular velocity of frame F_2 relative to frame F_1.

Let $\mathbf{r}$ denote the position vector of a differential mass dm in a continuum relative to F_1, and let $\mathbf{x}$ denote the position vector relative to F_2 (*see* Fig. 7.6). Then the velocity of (dm) relative to F_1 is $(d\mathbf{r}/dt)_{F_1}$ and the velocity relative to F_2 is $(d\mathbf{x}/dt)_{F_2}$. Since

$$\mathbf{r} = \mathbf{R}_0 + \mathbf{x}, \tag{7.31}$$

thus

$$\left(\frac{d\mathbf{r}}{dt}\right)_{F_1} = \left(\frac{d\mathbf{R}_0}{dt}\right)_{F_1} + \left(\frac{d\mathbf{x}}{dt}\right)_{F_1}. \tag{7.32}$$

Thus

$$\mathbf{v}_{F_1} = (\mathbf{v}_0)_{F_1} + (\mathbf{v})_{F_2} + \boldsymbol{\omega} \times \mathbf{x}. \tag{7.33}$$

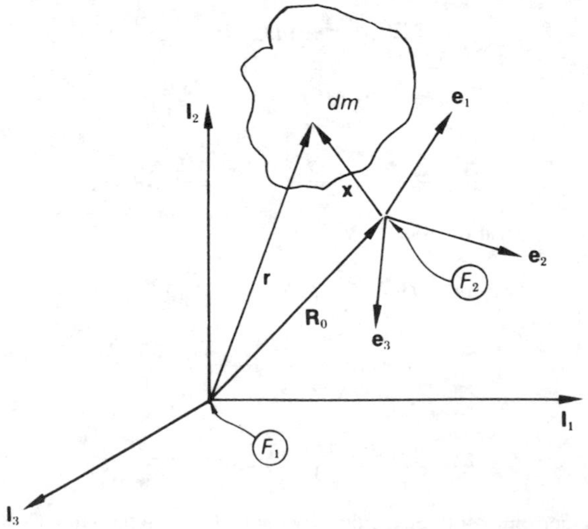

Fig. 7.6

Now, the linear momentum relative to F_1 is $\int \mathbf{v}_{F_1}\, dm$ and that relative to F_2 is $\int \mathbf{v}_{F_2}\, dm$. The rates of change of linear momentum are related in the following way:†

$$\left(\frac{D}{Dt}\right)_{F_1} \int \mathbf{v}_{F_1}\, dm = \left(\frac{D}{Dt}\right)_{F_1} \left[(\mathbf{v}_0)_{F_1} \int dm + \int \mathbf{v}_{F_2}\, dm + \boldsymbol{\omega} \times \int \mathbf{x}\, dm \right]$$
$$= (\mathbf{a}_0)_{F_1} \int dm + \left(\frac{D}{Dt}\right)_{F_1} \int \mathbf{v}_{F_2}\, dm + \left(\frac{D}{Dt}\right)_{F_1} \left\{ \boldsymbol{\omega} \times \int \mathbf{x}\, dm \right\}. \tag{7.34}$$

Now

$$\left(\frac{D}{Dt}\right)_{F_1} \int \mathbf{v}_{F_2}\, dm = \left(\frac{D}{Dt}\right)_{F_2} \int \mathbf{v}_{F_2}\, dm + \boldsymbol{\omega} \times \int \mathbf{v}_{F_2}\, dm \tag{7.35}$$

and

$$\left(\frac{D}{Dt}\right)_{F_1} \left\{ \boldsymbol{\omega} \times \int \mathbf{x}\, dm \right\} = \dot{\boldsymbol{\omega}} \times \int \mathbf{x}\, dm + \boldsymbol{\omega} \times \left(\frac{D}{Dt}\right)_{F_1} \int \mathbf{x}\, dm$$
$$= \dot{\boldsymbol{\omega}} \times \int \mathbf{x}\, dm + \boldsymbol{\omega} \times \int \mathbf{v}_{F_2}\, dm + \boldsymbol{\omega} \times \left(\boldsymbol{\omega} \times \int \mathbf{x}\, dm \right). \tag{7.36}$$

Thus

$$\left(\frac{D}{Dt}\right)_{F_1} \int \mathbf{v}_{F_1}\, dm = \left(\frac{D}{Dt}\right)_{F_2} \int \mathbf{v}_{F_2}\, dm + (\mathbf{a}_0)_{F_1} \int dm + 2\boldsymbol{\omega} \times \int \mathbf{v}_{F_2}\, dm$$
$$+ \dot{\boldsymbol{\omega}} \times \int \mathbf{x}\, dm + \boldsymbol{\omega} \times \left(\boldsymbol{\omega} \times \int \mathbf{x}\, dm \right). \tag{7.37}$$

Now, let F_1 be an inertial frame so that momentum principle reads

$$\left(\frac{D}{Dt}\right)_{F_1} \int \mathbf{v}_{F_1}\, dm = \int \mathbf{t}\, dS + \int \rho \mathbf{B}\, dV. \tag{7.38}$$

Using Eq. (7.37), the momentum principle [Eq. (7.38)] becomes

$$\left(\frac{D}{Dt}\right)_{F_2} \int_{Vm} \mathbf{v}_{F_2} dm = \int_{Sc} \mathbf{t} dS + \int_{Vc} \rho \mathbf{B} dV + \left[-m\mathbf{a}_0 - 2\boldsymbol{\omega} \times \int \mathbf{v}_{F_2} dm - \dot{\boldsymbol{\omega}} \right.$$
$$\left. \times \int \mathbf{x} dm - \boldsymbol{\omega} \times (\boldsymbol{\omega} \times \int \mathbf{x} dm) \right], \tag{7.39}$$

where $m = \int dm$.

Equation (7.39) shows that when a moving frame is used to compute

†For simplicity, we drop the subscript of the integrals V_m.

momentum and its time rate of change, the same momentum principle for an inertial frame can be used provided we add to the surface and body force terms those terms given inside the bracket in the right-hand side of Eq. (7.39). Thus, if we regard a (rigid) moving control volume as a moving frame, we need to modify the force terms accordingly. In particular, if the control volume has only translative motion (no rotation) with absolute acceleration $\mathbf{a}_0$, we have, with $\mathbf{v}$, $\partial/\partial t$ denoting relative velocity and relative time rate, respectively

$$\int_{V_c} \frac{\partial}{\partial t}(\rho \mathbf{v})\,dV + \int_{S_c} \rho \mathbf{v}(\mathbf{v}\cdot\mathbf{n})\,dS = \int_{S_c} \mathbf{t}\,dS + \int_{V_c} \rho \mathbf{B}\,dV - m\mathbf{a}_0. \quad (7.40)$$

EXAMPLE 7.10

A rocket of initial total mass M_0 moves upward while ejecting a jet of gases at the rate of γ slug/sec. The exhaust velocity of the jet relative to the rocket is v_e and the gage pressure in the jet of area A is p. Derive the differential equation governing the motion of the rocket and find the velocity as a function of time. Neglect drag forces.

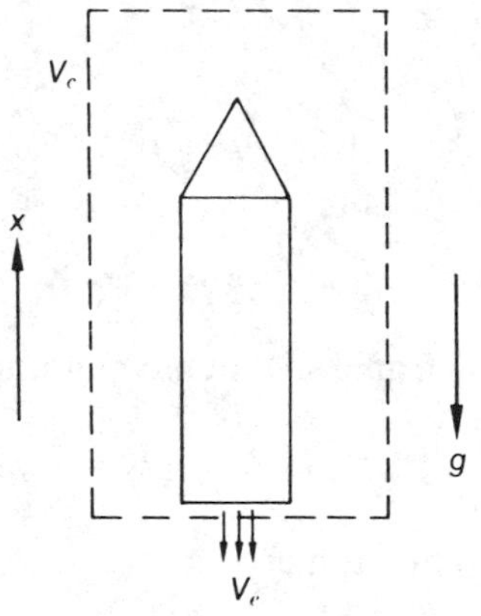

Fig. 7.7

Solution. Let V_c be a control volume which moves upward with the rocket. Then, relative to V_c, the net x-momentum outflux is $-\gamma v_e$. The motion of gases due to internal combustion does not produce any net momentum change relative to the rocket, therefore there is no rate of change of momentum inside the (moving) control volume. The net surface force on the control volume is an upward force of pA, and the body force is $(M_0 - \gamma t)g$ downward. However, since the control volume is moving with the rocket with acceleration $\ddot{x}$, therefore the term $-\ddot{x}(M - \gamma t)$ is to be added to the other force terms [*see* Eq. (7.40)]. Thus we have

$$-\gamma v_e = pA - (M_0 - \gamma t)g - (M_0 - \gamma t)\ddot{x},$$

i.e.,

$$(M_0-\gamma t)\ddot{x}=\gamma v_e+pA-(M_0-\gamma t)g.$$

This equation can be written

$$d\dot{x}=\frac{\gamma v_e+pA}{M_0-\gamma t}dt-gdt.$$

If at $t=0$, $\dot{x}=0$, then

$$\dot{x}=\frac{(\gamma v_e+pA)}{\gamma}\ \ln\left(\frac{M_0}{M_0-\gamma t}\right)-gt.$$

7.7 PRINCIPLE OF MOMENT OF MOMENTUM†

The **global principle of moment of momentum** states that the total moment of surface and body forces on a fixed part of material about a fixed point is equal to the time rate of change of total moment of momentum of the part about the same point. That is,

$$\frac{D}{Dt}\int_{V_m}\mathbf{x}\times\rho\mathbf{v}dV=\int_{S_c}\mathbf{x}\times\mathbf{t}dS+\int_{V_c}\mathbf{x}\times\rho\mathbf{B}dV, \tag{7.41}$$

where $\mathbf{x}$ is the position vector for a typical particle. Now, since $(D/Dt)(\rho dV)=0$,

$$\frac{D}{Dt}\int_{V_m}\mathbf{x}\times\rho\mathbf{v}dV=\int_{V_m=V_c}\left\{\mathbf{v}\times\rho\mathbf{v}dV+\mathbf{x}\times\frac{D\mathbf{v}}{Dt}\rho dV\right\}$$

$$=\int_{V_c}\mathbf{x}\times\rho\frac{D\mathbf{v}}{Dt}dV, \tag{7.42}$$

therefore, making use of Eq. (7.42) and converting the surface integral in the right-hand side of Eq. (7.41) [*see* Example 7.3, Eq. (7.7)], we obtain

$$\int_{V_c}\mathbf{x}\times\left\{\rho\frac{D\mathbf{v}}{Dt}-\operatorname{div}\mathbf{T}-\rho\mathbf{B}\right\}dV-2\int_{V_c}\mathbf{t}^A dV=0, \tag{7.43}$$

where $\mathbf{t}^A$ is the vector of the antisymmetric part of the stress tensor T. Now the first term in Eq. (7.43) vanishes because of Eq. (7.25), therefore, $\mathbf{t}^A=0$ and the symmetry of the stress tensor

$$\mathbf{T}=\mathbf{T}^T \tag{7.44}$$

is rediscovered.

It will be left as an exercise for the student to demonstrate from $(D/Dt)\int\mathbf{x}\times\rho\mathbf{v}\,dV$ the **Reynolds transport theorem for moment of**

†We assume that all torques are moments of forces and that the traction is simple.

momenta. Using the theorem, the moment of momentum principle [Eq. (7.43)], takes the form of the following balance equation

> **The total moment about a fixed point due to surface and body forces acting on the material instantaneously inside a control volume = total rate of change of moment of momenta inside the control volume + total net rate of outflow of moment of momenta across the control surface** (7.45)

Furthermore, if a moving control volume is used, then the following terms should be added to the left side of Eq. (7.45):

$$-\left(\int \mathbf{x}\,dm\right)\times \mathbf{a}_0 - \int \mathbf{x}\times(\dot{\boldsymbol{\omega}}\times\mathbf{x})\,dm - \int \mathbf{x}\times\{\boldsymbol{\omega}\times(\boldsymbol{\omega}\times\mathbf{x})\}\,\mathrm{dm}$$
$$-2\int \mathbf{x}\times(\boldsymbol{\omega}\times\mathbf{v})\,dm, \qquad (7.46)$$

where $\boldsymbol{\omega}$ and $\dot{\boldsymbol{\omega}}$ are absolute angular velocity and acceleration of the control volume, the vector $\mathbf{x}$ of (dm) is measured from the arbitrary chosen point O in the control volume, $\mathbf{a}_0$ is the absolute acceleration of point O and $\mathbf{v}$ is the velocity of (dm) *relative* to the control volume.

EXAMPLE 7.11

Each sprinkler arm in Fig. 7.8 discharges a constant volume of water Q and is free to rotate around the vertical center axis. Determine its constant speed of rotation.

Solution. Let V_c be a control volume that rotates with the sprinkler arms. The velocity of water particles relative to the sprinkler is $(Q/A)\mathbf{e}_1$ in the right arm and $(Q/A)(-\mathbf{e}_1)$ in the left arm. If ρ is density, then the total net outflux of moment of momentum about point O is $2\rho Q(Q/A)\sin\theta\, r_0\mathbf{e}_3$. The moment of momentum about O due to weight is zero, and assum-

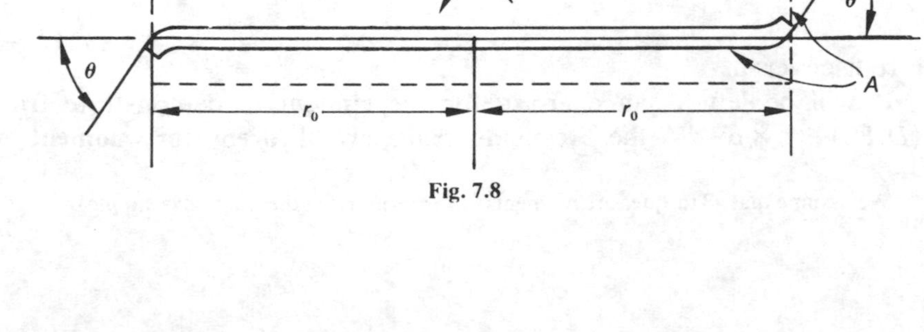

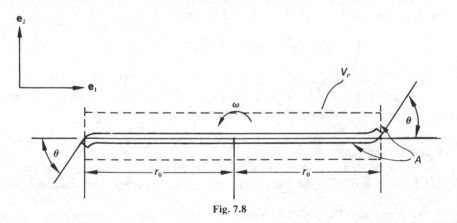

Fig. 7.8

ing the pressure in the water jet is the same as the atmospheric pressure, taken to be zero gage pressure, there is no contribution due to surface force on the control volume. Now, since the control volume is rotating with the sprinkler, therefore we need to add terms in Eq. (7.46) to the moment of forces. However, if **x** is measured from O, the only nonzero term is

$$-2\int \mathbf{x}\times(\boldsymbol{\omega}\times\mathbf{v})\,dm,$$

which is the moment due to Coriolis forces. Now, for the right arm, $\mathbf{v}=(Q/A)\,\mathbf{e}_1$, therefore

$$\boldsymbol{\omega}\times\mathbf{v}=\omega\mathbf{e}_3\times\frac{Q}{A}\mathbf{e}_1=\frac{\omega Q}{A}\mathbf{e}_2$$

and

$$\mathbf{x}\times(\boldsymbol{\omega}\times\mathbf{v})=x\mathbf{e}_1\times\frac{\omega Q}{A}\mathbf{e}_2=\frac{x\omega Q}{A}\mathbf{e}_3.$$

Thus, the contribution from the fluid in the right arm to the integral is

$$-\frac{2\omega Q}{A}\mathbf{e}_3\int_0^{r_0}x(\rho A\,dx)=-\omega Q\rho r_0^{2}\mathbf{e}_3.$$

Including that due to the left arm, the integral has the value $-2\omega Q\rho r_0^2\mathbf{e}_3$. Therefore, from the moment of momentum principle for a moving control volume, we have

$$2\rho Q\left(\frac{Q}{A}\right)\sin\theta r_0=-2\omega Q\rho r_0^{2}.$$

From which

$$\omega=-\frac{Q}{A}\frac{\sin\theta}{r_0}.$$

7.8 PRINCIPLE OF CONSERVATION OF ENERGY

The principle of conservation of energy states that the time rate of change of the kinetic energy and internal energy for a fixed part of material is equal to the sum of the rate of work done by the surface and body forces and the heat energy entering the boundary surface.† That is, if v^2 denotes $\mathbf{v}\cdot\mathbf{v}$, u the internal energy per unit mass, and **q** the rate of heat flow vector across a unit area, then the principle states:

$$\frac{D}{Dt}\int_{Vm}\left(\rho\frac{v^2}{2}+\rho u\right)dV=\int_{S_c}\mathbf{t}\cdot\mathbf{v}\,dS+\int_{S_c}\rho\mathbf{B}\cdot\mathbf{v}\,dV-\int_{S_c}\mathbf{q}\cdot\mathbf{n}\,dS,\quad(7.47)$$

the minus sign in the last term is again due to the convention that **n** is an outward unit normal vector.

Again, since $(D/Dt)\,(\rho\,dV)=0$, therefore

$$\frac{D}{Dt}\int_{V_m}\rho\left(\frac{v^2}{2}+u\right)dV=\int_{V_c}\left\{\frac{D}{Dt}\left(\frac{v^2}{2}+u\right)\right\}\rho\,dV.\quad(7.48)$$

†If there are other energies entering the boundary or if there are energy sources, they should be included in the right-hand side of Eq. (7.47).

Making use of Eq. (7.48) and converting the surface integrals in the right-hand side of Eq. (7.47) [*see* Example 7.4, Eq. (7.8)], we have

$$\int_{V_c} \rho \frac{D}{Dt}\left(\frac{v^2}{2}+u\right)dV = \int_{V_c} \{(\text{div}\,\mathbf{T}+\rho\mathbf{B})\cdot\mathbf{v}+\text{tr}\,(\mathbf{T}^T\nabla\mathbf{v})-\text{div}\,\mathbf{q}\}dV. \quad (7.49)$$

Since $\mathbf{T}=\mathbf{T}^T$ and

$$(\text{div}\,\mathbf{T}+\rho\mathbf{B})\cdot\mathbf{v}=\rho\frac{D\mathbf{v}}{Dt}\cdot\mathbf{v}=\frac{1}{2}\rho\frac{Dv^2}{Dt}.$$

Therefore, Eq. (7.49) becomes

$$\int_{V_c} \rho \frac{Du}{Dt}\,dV = \int_{V_c} \{\text{tr}\,(\mathbf{T}\nabla\mathbf{v})-\text{div}\,\mathbf{q}\}dV. \quad (7.50)$$

Thus, at every point, we have

$$\rho\frac{Du}{Dt}=\text{tr}\,(\mathbf{T}\nabla\mathbf{v})-\text{div}\,\mathbf{q}. \quad (7.51)$$

This is the energy equation. A slightly different form of Eq. (7.51) can be obtained if we recall that $\nabla\mathbf{v}=\mathbf{D}+\mathbf{W}$, where $\mathbf{D}$, the symmetric part of $\nabla\mathbf{v}$ is the rate of deformation tensor and $\mathbf{W}$, the antisymmetric part of $\nabla\mathbf{v}$, is spin tensor, we have

$$\text{tr}\,(\mathbf{T}\nabla\mathbf{v})=\text{tr}\,(\mathbf{TD}+\mathbf{TW})=\text{tr}\,(\mathbf{TD})+\text{tr}\,(\mathbf{TW}). \quad (7.52)$$

But $\text{tr}\,(\mathbf{TW})=T_{ij}W_{ji}=T_{ji}W_{ij}=T_{ij}(-W_{ji})=-T_{ij}W_{ji}$ so that

$$\text{tr}\,(\mathbf{TW})=0, \quad (7.53)$$

therefore, we rediscover the energy equation in the following form:

$$\rho\frac{Du}{Dt}=\text{tr}\,(\mathbf{TD})-\text{div}\,\mathbf{q}. \quad (7.54)$$

Again, **the global conservation of energy principle** can be stated in the following form of balance equation:

> **The time rate of work done by surface and body forces in a control volume + rate of heat input = total rate of increase of internal and kinetic energy of the material inside the control volume + rate of outflow of the internal and kinetic energy across the control surface**

(7.55)

EXAMPLE 7.12

A supersonic one-dimensional flow in an insulating duct suffers a normal compression shock. Assuming ideal gas, find the pressure after the shock in terms of the pressure and velocity before the shock.

Solution. For the control volume shown, we have, for steady flow:

(1) Mass outflux = mass influx†, that is,

$$\rho_1 A v_1 = \rho_2 A v_2,$$

i.e.,

$$\rho_1 v_1 = \rho_2 v_2. \tag{i}$$

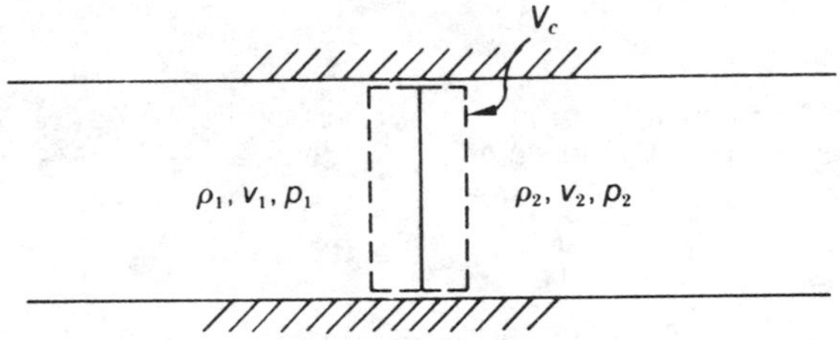

Fig. 7.9

(2) Force in x-direction = net momentum outflux in x-direction

$$p_1 A - p_2 A = (\rho_2 A v_2) v_2 - (\rho_1 A v_1) v_1$$

i.e.,

$$(p_1 - p_2) = \rho_2 v_2^2 - \rho_1 v_1^2. \tag{ii}$$

(3) Rate of work done by surface force = net energy (internal and kinetic) outflux.† That is,

$$p_1 A v_1 - p_2 A v_2 = [(\rho_2 A v_2) u_2 - (\rho_1 A v_1) u_1] + [\tfrac{1}{2} (\rho_2 A v_2) v_2^2 - \tfrac{1}{2} (\rho_1 A v_1) v_1^2]. \tag{iii}$$

For ideal gas the internal energy per unit mass is given by

$$u = \frac{p}{\rho}\left(\frac{1}{\gamma - 1}\right),$$

where $\gamma = c_p/c_v$ is ratio of specific heat. Thus Eq. (iii) becomes

$$p_1 v_1 - p_2 v_2 = p_2 v_2 \left(\frac{1}{\gamma - 1}\right) - p_1 v_1 \left(\frac{1}{\gamma - 1}\right) + \frac{1}{2} \rho_2 v_2^3 - \frac{1}{2} \rho_1 v_1^3$$

or

$$\frac{\gamma}{\gamma - 1} p_1 v_1 + \frac{1}{2} \rho_1 v_1^3 = \frac{\gamma}{\gamma - 1} p_2 v_2 + \frac{1}{2} \rho_2 v_2^3,$$

that is,

$$\rho_1 v_1 \left[\frac{\gamma}{\gamma - 1} \frac{p_1}{\rho_1} + \frac{1}{2} v_1^2\right] = \rho_2 v_2 \left[\frac{\gamma}{\gamma - 1} \frac{p_2}{\rho_2} + \frac{1}{2} v_2^2\right].$$

†Flux ≡ rate of flow across a surface.

In view of Eq. (i), this equation becomes

$$\frac{\gamma}{\gamma-1}\frac{p_1}{\rho_1}+\frac{1}{2}v_1^2=\frac{\gamma}{\gamma-1}\frac{p_2}{\rho_2}+\frac{1}{2}v_2^2. \tag{iv}$$

We note here that this is the same energy equation derived in Chapter 6 (Example 6.20) using differential forms of energy equations for inviscid nonheat-conducting fluid. From Eqs. (i), (ii), and (iii) we obtain

$$p_2=\frac{1}{\gamma+1}[2\rho_1 v_1^2-(\gamma-1)p_1].$$

PROBLEMS

7.1. Verify the divergence theorem for the vector field $\mathbf{v}=2x\mathbf{e}_1+z\mathbf{e}_2$, by considering the region bounded by $x=0, x=2, y=0, y=2, z=0, z=2$.

7.2. Show that

$$\int_S \mathbf{x}\cdot\mathbf{n}\,dS=3V,$$

where V is the volume enclosed by the boundary S.

7.3. (a) Consider the vector field $\mathbf{v}=\phi\mathbf{a}$, where ϕ is a given scalar field and $\mathbf{a}$ is an arbitrary constant vector (independent of position). Using the divergence theorem, prove that

$$\int_V \nabla\phi\,dV=\int_S \phi\mathbf{n}\,dS.$$

(b) Show that for any closed surface S that

$$\int_S \mathbf{n}\,dS=0.$$

7.4. A stress field $\mathbf{T}$ is in equilibrium with a body force $\rho\mathbf{B}$. Using the divergence theorem, show that for any volume V, and boundary surface S, that

$$\int_S \mathbf{t}\,dS+\int_V \rho\mathbf{B}\,dV=0.$$

That is, the total resultant force is equipollent to zero.

7.5. (a) Let $\mathbf{u}^*$ define a strain field $\mathbf{E}^*=\frac{1}{2}(\nabla\mathbf{u}^*+(\nabla\mathbf{u}^*)^T)$ and let $\mathbf{T}^{**}$ be in equilibrium with a body force $\rho\mathbf{B}^{**}$ and a surface traction $\mathbf{t}^{**}$. Using the divergence theorem, verify the identity (theorem of virtual work)

$$\int_S \mathbf{t}^{**}\cdot\mathbf{u}^*\,dS+\int_V \rho\mathbf{B}^{**}\cdot\mathbf{u}^*\,dV=\int_V T_{ij}^{**}E_{ij}^*\,dV.$$

(b) Let $\mathbf{u}^*=X_1\mathbf{e}_1$ and $\mathbf{T}^{**}$ be in equilibrium with zero body force and zero surface traction on a specific boundary. Using the identity of part (a), show that

$$\int_V T_{11}^{**}\,dV=0.$$

7.6. (a) Using the equations of motion and the divergence theorem, verify the following rate of work identity

$$\int_V \rho \mathbf{B} \cdot \mathbf{v}\, dV + \int_S \mathbf{t} \cdot \mathbf{v}\, dS = \int_V \rho \frac{D}{Dt}\left(\frac{v^2}{2}\right) dV + \int_V T_{ij} D_{ij}\, dV.$$

(b) How will the identity of part (a) be altered if the material is rigid?

7.7. Consider the velocity and density fields

$$\mathbf{v} = x_1 \mathbf{e}_1 \quad \text{and} \quad \rho = \rho_0 e^{-t}.$$

(a) Check the equation of mass conservation.
(b) Compute the mass and rate of increase of mass in the cylindrical control volume of cross-section A and bounded by $x_1 = 0$ and $x_1 = 3$.
(c) Compute the mass inflow into the control volume of part (b). Does mass inflow equal the rate of mass increase?

7.8. (a) Check that the motion

$$x_1 = X_1 e^{t-t_0}, \quad x_2 = X_2, \quad x_3 = X_3$$

corresponds to the velocity field of the previous problem.
(b) For a density field $\rho = \rho_0 e^{-(t-t_0)}$, verify that the mass contained in the material volume that is coincident with the control volume of Problem 7.7 at time t_0, remains constant.
(c) Compute the total linear momentum for the material volume of part (b). Why in this case, as compared to Example 7.6, is it not constant?

7.9. Do Problem 7.7 for the velocity field $\mathbf{v} = x_1 \mathbf{e}_1$ and the density field $\rho = \rho_0 / x_1$ and for the cylindrical control volume bounded by $x_1 = 1$ and $x_1 = 3$.

7.10. The center of mass $\mathbf{x}_{\text{c.m.}}$ of a material volume is defined by the equation

$$m\mathbf{x}_{\text{c.m.}} = \int_{V_m} \mathbf{x} \rho\, dV, \qquad \text{where } m = \int_{V_m} \rho\, dV.$$

Demonstrate that the linear momentum principle may be written in the form

$$\int_{S_C} \mathbf{t}\, dS + \int_{V_C} \rho \mathbf{B}\, dV = m\, \mathbf{a}_{\text{c.m.}}, \qquad \text{where } \mathbf{a}_{\text{c.m.}} = \frac{D^2}{Dt^2} \mathbf{x}_{\text{c.m.}}$$

is the acceleration of the mass center.

7.11. Consider the velocity field and density field of Example 7.6, i.e.,

$$\mathbf{v} = \left(\frac{x_1}{1+t}\right) \mathbf{e}_1, \qquad \rho = \frac{\rho_0}{1+t},$$

(a) Compute the total linear momentum and rate of increase of linear momentum in a cylindrical control volume of cross-sectional area A and bounded by the plane $x_1 = 1$ and $x_1 = 3$.
(b) Compute the net rate of outflow of linear momentum from the control volume of part (a).
(c) Compute the total force on the material in the control volume.

(d) Compute the total kinetic energy and rate of increase of kinetic energy for the control volume of part (a).

(e) Compute the net rate of outflow of kinetic energy from the control volume.

7.12. Consider the velocity and density fields of Problem 7.7, i.e.,

$$\mathbf{v} = x_1\mathbf{e}_1, \quad \rho = \rho_0 e^{-t}.$$

For an arbitrary time t, consider the material contained in the cylindrical control volume bounded by $x_1 = 0$ and $x_1 = 3$,

(a) Determine the linear momentum and rate of increase of linear momentum in this control volume.

(b) Determine the outflux of linear momentum.

(c) Determine the net resultant force that is acting on the material contained in the control volume.

7.13. Do the previous problem for the same velocity field, with $\rho = \rho_0/x_1$ and the cylindrical control volume bounded by $x_1 = 1$ and $x_1 = 3$.

7.14. Consider the flow field $\mathbf{v} = x\mathbf{e}_1 - y\mathbf{e}_2$ with $\rho =$ constant. For a control volume defined by $x = 0$, $x = 2$, $y = 0$, $y = 2$, $z = 0$, $z = 2$, determine the net resultant force and couple that is acting on the material contained in this volume.

7.15. Do the previous problem for the control volume defined by $x = 2$, $y = 2$, $xy = 2$, $z = 0$, $z = 2$.

7.16. For Hagen–Poiseuille flow in a pipe

$$\mathbf{v} = C\,(r_0^2 - r^2)\mathbf{e}_1,$$

calculate the momentum flux across a cross-section. For the same flow rate, if the velocity is assumed to be uniform, what is the momentum flux across a cross-section? Compare the two results.

7.17. A pile of chain on a table falls through a hole from the table under the action of gravity. Show that the hanging length x satisfies the equation

$$gx = x\ddot{x} + (\dot{x})^2 = \frac{d}{dt}(x\dot{x}).$$

7.18. A water jet of 5 cm. diameter moves at 12 m/sec, impinges on a curved vane which deflects it 60° from its direction. Neglect the weight. Obtain the force exerted by the liquid to the vane.

7.19. A horizontal pipeline of 10 cm. diameter bends through 90°, and while bending, changes its diameter to 5 cm. The pressure in the 10 cm. pipe is 140 kPa. Estimate the resultant force on the bend when 0.05 m³/sec of water is flowing in the pipeline.

7.20. Figure P7.1 shows a steady water jet of area A impinging onto the flat wall. Find the force exerted on the wall. Neglect weight and viscosity of water.

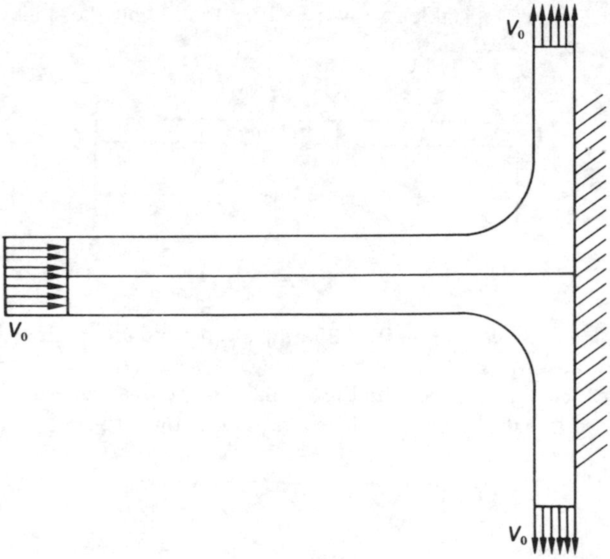

Fig. P7.1

7.21. Frequently in open channel flow, when water at high velocity discharges into a region of lower velocity, an abrupt rise occurs in the water surface. This is known as a "hydraulic jump." Referring to Fig. P7.2, if the flow rate is Q per unit width, find the relation between y_1 and y_2. Assume the flow before and after the jump is uniform and the pressure distribution there is hydrostatic.

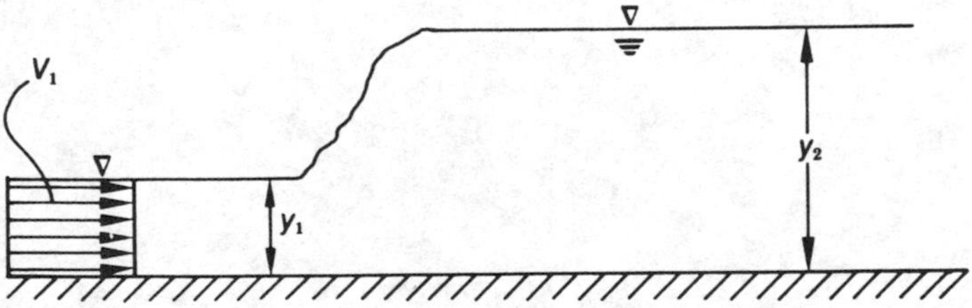

Fig. P7.2

7.22. If the curved vane of Example 7.8 moves with velocity $v < v_0$ in the same direction as the oncoming jet, find the resultant force exerted on the vane by the jet.

7.23. Write out the verbal moment of momentum Eq. (7.45).

7.24. For the half-arm sprinkler shown in Fig. P7.3, find the angular speed if $Q = 0.566 \text{m}^3/\text{sec}$. Neglect friction.

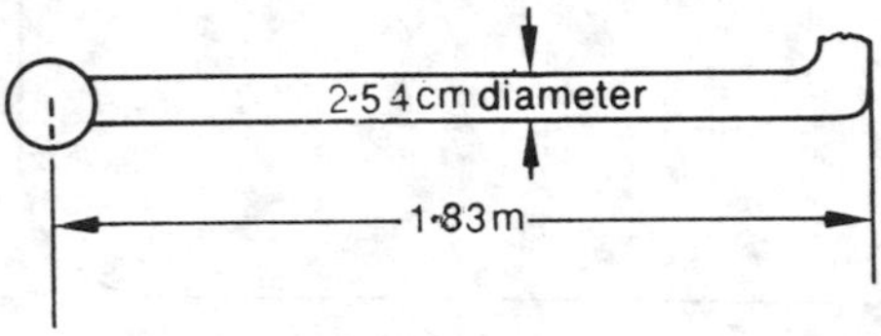

Fig. P7.3

7.25. The tank car shown in Fig. P7.4 contains water and compressed air which is regulated to force a water jet out of the nozzle at a constant rate of Q m³/sec. The diameter of the jet is d cm. The initial total mass of the tank car is M_0. Neglecting frictional forces, find the velocity of the car as a function of time.

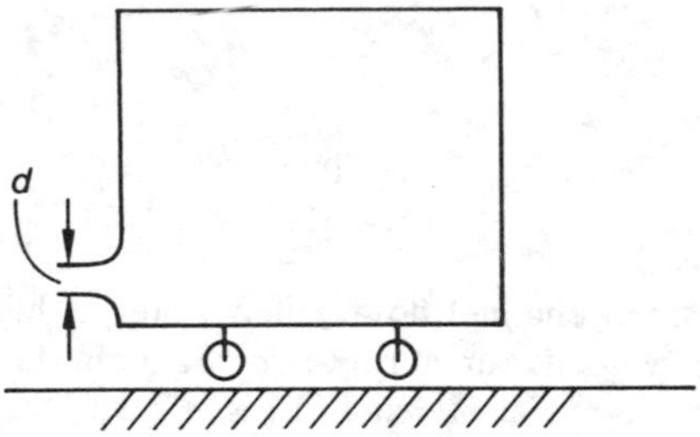

Fig. P7.4

7.26. Write out the verbal energy equation [Eq. (7.55)].

8

Incompressible Simple Fluid

In Chapter 6, the linear viscous fluid was discussed as an example of a constitutive equation of an idealized fluid. The mechanical behavior of many real fluids appears to be adequately described under a wide range of circumstances by this constitutive equation which is often referred to as the constitutive equation of a Newtonian fluid. Many real fluids exhibit behaviors which are not accounted for by the theory of Newtonian fluid. Examples of such substances are polymeric solution, paints, molasses, etc.

For a steady unidirectional laminar flow of water in a circular pipe, the theory of Newtonian fluid gives the result, agreeing with the experimental observation first made by Hagen and Poiseuille, that the volume discharge Q is proportional to the (constant) pressure gradient in the axial direction and to the fourth power of the diameter d of the pipe, that is, [*see* Eq. (6.27)]

$$Q = \frac{\rho \pi d^4}{128\mu} \left| \frac{dp}{dx} \right| . \tag{i}$$

However, for high polymeric solutions, it was observed that the above equation does not hold. For fixed d, the Q versus $|dp/dx|$ relation is nonlinear, as sketched in Fig. 8.1.

For a steady laminar flow of water placed between two very long coaxial cylinders of radii r_1 and r_2, if one of the cylinders is at rest while the other is rotating with an angular velocity of Ω, the theory of Newtonian fluid gives the result agreeing with experimental observations that the torque per unit height which must be applied to the cylinders to maintain

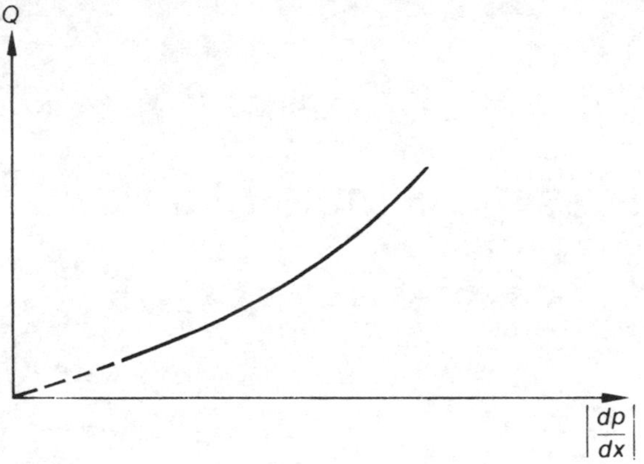

Fig. 8.1

the flow is proportional to Ω. In fact, [*see* Eq. (6.30)]

$$M = \frac{4\pi\mu r_1^2 r_2^2 \Omega}{r_2^2 - r_1^2}. \tag{ii}$$

However, for those fluids which do not obey Eq. (6.27), it is found that they do not obey Eq. (6.30) either.

Furthermore, for a substance such as water in the steady Couette flow, the normal stress at the outer cylinder is always larger than that at the inner cylinder, the difference being due to the centrifugal forces. However, for the fluid which does not obey Eq. (i), the compressive normal stress at the inner cylinder can be larger than that at the outer cylinder.

In this chapter, we shall discuss a class of flows, known as the viscometric flows, of an idealized material known as a simple fluid, which is capable of exhibiting many of the behaviors that have been observed for many real fluids. We begin by considering some more kinematics of deformation needed later in the chapter.

8.1 CURRENT CONFIGURATION AS REFERENCE CONFIGURATION

Let $\mathbf{x}$ be the position vector of a particle at current time t, and let $\boldsymbol{\xi}$ be the position vector of the same particle at some time τ. Then the equation

$$\boldsymbol{\xi} = \boldsymbol{\xi}(\mathbf{x}, \tau), \qquad \text{with } \mathbf{x} = \boldsymbol{\xi}(\mathbf{x}, t) \tag{8.1}$$

defines the motion of a continuum using the current time as the reference time.

For a given velocity field $\mathbf{v} = \mathbf{v}(\mathbf{x}, t)$, the velocity at the position $\boldsymbol{\xi}$ at time τ is $\mathbf{v} = \mathbf{v}(\boldsymbol{\xi}, \tau)$. On the other hand, for a particular particle (i.e., for fixed $\mathbf{x}$ and t), the velocity at time τ is given by $(\partial\boldsymbol{\xi}/\partial\tau)_{\text{fixed } \mathbf{x}, t}$.

Thus

$$\frac{\partial \boldsymbol{\xi}}{\partial \tau} = \mathbf{v}(\boldsymbol{\xi}, \tau). \tag{8.2}$$

EXAMPLE 8.1

Given the velocity field of the steady unidirectional flow

$$v_1 = v(x_2), \quad v_2 = 0, \quad v_3 = 0,$$

describe the motion of the particles by using the current time t as the reference time.

Solution. From the given velocity field, we have the velocity components at the position (ξ_1, ξ_2, ξ_3) at time τ:

$$v_1 = v(\xi_2), \quad v_2 = 0, \quad v_3 = 0.$$

Thus, with $\boldsymbol{\xi} = \xi_i \mathbf{e}_i$, Eq. (8.2) gives

$$\frac{\partial \xi_1}{\partial \tau} = v(\xi_2), \quad \frac{\partial \xi_2}{\partial \tau} = 0, \quad \frac{\partial \xi_3}{\partial \tau} = 0.$$

From $\partial\xi_2/\partial\tau = 0$, we have

$$\xi_2 = f(x_1, x_2, x_3, t).$$

But at $\tau = t$, $\xi_2 = x_2$. Therefore,

$$\xi_2 = x_2 \qquad \text{for all } \tau.$$

Similarly,

$$\xi_3 = x_3 \qquad \text{for all } \tau.$$

From $\partial\xi_1/\partial\tau = v(\xi_2)$, since $\xi_2 = x_2$ for all τ, we have

$$\frac{\partial \xi_1}{\partial \tau} = v(x_2),$$

so that

$$\xi_1 = v(x_2)\tau + g(x_1, x_2, x_3, t).$$

At $\tau = t$, $\xi_1 = x_1$, therefore

$$x_1 = v(x_2)t + g(x_1, x_2, x_3, t)$$

from which

$$g(x_1, x_2, x_3, t) = x_1 - v(x_2)t$$

and

$$\xi_1 = v(x_2)\tau + x_1 - v(x_2)t.$$

That is,

$$\xi_1 = x_1 + v(x_2)(\tau - t),$$

$$\xi_2 = x_2,$$

$$\xi_3 = x_3.$$

8.2 RELATIVE DEFORMATION TENSOR

Let $d\mathbf{x}$ and $d\boldsymbol{\xi}$ be the (infinitesimal) vectors representing the *same* material element at time t and τ, respectively. Then $d\mathbf{x}$ and $d\boldsymbol{\xi}$ are related by

$$d\boldsymbol{\xi} = \boldsymbol{\xi}(\mathbf{x}+d\mathbf{x}, \tau) - \boldsymbol{\xi}(\mathbf{x}, \tau) = (\nabla\boldsymbol{\xi})\, d\mathbf{x}, \tag{8.3}$$

where in Cartesian coordinates

$$[\nabla\boldsymbol{\xi}] = \begin{bmatrix} \frac{\partial\xi_1}{\partial x_1} & \frac{\partial\xi_1}{\partial x_2} & \frac{\partial\xi_1}{\partial x_3} \\ \frac{\partial\xi_2}{\partial x_1} & \frac{\partial\xi_2}{\partial x_2} & \frac{\partial\xi_2}{\partial x_3} \\ \frac{\partial\xi_3}{\partial x_1} & \frac{\partial\xi_3}{\partial x_2} & \frac{\partial\xi_3}{\partial x_3} \end{bmatrix}. \tag{8.4}$$

The tensor $\nabla\boldsymbol{\xi}$ is known as the relative deformation gradient. Note for $\tau = t$, $d\boldsymbol{\xi} = \boldsymbol{\xi}(\mathbf{x}+d\mathbf{x}, t) - \boldsymbol{\xi}(\mathbf{x}, t) = (\mathbf{x}+d\mathbf{x}) - \mathbf{x} = d\mathbf{x}$. Thus, $(\nabla\boldsymbol{\xi})_{\tau=t} = \mathbf{I}$ (identity tensor).

Consider two elements $d\mathbf{x}^{(1)}$ and $d\mathbf{x}^{(2)}$ emanating from a point P at time t and $d\boldsymbol{\xi}^{(1)}$ and $d\boldsymbol{\xi}^{(2)}$, the corresponding elements at time τ. Then

$$d\boldsymbol{\xi}^{(1)} \cdot d\boldsymbol{\xi}^{(2)} = [(\nabla\boldsymbol{\xi})\, d\mathbf{x}^{(1)}] \cdot [(\nabla\boldsymbol{\xi})\, d\mathbf{x}^{(2)}]. \tag{8.5}$$

Then, by definition of the transpose,

$$d\boldsymbol{\xi}^{(1)} \cdot d\boldsymbol{\xi}^{(2)} = d\mathbf{x}^{(1)} \cdot (\nabla\boldsymbol{\xi})^T(\nabla\boldsymbol{\xi})\, d\mathbf{x}^{(2)}. \tag{8.6}$$

Let

$$\mathbf{C}_t(\mathbf{x}, \tau) \equiv (\nabla\boldsymbol{\xi})^T(\nabla\boldsymbol{\xi}), \tag{8.7}$$

then

$$d\boldsymbol{\xi}^{(1)} \cdot d\boldsymbol{\xi}^{(2)} = d\mathbf{x}^{(1)} \cdot \mathbf{C}_t d\mathbf{x}^{(2)}. \tag{8.8}$$

If $\mathbf{C}_t = \mathbf{I}$ (identity tensor), then $d\boldsymbol{\xi}^{(1)} \cdot d\boldsymbol{\xi}^{(2)} = d\mathbf{x}^{(1)} \cdot d\mathbf{x}^{(2)}$ and no deformation takes place. The symmetric tensor $\mathbf{C}_t(\mathbf{x}, \tau)$ is called the relative deformation tensor (or relative Cauchy–Green tensor). Note:

$$\mathbf{C}_t(\mathbf{x}, t) = (\nabla\boldsymbol{\xi})^T_{\tau=t}(\nabla\boldsymbol{\xi})_{\tau=t} = \mathbf{I}.$$

EXAMPLE 8.2

Find the relative deformation tensor for the velocity field given in Example 8.1.

Solution. Since

$$\xi_1 = x_1 + v(x_2)\,(\tau - t), \quad \xi_2 = x_2, \quad \xi_3 = x_3.$$

We have, with $k \equiv dv/dx_2$,

$$[\nabla\xi] = \begin{bmatrix} 1 & k(\tau-t) & 0 \\ 0 & 1 & 0 \\ 0 & 0 & 1 \end{bmatrix},$$

and $\mathbf{C}_t(\mathbf{x}, \tau) = (\nabla\xi)^T(\nabla\xi)$ gives

$$[\mathbf{C}_t] = \begin{bmatrix} 1 & 0 & 0 \\ k(\tau-t) & 1 & 0 \\ 0 & 0 & 1 \end{bmatrix}\begin{bmatrix} 1 & k(\tau-t) & 0 \\ 0 & 1 & 0 \\ 0 & 0 & 1 \end{bmatrix} = \begin{bmatrix} 1 & k(\tau-t) & 0 \\ k(\tau-t) & k^2(\tau-t)^2+1 & 0 \\ 0 & 0 & 1 \end{bmatrix}.$$

8.3 HISTORY OF DEFORMATION TENSOR. RIVLIN–ERICKSEN TENSORS

The tensor $\mathbf{C}_t(\mathbf{x}, \tau)$ describes the deformation at time τ of the element which is at $\mathbf{x}$ at time t. Thus, as one varies τ from $\tau = -\infty$ to $\tau = t$ in the function $\mathbf{C}_t(\mathbf{x}, \tau)$, one gets the whole history of the deformation from infinitely long time ago to the present time t.

If we assume that we can expand the components of $\mathbf{C}_t$ in Taylor series about $\tau = t$, we have, with $(\partial\mathbf{C}_t/\partial\tau)_{ij} \equiv (\partial/\partial\tau)(\mathbf{C}_t)_{ij}$,

$$\mathbf{C}_t(\mathbf{x}, \tau) = \mathbf{C}_t(\mathbf{x}, t) + \left(\frac{\partial\mathbf{C}_t}{\partial\tau}\right)_{\tau=t}(\tau-t) + \frac{1}{2}\left(\frac{\partial^2\mathbf{C}_t}{\partial\tau^2}\right)_{\tau=t}(\tau-t)^2 + \cdots.$$

Let

$$\mathbf{A}_1 = \left(\frac{\partial\mathbf{C}_t}{\partial\tau}\right)_{\tau=t},$$

$$\mathbf{A}_2 = \left(\frac{\partial^2\mathbf{C}_t}{\partial\tau^2}\right)_{\tau=t}, \quad \text{etc.} \tag{8.9}$$

We have (note, $\mathbf{C}_t(\mathbf{x}, t) = \mathbf{I}$),

$$\mathbf{C}_t(\mathbf{x}, \tau) = \mathbf{I} + (\tau-t)\mathbf{A}_1 + \frac{(\tau-t)^2}{2}\mathbf{A}_2 + \cdots. \tag{8.10}$$

The tensors $\mathbf{A}_1, \mathbf{A}_2, \ldots, \mathbf{A}_n$ are known as Rivlin–Ericksen tensors.

We see from the above equation that the $\mathbf{A}_n$'s determine the history of relative deformation.

EXAMPLE 8.3

Find the Rivlin–Ericksen tensor for the unidirectional flows of Example 8.1.

Solution. We have, from Example 8.2,

$$[\mathbf{C}_t(\mathbf{x},\tau)] = \begin{bmatrix} 1 & k(\tau-t) & 0 \\ k(\tau-t) & 1+k^2(\tau-t)^2 & 0 \\ 0 & 0 & 1 \end{bmatrix}$$

$$= \begin{bmatrix} 1 & 0 & 0 \\ 0 & 1 & 0 \\ 0 & 0 & 1 \end{bmatrix} + \begin{bmatrix} 0 & k & 0 \\ k & 0 & 0 \\ 0 & 0 & 0 \end{bmatrix}(\tau-t) + \begin{bmatrix} 0 & 0 & 0 \\ 0 & 2k^2 & 0 \\ 0 & 0 & 0 \end{bmatrix}\frac{(\tau-t)^2}{2}.$$

Thus (*see* Eq. 8.10)

$$[\mathbf{A}_1] = \begin{bmatrix} 0 & k & 0 \\ k & 0 & 0 \\ 0 & 0 & 0 \end{bmatrix},$$

$$[\mathbf{A}_2] = \begin{bmatrix} 0 & 0 & 0 \\ 0 & 2k^2 & 0 \\ 0 & 0 & 0 \end{bmatrix},$$

$$[\mathbf{A}_3] = [\mathbf{A}_4] = \ldots, [\mathbf{A}_n] = 0 \qquad \text{for all } n \geq 3,$$

where $k = dv/dx_2$. We note that the same results can be obtained by using Eq. (8.9) directly.

EXAMPLE 8.4

Given the Cartesian components of an axisymmetric velocity field:

$$v_1 = v(r), \quad v_2 = 0, \quad v_3 = 0, \qquad \text{where } r^2 = x_2{}^2 + x_3{}^2$$

(a) find the motion using current time t as reference;
(b) compute the relative deformation tensor; and
(c) find the Rivlin–Ericksen tensors.

Solution. (a) Let ξ_1, ξ_2, ξ_3 be Cartesian components of the position vector of a particle $(\mathbf{x}, t)$ at time τ. Then, following Example 8.1, we see clearly since $v_2 = 0$ and $v_3 = 0$, we have for all time τ

$$\xi_2 = x_2$$

and

$$\xi_3 = x_3.$$

Thus

$$\xi_2{}^2 + \xi_3{}^2 = x_2{}^2 + x_3{}^2 = r^2.$$

Now

$$\left(\frac{\partial \xi_1}{\partial \tau}\right)_{x_i \text{ and } t \text{ fixed}} = v(r'),$$

where $r'^2 = \xi_2{}^2 + \xi_3{}^2 = r^2$ (independent of τ). Thus

$$\xi_1 = v(r)\tau + f(x_1, x_2, x_3, t).$$

Since, $\xi_1 = x_1$ at $\tau = t$, we have

$$f = x_1 - v(r)t.$$

Therefore,

$$\xi_1 = x_1 + v(r)(\tau - t),$$

$$\xi_2 = x_2,$$

$$\xi_3 = x_3.$$

(b) Since

$$\frac{\partial \xi_1}{\partial x_1} = 1, \quad \frac{\partial \xi_1}{\partial x_2} = (\tau - t)\frac{dv}{dr}\frac{\partial r}{\partial x_2},$$

and

$$\frac{\partial \xi_1}{\partial x_3} = (\tau - t)\frac{dv}{dr}\frac{\partial r}{\partial x_3}, \text{ etc.,}$$

where

$$\frac{\partial r}{\partial x_2} = \frac{x_2}{r} = \cos\theta,$$

and

$$\frac{\partial r}{\partial x_3} = \frac{x_3}{r} = \sin\theta,$$

we have, therefore, denoting dv/dr by $k(r)$,

$$[\nabla \boldsymbol{\xi}] = \begin{bmatrix} 1 & (\tau - t)k(r)\cos\theta & (\tau - t)k(r)\sin\theta \\ 0 & 1 & 0 \\ 0 & 0 & 1 \end{bmatrix}.$$

Thus

$$[\mathbf{C}_t] = [\nabla \boldsymbol{\xi}]^T [\nabla \boldsymbol{\xi}] = \begin{bmatrix} 1 & (\tau - t)k(r)\cos\theta & (\tau - t)k(r)\sin\theta \\ (\tau - t)k(r)\cos\theta & 1 + (\tau - t)^2k^2(r)\cos^2\theta & (\tau - t)^2k^2(r)\sin\theta\cos\theta \\ (\tau - t)k(r)\sin\theta & (\tau - t)^2k^2(r)\sin\theta\cos\theta & 1 + (\tau - t)^2k^2(r)\sin^2\theta \end{bmatrix}.$$

(c) From the result of part (b), we have

$$[\mathbf{C}_t] = \begin{bmatrix} 1 & 0 & 0 \\ 0 & 1 & 0 \\ 0 & 0 & 1 \end{bmatrix} + k(r)\begin{bmatrix} 0 & \cos\theta & \sin\theta \\ \cos\theta & 0 & 0 \\ \sin\theta & 0 & 0 \end{bmatrix}(\tau - t)$$

$$+ k^2(r)\begin{bmatrix} 0 & 0 & 0 \\ 0 & \cos^2\theta & \sin\theta\cos\theta \\ 0 & \sin\theta\cos\theta & \sin^2\theta \end{bmatrix}(\tau - t)^2.$$

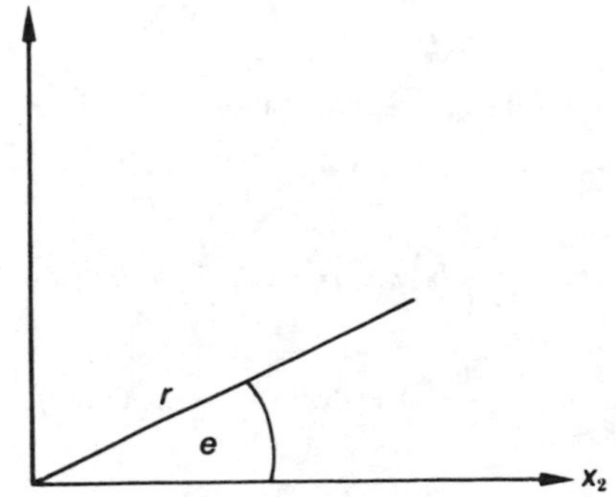

Fig. 8.2

Thus, with respect to Cartesian basis $\mathbf{e}_1, \mathbf{e}_2, \mathbf{e}_3$, the Rivlin–Ericksen tensors are (*see* Eq. 8.10)

$$[\mathbf{A}_1] = k(r)\begin{bmatrix} 0 & \cos\theta & \sin\theta \\ \cos\theta & 0 & 0 \\ \sin\theta & 0 & 0 \end{bmatrix}.$$

$$[\mathbf{A}_2] = 2k^2(r)\begin{bmatrix} 0 & 0 & 0 \\ 0 & \cos^2\theta & \sin\theta\cos\theta \\ 0 & \sin\theta\cos\theta & \sin^2\theta \end{bmatrix}.$$

$$\mathbf{A}_3 = \mathbf{A}_4 = \cdots = \mathbf{A}_n = 0.$$

EXAMPLE 8.5

Show that

(a)

$$\frac{D}{Dt}(ds^2) = d\mathbf{x}\cdot\mathbf{A}_1 d\mathbf{x} \quad \text{and} \quad \frac{D^2}{Dt^2}(ds^2) = d\mathbf{x}\cdot\mathbf{A}_2 d\mathbf{x}\cdot$$

(b)

$$\mathbf{A}_1 = 2\mathbf{D} = \nabla\mathbf{v} + (\nabla\mathbf{v})^T, \quad \mathbf{A}_2 = \frac{D\mathbf{A}_1}{Dt} + \mathbf{A}_1(\nabla\mathbf{v}) + (\nabla\mathbf{v})^T\mathbf{A}_1\,. \qquad (8.11a)$$

Solution. (a) We have, at any time τ

$$ds^2 = d\boldsymbol{\xi}\cdot d\boldsymbol{\xi} = d\mathbf{x}\cdot\mathbf{C}_t d\mathbf{x},$$

$$\left[\frac{D}{Dt}(ds^2)\right]_{\text{at time } t} \equiv \left[\frac{\partial}{\partial\tau}(ds^2)\right]_{\substack{x_i\text{-fixed}\\ \tau=t}} = d\mathbf{x}\cdot\left(\frac{\partial\mathbf{C}_t}{\partial\tau}\right) d\mathbf{x} = d\mathbf{x}\cdot\mathbf{A}_1 d\mathbf{x}.$$

Similarly,

$$\left[\frac{D^2}{Dt^2}(ds^2)\right]_{\text{at time } t} = d\mathbf{x}\cdot\left(\frac{\partial^2\mathbf{C}_t}{\partial\tau^2}\right)_{\tau=t} d\mathbf{x} = d\mathbf{x}\cdot\mathbf{A}_2 d\mathbf{x}.$$

(b) From Chapter 3, we have, at time t

$$\frac{D}{Dt}(ds^2) = 2d\mathbf{x}\cdot\mathbf{D}d\mathbf{x},$$

where $\mathbf{D} = \frac{1}{2}[\nabla\mathbf{v} + (\nabla\mathbf{v})^T]$ is the rate of deformation tensor. Therefore,

$$\mathbf{A}_1 = 2\mathbf{D} = (\nabla\mathbf{v}) + (\nabla\mathbf{v})^T.$$

Next

$$\frac{D^2}{Dt^2}(ds^2) = \left(\frac{D}{Dt}d\mathbf{x}\right)\cdot\mathbf{A}_1 d\mathbf{x} + d\mathbf{x}\cdot\frac{D}{Dt}\mathbf{A}_1 d\mathbf{x} + d\mathbf{x}\cdot\mathbf{A}_1\frac{D}{Dt}d\mathbf{x}.$$

But

$$\frac{D}{Dt}d\mathbf{x} = (\nabla\mathbf{v})d\mathbf{x} \qquad [\textit{see } \text{Eq. } (3.20)].$$

Therefore,

$$\frac{D^2}{Dt^2}(ds^2) = (\nabla\mathbf{v})d\mathbf{x}\cdot\mathbf{A}_1 d\mathbf{x} + d\mathbf{x}\cdot\frac{D\mathbf{A}_1}{Dt}d\mathbf{x} + d\mathbf{x}\cdot\mathbf{A}_1(\nabla\mathbf{v})d\mathbf{x}.$$

From the definition of transpose

$$(\nabla \mathbf{v})d\mathbf{x} \cdot \mathbf{A}_1 d\mathbf{x} = d\mathbf{x} \cdot (\nabla \mathbf{v})^T \mathbf{A}_1 d\mathbf{x},$$

$$\frac{D^2}{Dt^2}(ds^2) = d\mathbf{x} \cdot \left[\frac{D\,\mathbf{A}_1}{Dt} + (\nabla \mathbf{v})^T \mathbf{A}_1 + \mathbf{A}_1(\nabla \mathbf{v})\right] d\mathbf{x}.$$

Thus†

$$\mathbf{A}_2 = \frac{D\mathbf{A}_1}{Dt} + \mathbf{A}_1(\nabla \mathbf{v}) + (\nabla \mathbf{v})^T \mathbf{A}_1.$$

We note that in Cartesian coordinates

$$\frac{DA_{ij}}{Dt} = \frac{\partial A_{ij}}{\partial t} + v_m \frac{\partial A_{ij}}{\partial x_m} \tag{8.11b}$$

8.4 INCOMPRESSIBLE SIMPLE FLUID

An incompressible simple fluid is an isotropic ideal material having the following constitutive equation

$$\mathbf{T} = -p\mathbf{I} + \mathbf{T}', \tag{8.12}$$

where $\mathbf{T}'$ depends on the past histories up to the current time t of the relative deformation tensor $\mathbf{C}_t$ and p is an indeterminate hydrostatic pressure. In other words,

$$\mathbf{T} = -p\mathbf{I} + \underset{\tau=-\infty}{\overset{\tau=t}{\mathbf{H}}} [\mathbf{C}_t(\mathbf{x}, \tau)], \tag{8.13}$$

where $\underset{\tau=-\infty}{\overset{\tau=t}{\mathbf{H}}}$ indicates that the values of $\mathbf{H}$ depend on all $\mathbf{C}_t$ from $\mathbf{C}_t(\mathbf{x}, -\infty)$ to $\mathbf{C}_t(\mathbf{x}, t)$.

We see from Section 8.3 that under the assumption that the Taylor series expansion of $\mathbf{C}_t(\mathbf{x},t)$ is justified,† the Rivlin-Ericksen tensor $\mathbf{A}_n (n = 1, 2, \ldots, \infty)$ determines the history of $\mathbf{C}_t(\mathbf{x},\tau)$. Thus, we may write

$$\mathbf{T} = -p\mathbf{I} + \mathbf{f}(\mathbf{A}_1, \mathbf{A}_2, \ldots, \mathbf{A}_n, \ldots), \tag{8.14a}$$

where tr $\mathbf{A}_1 = 0$, which follows from the equation of conservation of mass for an incompressible fluid‡.

†Note: The remaining Rivlin–Ericksen tensors satisfy the recursion relation

$$\mathbf{A}_{n+1} = \frac{D}{Dt}\mathbf{A}_n + \mathbf{A}_n(\nabla \mathbf{v}) + (\nabla \mathbf{v})^T \mathbf{A}_n.$$

†This is a very severe restriction. For example, stress relaxation, which has a discontinuity in its history of deformation, is ruled out.

‡Or, one may include tr $\mathbf{A}_1 = 0$ as part of the definition of incompressible simple fluid.

We further require that the function **f** cannot be arbitrary but must satisfy the relation

$$\mathbf{Q}^T\mathbf{f}(\mathbf{A}_1, \mathbf{A}_2, \ldots, \mathbf{A}_n)\mathbf{Q} = \mathbf{f}(\mathbf{Q}^T\mathbf{A}_1\mathbf{Q}, \mathbf{Q}^T\mathbf{A}_2\mathbf{Q}, \ldots, \mathbf{Q}^T\mathbf{A}_n\mathbf{Q}) \qquad (8.14b)$$

for any orthogonal tensor **Q**.

Equation (8.14b) makes "isotropy of material" as part of the definition of a simple fluid. It may be remarked that according to the principle of material frame indifference,† under the assumption that **f** depends only on the $\mathbf{A}_n$'s, the only admissible form of **f** is that satisfying Eq. (8.14b).

8.5 RIVLIN–ERICKSEN FLUID

If, in Eq. (8.14) only a finite number of the Rivlin–Ericksen tensors is included as arguments of **f**, the fluid is then known as the Rivlin–Ericksen fluid. In fact, the **Rivlin–Ericksen incompressible fluid of complexity** n is defined by the constitutive equation

$$\mathbf{T} = -p\mathbf{I} + \mathbf{f}(\mathbf{A}_1, \mathbf{A}_2, \ldots, \mathbf{A}_n), \qquad (8.15)$$

In particular, a **Rivlin–Ericksen liquid of complexity 2** is given by‡ (with $\mathbf{A}_1{}^2 \equiv \mathbf{A}_1\mathbf{A}_1$, etc.)

$$\begin{aligned}\mathbf{T} = -p\mathbf{I} &+ \mu_1\mathbf{A}_1 + \mu_2\mathbf{A}_1{}^2 + \mu_3\mathbf{A}_2 + \mu_4\mathbf{A}_2{}^2 \\ &+ \mu_5(\mathbf{A}_1\mathbf{A}_2 + \mathbf{A}_2\mathbf{A}_1) + \mu_6(\mathbf{A}_1\mathbf{A}_2{}^2 + \mathbf{A}_2{}^2\mathbf{A}_1) \\ &+ \mu_7(\mathbf{A}_1{}^2\mathbf{A}_2 + \mathbf{A}_2\mathbf{A}_1{}^2) + \mu_8(\mathbf{A}_1{}^2\mathbf{A}_2{}^2 + \mathbf{A}_2{}^2\mathbf{A}_1{}^2), \end{aligned} \qquad (8.16a)$$

where $\mu_1, \mu_2, \ldots, \mu_8$ are scalar material functions of the following scalar invariants:

$$\begin{aligned}&\operatorname{tr}\mathbf{A}_1{}^2, \operatorname{tr}\mathbf{A}_1{}^3, \operatorname{tr}\mathbf{A}_2, \operatorname{tr}\mathbf{A}_2{}^2, \operatorname{tr}\mathbf{A}_2{}^3, \\ &\operatorname{tr}\mathbf{A}_1\mathbf{A}_2, \operatorname{tr}\mathbf{A}_1{}^2\mathbf{A}_2, \operatorname{tr}\mathbf{A}_1\mathbf{A}_2{}^2, \operatorname{tr}\mathbf{A}_1{}^2\mathbf{A}_2{}^2.\end{aligned} \qquad (8.16b)$$

We note that if $\mu_2 = \mu_3 = \cdots \mu_8 = 0$ and $\mu_1 =$ a constant, Eq. (8.16a) reduces to the constitutive equation for a Newtonian liquid with viscosity μ_1.

†For a discussion of the principle of material frame-indifference, *see* p. 44, Encyclopedia of Physics, Vol. III/3.

‡Actually, Eq. (8.16) is the result of a "representation theorem" which states that **f**, $\mathbf{A}_1$, $\mathbf{A}_2$ being symmetric, the most general polynomial form of $\mathbf{f}(\mathbf{A}_1, \mathbf{A}_2)$ under the restriction Eq. (8.14b) is that given in Eqs. (8.16a) and (8.16b).

Another important special case is the so-called "**second-order fluid**" defined by the constitutive equation

$$\mathbf{T} = -p\mathbf{I} + \mu_1\mathbf{A}_1 + \mu_2\mathbf{A}_1^2 + \mu_3\mathbf{A}_2, \tag{8.17}$$

where μ_1, μ_2, and μ_3 are material constants.

EXAMPLE 8.6

For a second-order fluid, compute the stress components in a simple shearing flow given by the velocity field

$$v_1 = kx_2, \quad v_2 = 0, \quad v_3 = 0.$$

Solution. From Example 8.3, we have, for the simple shearing flow,

$$[\mathbf{A}_1] = \begin{bmatrix} 0 & k & 0 \\ k & 0 & 0 \\ 0 & 0 & 0 \end{bmatrix},$$

$$[\mathbf{A}_2] = \begin{bmatrix} 0 & 0 & 0 \\ 0 & 2k^2 & 0 \\ 0 & 0 & 0 \end{bmatrix},$$

and

$$\mathbf{A}_3 = \mathbf{A}_4 = \cdots = \mathbf{0}.$$

Now

$$[\mathbf{A}_1^2] = [\mathbf{A}_1][\mathbf{A}_1] = \begin{bmatrix} k^2 & 0 & 0 \\ 0 & k^2 & 0 \\ 0 & 0 & 0 \end{bmatrix},$$

therefore, Eq. (8.17) gives

$$T_{11} = -p + \mu_2 k^2,$$

$$T_{22} = -p + \mu_2 k^2 + 2\mu_3 k^2,$$

$$T_{33} = -p,$$

$$T_{12} = \mu_1 k,$$

$$T_{13} = T_{23} = 0.$$

We see that because of the presence of μ_2 and μ_3, normal stresses, in excess of p, on the planes x_1 = constant and x_2 = constant are necessary to maintain the shearing flow. Furthermore, these normal stress components are unequal. The normal stress differences $\sigma_1(k) \equiv T_{22} - T_{33}$ and $\sigma_2(k) \equiv T_{11} - T_{33}$ are known as normal stress functions. For a Newtonian fluid they are zero. For a second-order fluid,

$$\sigma_2(k) = \mu_2 k^2,$$

$$\sigma_1(k) = \mu_2 k^2 + 2\mu_3 k^2.$$

By measuring the normal stress differences and the shearing stress components T_{12}, the three material constants are determined. These three constants constitute an adequate characterization of a simple fluid if k is not too large.

EXAMPLE 8.7

For the simple shearing flow of the previous example, compute the scalar invariants of Eq. (8.16b).

Solution. Since

$$[\mathbf{A}_1] = \begin{bmatrix} 0 & k & 0 \\ k & 0 & 0 \\ 0 & 0 & 0 \end{bmatrix}, \qquad [\mathbf{A}_1]^2 = \begin{bmatrix} k^2 & 0 & 0 \\ 0 & k^2 & 0 \\ 0 & 0 & 0 \end{bmatrix},$$

$$[\mathbf{A}_1{}^3] = \begin{bmatrix} 0 & k^3 & 0 \\ k^3 & 0 & 0 \\ 0 & 0 & 0 \end{bmatrix}, \qquad [\mathbf{A}_2] = \begin{bmatrix} 0 & 0 & 0 \\ 0 & 2k^2 & 0 \\ 0 & 0 & 0 \end{bmatrix},$$

$$[\mathbf{A}_2]^2 = \begin{bmatrix} 0 & 0 & 0 \\ 0 & 4k^4 & 0 \\ 0 & 0 & 0 \end{bmatrix}, \qquad [\mathbf{A}_2]^3 = \begin{bmatrix} 0 & 0 & 0 \\ 0 & 8k^6 & 0 \\ 0 & 0 & 0 \end{bmatrix},$$

$$[\mathbf{A}_1][\mathbf{A}_2] = \begin{bmatrix} 0 & 2k^3 & 0 \\ 0 & 0 & 0 \\ 0 & 0 & 0 \end{bmatrix}, \quad [\mathbf{A}_1{}^2\mathbf{A}_2] = \begin{bmatrix} 0 & 0 & 0 \\ 0 & 2k^4 & 0 \\ 0 & 0 & 0 \end{bmatrix},$$

$$[\mathbf{A}_1\mathbf{A}_2{}^2] = \begin{bmatrix} 0 & 4k^5 & 0 \\ 0 & 0 & 0 \\ 0 & 0 & 0 \end{bmatrix}, \quad [\mathbf{A}_1{}^2\mathbf{A}_2{}^2] = \begin{bmatrix} 0 & 0 & 0 \\ 0 & 4k^6 & 0 \\ 0 & 0 & 0 \end{bmatrix}.$$

Therefore,

$$\text{tr}\,\mathbf{A}_1{}^2 = 2k^2,\ \text{tr}\,\mathbf{A}_1{}^3 = 0,\ \text{tr}\,\mathbf{A}_2 = 2k^2,$$
$$\text{tr}\,\mathbf{A}_2{}^2 = 4k^4,\ \text{tr}\,\mathbf{A}_2{}^3 = 8k^6,\ \text{tr}\,\mathbf{A}_1\mathbf{A}_2 = 0,$$
$$\text{tr}\,\mathbf{A}_1{}^2\mathbf{A}_2 = 2k^4,\ \text{tr}\,\mathbf{A}_1\mathbf{A}_2{}^2 = 0,\ \text{tr}\,\mathbf{A}_1{}^2\mathbf{A}_2{}^2 = 4k^6,$$

EXAMPLE 8.8

In a simple shearing flow, compute the stress components for the Rivlin–Ericksen liquid.

Solution. From Eqs. (8.16) and the results of the previous example, we have (note: $\mathbf{A}_3 = \mathbf{A}_4 = \cdots = 0$)

$$T_{11} = -p + k^2[\mu_2(k^2)] \equiv -p + \sigma_2(k^2),$$
$$T_{22} = -p + 2k^2[\mu_3(k^2) + \tfrac{1}{2}\mu_2(k^2) + 2\{\mu_4(k^2) + \mu_7(k^2)\}k^2 + 4k^4\mu_8(k^2)] = -p + \sigma_1(k^2),$$
$$T_{33} = -p,$$
$$T_{12} = k[\mu_1(k^2) + 2k^2\mu_5(k^2) + 4k^4\mu_6(k^2)] = k\eta(k^2) \equiv s(k),$$
$$T_{23} = T_{13} = 0,$$

where $\mu_1(k^2)$ indicates μ_1 is a function of k^2, etc. The normal stress differences are

$$T_{22} - T_{33} = \sigma_1(k^2),$$
$$T_{11} - T_{33} = \sigma_2(k^2).$$

The above results indicate that the normal stress functions σ_1 and σ_2 are even functions of k (= rate of shear), whereas the shear stress function $s(k)$ is an odd function of k. Figures 8.3 and 8.4 show some experimental results of σ_1, σ_2, and $\eta(k)$, where $\eta(k) \equiv s(k)/k$ is

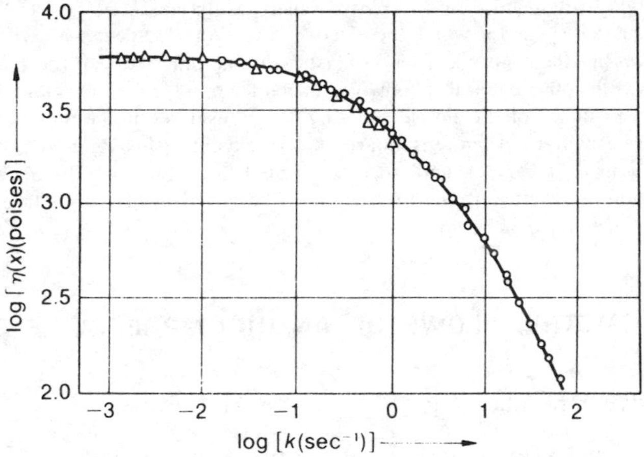

Fig. 8.3 Graph of η for a 13% solution of polyisobutylene in decalin. The data were obtained using cone and plate viscometers with $\alpha = 2°$ (Δ) and $\frac{1}{2}°$ (○). Markovitz, Elyash, Padden, and DeWitt [1955]. From "Viscometric Flows of Non-Newtonian Fluids" by Coleman, Markovitz, and Noll, New York: *Springer Tracts in Natural Philosophy*, Vol. 5, 1966.

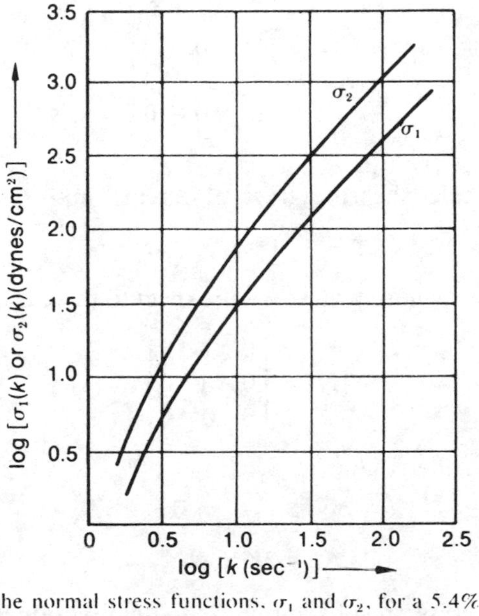

Fig. 8.4 Graph of the normal stress functions, σ_1 and σ_2, for a 5.4% solution of polyisobutylene in cetane. Markovitz and Brown [1965]. From "Viscometric Flows of Non-Newtonian Fluids" by Coleman, Markovitz, and Noll, New York: *Springer Tracts in Natural Philosophy*, Vol. 5, 1966.

known as the viscosity function or apparent viscosity. Materials such as the 13% solution of polyisobutylene in decalin, for which the viscosity function $\eta(k)$ decreases with the rate of shear, is known by Rheologists as pseudo-plastic. On the other hand, if the viscosity function increases with k, the material is known as dilatant. For a Newtonian liquid the viscosity function is a constant. For the simple shearing flow considered in the previous examples, we see that the Rivlin–Ericksen tensors are all zero except $\mathbf{A}_1$ and $\mathbf{A}_2$, and that only three material functions $\sigma_1(k)$, $\sigma_2(k)$, and $\eta(k)$ are needed to characterize the material. In the following section, we shall consider a whole class of flows for which this is true. This class of flows is known as viscometric flow.

8.6 VISCOMETRIC FLOWS OF AN INCOMPRESSIBLE SIMPLE FLUID

For steady simple shearing flow,

$$v_1 = kx_2, \quad v_2 = 0, \quad v_3 = 0, \quad k = \text{constant}.$$

The result of Example 8.3 shows that the Rivlin–Ericksen tensors are

$$[\mathbf{A}_1] = \begin{bmatrix} 0 & k & 0 \\ k & 0 & 0 \\ 0 & 0 & 0 \end{bmatrix},$$

$$[\mathbf{A}_2] = \begin{bmatrix} 0 & 0 & 0 \\ 0 & 2k^2 & 0 \\ 0 & 0 & 0 \end{bmatrix},$$

and all $\mathbf{A}_3, \mathbf{A}_4, \mathbf{A}_5 \ldots$ vanish.

Thus in a simple shearing flow of an incompressible simple fluid

$$\mathbf{T} = -p\mathbf{I} + \mathbf{f}(\mathbf{A}_1, \mathbf{A}_2). \tag{8.18}$$

Let $\mathbf{N}$ be a tensor whose matrix with respect to the Cartesian bases is

$$[\mathbf{N}] = \begin{bmatrix} 0 & 1 & 0 \\ 0 & 0 & 0 \\ 0 & 0 & 0 \end{bmatrix}, \tag{8.19}$$

then, it is easily verified that,

$$\mathbf{A}_1 = k(\mathbf{N} + \mathbf{N}^T),$$
$$\mathbf{A}_2 = 2k^2\mathbf{N}^T\mathbf{N}. \tag{8.20}$$

We note that the above results apply to the unidirectional flow $v_1 = v(x_2)$, $v_2 = 0$, $v_3 = 0$ as well. (However, $k \equiv dv/dx_2$.) Also, we have from

Example 8.4 that for the axisymmetric flow field with the Cartesian velocity components given by

$$v_1 = v(r), \quad v_2 = 0, \quad v_3 = 0, \quad r^2 = x_2^2 + x_3^2, \tag{8.21}$$

the Rivlin–Ericksen tensors are, with $\sin\theta = x_2/r$, $\cos\theta = x_2/r$,

$$[\mathbf{A}_1] = k(r)\begin{bmatrix} 0 & \cos\theta & \sin\theta \\ \cos\theta & 0 & 0 \\ \sin\theta & 0 & 0 \end{bmatrix},$$
$$[\mathbf{A}_2] = 2k^2(r)\begin{bmatrix} 0 & 0 & 0 \\ 0 & \cos^2\theta & \sin\theta\cos\theta \\ 0 & \sin\theta\cos\theta & \sin^2\theta \end{bmatrix}, \tag{8.22}$$

and all other $\mathbf{A}_n = 0$, $n = 3, 4, 5 \ldots$.

Thus, in this axially symmetric flow field of an incompressible simple fluid, we also have

$$\mathbf{T} = -p\mathbf{I} + \mathbf{f}(\mathbf{A}_1, \mathbf{A}_2).$$

Let $\mathbf{N}$ be a tensor whose matrix with respect to the Cartesian basis $(\mathbf{e}_1, \mathbf{e}_2, \mathbf{e}_3)$ is

$$[\mathbf{N}] = \begin{bmatrix} 0 & \cos\theta & \sin\theta \\ 0 & 0 & 0 \\ 0 & 0 & 0 \end{bmatrix}. \tag{8.23}$$

We easily verify that

$$\mathbf{A}_1 = k(r)(\mathbf{N} + \mathbf{N}^T),$$
$$\mathbf{A}_2 = 2k^2(r)\mathbf{N}^T\mathbf{N}. \tag{8.24}$$

We see here $\mathbf{A}_1$ and $\mathbf{A}_2$ are of the same form as that of the simple shearing

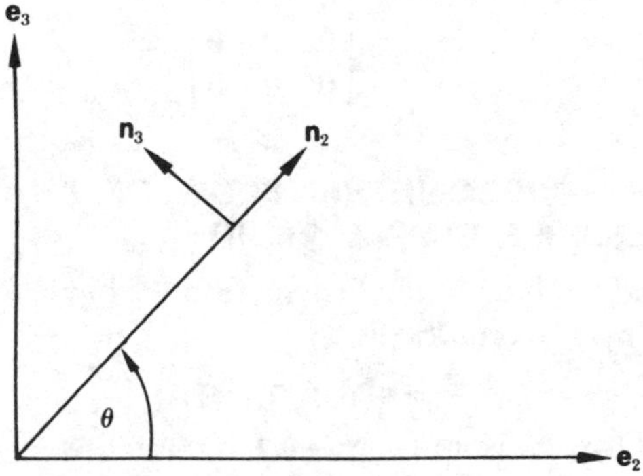

Fig. 8.5

flow. In fact, if we let $\mathbf{n}_1, \mathbf{n}_2, \mathbf{n}_3$ be bases related to $\mathbf{e}_1, \mathbf{e}_2, \mathbf{e}_3$ by

$$\begin{aligned} \mathbf{n}_1 &= \mathbf{e}_1, \\ \mathbf{n}_2 &= \cos\theta\mathbf{e}_2 + \sin\theta\mathbf{e}_3, \\ \mathbf{n}_3 &= -\sin\theta\mathbf{e}_2 + \cos\theta\mathbf{e}_3, \end{aligned} \tag{8.25}$$

then, with respect to this bases, $\mathbf{n}_1, \mathbf{n}_2, \mathbf{n}_3$, the matrix of $\mathbf{N}$ is

$$[\mathbf{N}]_{\mathbf{n}_1,\mathbf{n}_2,\mathbf{n}_3} = \begin{bmatrix} 1 & 0 & 0 \\ 0 & \cos\theta & \sin\theta \\ 0 & -\sin\theta & \cos\theta \end{bmatrix} \begin{bmatrix} 0 & \cos\theta & \sin\theta \\ 0 & 0 & 0 \\ 0 & 0 & 0 \end{bmatrix} \begin{bmatrix} 1 & 0 & 0 \\ 0 & \cos\theta & -\sin\theta \\ 0 & \sin\theta & \cos\theta \end{bmatrix},$$

i.e.,

$$[\mathbf{N}]_{\mathbf{n}_1,\mathbf{n}_2,\mathbf{n}_3} = \begin{bmatrix} 0 & 1 & 0 \\ 0 & 0 & 0 \\ 0 & 0 & 0 \end{bmatrix}.$$

This is the same as Eq. (8.19) of the simple shearing flow. In other words, the axisymmetric flow we are considering is essentially a simple shearing flow, that is, for a given element if one orientates oneself properly [according to Eq. (8.25)], one sees that the element is under the state of simple shearing.

The above two examples are flows belonging to a class of flows known as the **steady viscometric flow** which can be defined as one for which all Rivlin–Ericksen tensors vanish except $\mathbf{A}_1$ and $\mathbf{A}_2$, and

$$\begin{aligned} \mathbf{A}_1 &= k(\mathbf{N}+\mathbf{N}^T), \\ \mathbf{A}_2 &= 2k^2\,\mathbf{N}^T\mathbf{N}, \end{aligned} \tag{8.26}$$

where k is in general a function of position and the matrix of $\mathbf{N}$ relative to a proper choice of bases (in general, rotating) is†

$$[\mathbf{N}]_{\mathbf{n}_i} = \begin{bmatrix} 0 & 1 & 0 \\ 0 & 0 & 0 \\ 0 & 0 & 0 \end{bmatrix}. \tag{8.27}$$

8.7 STRESSES IN VISCOMETRIC FLOW OF AN INCOMPRESSIBLE SIMPLE FLUID

From the definition given in the last section we have, for an incompressible simple fluid in a viscometric flow,

$$\mathbf{T} = -p\mathbf{I} + \mathbf{f}(\mathbf{A}_1, \mathbf{A}_2), \tag{8.28}$$

†The tensor $\mathbf{N}$ must have the properties $\mathbf{N}^2 = 0$, $\operatorname{tr}\mathbf{N} = 0$, $\operatorname{tr}\mathbf{N}^T\mathbf{N} = 1$ for Eq. (8.27) to be possible.

where the Rivlin–Ericksen tensors $\mathbf{A}_1$ and $\mathbf{A}_2$ are expressible as

$$\mathbf{A}_1 = k(\mathbf{N} + \mathbf{N}^T),$$
$$\mathbf{A}_2 = 2k^2\mathbf{N}^T\mathbf{N},$$

and the matrix of $\mathbf{N}$ with respect to some choice of bases is

$$[\mathbf{N}] = \begin{bmatrix} 0 & 1 & 0 \\ 0 & 0 & 0 \\ 0 & 0 & 0 \end{bmatrix}_{\mathbf{n}_i}.$$

Furthermore, Eq. (8.14b) demands that for all orthogonal tensors $\mathbf{Q}$,

$$\mathbf{Q}^T\mathbf{f}(\mathbf{A}_1, \mathbf{A}_2)\mathbf{Q} = \mathbf{f}(\mathbf{Q}^T\mathbf{A}_1\mathbf{Q}, \mathbf{Q}^T\mathbf{A}_2\mathbf{Q}). \tag{8.29}$$

If we choose $\mathbf{Q}$, such that

$$[\mathbf{Q}]_{\mathbf{n}_i} = \begin{bmatrix} 1 & 0 & 0 \\ 0 & 1 & 0 \\ 0 & 0 & -1 \end{bmatrix}.$$

Then

$$[\mathbf{Q}]^T[\mathbf{N}]\,[\mathbf{Q}] = \begin{bmatrix} 1 & 0 & 0 \\ 0 & 1 & 0 \\ 0 & 0 & -1 \end{bmatrix} \begin{bmatrix} 0 & 1 & 0 \\ 0 & 0 & 0 \\ 0 & 0 & 0 \end{bmatrix} \begin{bmatrix} 1 & 0 & 0 \\ 0 & 1 & 0 \\ 0 & 0 & -1 \end{bmatrix}$$
$$= \begin{bmatrix} 0 & 1 & 0 \\ 0 & 0 & 0 \\ 0 & 0 & 0 \end{bmatrix} = [\mathbf{N}].$$

Also

$$[\mathbf{Q}^T]\,[\mathbf{N}^T\mathbf{N}]\,[\mathbf{Q}] = \begin{bmatrix} 1 & 0 & 0 \\ 0 & 1 & 0 \\ 0 & 0 & -1 \end{bmatrix} \begin{bmatrix} 0 & 0 & 0 \\ 0 & 1 & 0 \\ 0 & 0 & 0 \end{bmatrix} \begin{bmatrix} 1 & 0 & 0 \\ 0 & 1 & 0 \\ 0 & 0 & -1 \end{bmatrix}$$
$$= \begin{bmatrix} 0 & 0 & 0 \\ 0 & 1 & 0 \\ 0 & 0 & 0 \end{bmatrix} = [\mathbf{N}^T\mathbf{N}],$$

i.e., for the choice of $\mathbf{Q}$,

$$\mathbf{Q}^T\mathbf{N}\mathbf{Q} = \mathbf{N}$$

and

$$\mathbf{Q}^T(\mathbf{N}^T\mathbf{N})\mathbf{Q} = \mathbf{N}^T\mathbf{N}.$$

Thus

$$\mathbf{Q}^T\mathbf{A}_1\mathbf{Q} = k\mathbf{Q}^T(\mathbf{N} + \mathbf{N}^T)\mathbf{Q} = k(\mathbf{N} + \mathbf{N}^T) = \mathbf{A}_1$$

and

$$\mathbf{Q}^T\mathbf{A}_2\mathbf{Q} = 2k^2\mathbf{Q}^T(\mathbf{N}^T\mathbf{N})\mathbf{Q} = 2k^2\mathbf{N}^T\mathbf{N} = \mathbf{A}_2.$$

Now, from Eq. (8.29), we have, for this particular $\mathbf{Q}$,

$$\mathbf{Q}^T\mathbf{T}\mathbf{Q} = -p\mathbf{I} + \mathbf{f}(\mathbf{Q}^T\mathbf{A}_1\mathbf{Q}, \mathbf{Q}^T\mathbf{A}_2\mathbf{Q}),$$

$$\mathbf{Q}^T\mathbf{T}\mathbf{Q} = -p\mathbf{I} + \mathbf{f}(\mathbf{A}_1, \mathbf{A}_2),$$

i.e.,

$$\mathbf{Q}^T\mathbf{T}\mathbf{Q} = \mathbf{T}. \tag{i}$$

But

$$[\mathbf{Q}^T\mathbf{T}\mathbf{Q}]_{\mathbf{n}_i} = \begin{bmatrix} 1 & 0 & 0 \\ 0 & 1 & 0 \\ 0 & 0 & -1 \end{bmatrix} \begin{bmatrix} T_{11} & T_{12} & T_{13} \\ T_{21} & T_{22} & T_{23} \\ T_{31} & T_{32} & T_{33} \end{bmatrix} \begin{bmatrix} 1 & 0 & 0 \\ 0 & 1 & 0 \\ 0 & 0 & -1 \end{bmatrix}$$

$$= \begin{bmatrix} T_{11} & T_{12} & -T_{13} \\ T_{21} & T_{22} & -T_{23} \\ -T_{31} & -T_{32} & T_{33} \end{bmatrix}_{\mathbf{n}_i}.$$

Thus, from Eq. (i), with respect to the basis $\mathbf{n}_i$,

$$T_{13} = 0, \quad T_{23} = 0,$$

and with respect to this basis, $\mathbf{A}_1$ and $\mathbf{A}_2$ depend only on k, therefore, the nonzero stress components are $T_{12} = s(k)$, $T_{11} = -p + \alpha(k)$, $T_{22} = -p + \beta(k)$, and $T_{33} = -p + \gamma(k)$. Or, as customary, we let $\sigma_1(k) \equiv \beta - \gamma$ and $\sigma_2(k) \equiv \alpha - \gamma$, then the stress system in a viscometric flow of an incompressible simple fluid is, with respect to the basis $\mathbf{n}_i$, for which

$$[\mathbf{N}]_{\mathbf{n}_i} = \begin{bmatrix} 0 & 1 & 0 \\ 0 & 0 & 0 \\ 0 & 0 & 0 \end{bmatrix},$$

$$T_{12} = s(k), \qquad T_{13} = T_{23} = 0,$$

$$T_{22} - T_{33} = \sigma_1(k), \tag{8.30}$$

$$T_{11} - T_{33} = \sigma_2(k).$$

The function $s(k)$ is called the shear stress function and the function $\sigma_1(k)$, $\sigma_2(k)$ are called the normal stress functions. They are known also as the viscometric functions. These functions, when determined from the experiments on *one* viscometric flow of a fluid, determine completely the properties of the fluid in any other viscometric flow.

It can be shown (*see* Problems 8.12) that

$$s(-k) = -s(k),$$

$$\sigma_1(-k) = \sigma_1(k), \tag{8.31}$$

and

$$\sigma_2(-k) = \sigma_2(k).$$

That is, s is an odd function of k, while σ_1 and σ_2 are even functions of k.

8.8 SIMPLE SHEARING FLOW

With $v_1 = kx_2$, $v_2 = 0$, $v_3 = 0$, $k = \text{constant}$, the acceleration components are all zero. Thus, the equations of motion read, in the absence of body force,

$$0 = \frac{\partial T_{11}}{\partial x_1} + \frac{\partial T_{12}}{\partial x_2} + \frac{\partial T_{13}}{\partial x_3},$$

$$0 = \frac{\partial T_{21}}{\partial x_1} + \frac{\partial T_{22}}{\partial x_2} + \frac{\partial T_{23}}{\partial x_3},$$

$$0 = \frac{\partial T_{31}}{\partial x_1} + \frac{\partial T_{32}}{\partial x_2} + \frac{\partial T_{33}}{\partial x_3}.$$

For an incompressible simple fluid in this flow, the basis $\mathbf{n}_i$ for which

$$[\mathbf{N}]_{\mathbf{n}_i} = \begin{bmatrix} 0 & 1 & 0 \\ 0 & 0 & 0 \\ 0 & 0 & 0 \end{bmatrix}$$

is the basis $(\mathbf{e}_1, \mathbf{e}_2, \mathbf{e}_3)$, therefore with respect to $(\mathbf{e}_1, \mathbf{e}_2, \mathbf{e}_3)$,

$$T_{13} = T_{23} = 0, \quad T_{12} = s(k), \quad T_{11} = T_{33} + \sigma_2(k), \quad T_{22} = T_{33} + \sigma_1(k),$$

so that the equation of motion becomes

$$\frac{\partial T_{33}}{\partial x_1} = \frac{\partial T_{33}}{\partial x_2} = \frac{\partial T_{33}}{\partial x_3} = 0.$$

In other words, the flow is dynamically possible with $T_{33} = \text{constant}$.

If the fluid is between two plates of distance d apart, one at rest, the other moves with a constant velocity v in its own plane, then $k = v/d$ and the shear stress function $s(k)$ can be obtained by measuring the shearing stress needed to maintain the steady flow at different values of the constant k. Also, by measuring the differences in the normal stresses $(T_{11} - T_{33}$ and $T_{22} - T_{33})$ in the same experiment, the normal stress functions $\sigma_2(k)$ and $\sigma_1(k)$ can be determined.

For a Newtonian fluid such as water, the measurements discussed above give

$$s(k) = \mu k,$$
$$\sigma_1(k) = 0, \qquad (8.32)$$
$$\sigma_2(k) = 0.$$

That is, the shear stress function is a linear function of k and there are no normal stress differences. For a non-Newtonian fluid, such as a polymeric solution, for small k, the viscometric functions can be approximated by a few terms of their Taylor series expansions. Noting that s is an odd function of k, we have

$$s(k) = \mu k + \mu_1 k^3 + \cdots \qquad (8.33)$$

and noting that σ_1 and σ_2 are even functions of k, we have

$$\sigma_1(k) = s_1 k^2 + \ldots, \qquad (8.34)$$

$$\sigma_2(k) = s_2 k^2 + \ldots. \qquad (8.35)$$

Since the deviation from Newtonian behavior [*see* Eq. (8.32)] is of the order of k^2 for σ_1 and σ_2, and of k^3 for s, therefore, it is expected that the deviation of the normal stresses will manifest themselves within the range of k in which the response of the shear stress remains essentially the same as that of a Newtonian fluid.

8.9 CHANNEL FLOW

We now consider the steady shearing flow between two infinite parallel fixed plates. That is,

$$v_1 = v(x_2), \quad v_2 = 0, \quad v_3 = 0,$$

with

$$v\left(\pm\frac{h}{2}\right) = 0. \qquad (8.36)$$

Again, for an incompressible simple fluid in this flow, the basis $\mathbf{n}_i$ for which

$$[\mathbf{N}]_{\mathbf{n}_i} = \begin{bmatrix} 0 & 1 & 0 \\ 0 & 0 & 0 \\ 0 & 0 & 0 \end{bmatrix}$$

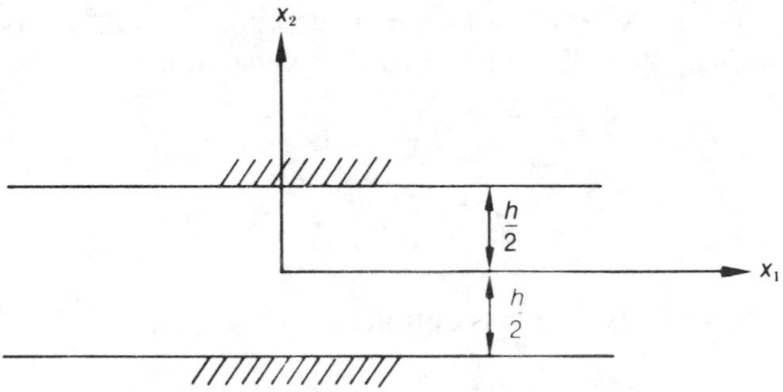

Fig. 8.6

is the Cartesian basis $(\mathbf{e}_1, \mathbf{e}_2, \mathbf{e}_3)$. Therefore, with respect to the basis $\mathbf{e}_i$, the stress components are, with $k(x_2) \equiv dv/dx_2$

$$\begin{aligned} T_{12} &= s(k), \quad T_{13} = T_{23} = 0, \\ T_{11} &= T_{33} + \sigma_2(k), \quad T_{22} = T_{33} + \sigma_1(k). \end{aligned} \tag{8.37}$$

Substituting the above equation in the equation of motion, we get, in the absence of body forces,

$$0 = \frac{\partial T_{11}}{\partial x_1} + \frac{\partial T_{12}}{\partial x_2} + \frac{\partial T_{13}}{\partial x_3} = \frac{\partial T_{33}}{\partial x_1} + \frac{\partial T_{12}}{\partial x_2}, \tag{8.38a}$$

$$0 = \frac{\partial T_{21}}{\partial x_1} + \frac{\partial T_{22}}{\partial x_2} + \frac{\partial T_{23}}{\partial x_3} = \frac{\partial T_{33}}{\partial x_2} + \frac{d\sigma_1}{dx_2} \tag{8.38b}$$

$$0 = \frac{\partial T_{31}}{\partial x_1} + \frac{\partial T_{32}}{\partial x_2} + \frac{\partial T_{33}}{\partial x_3} = \frac{\partial T_{33}}{\partial x_3}. \tag{8.38c}$$

From Eqs. (8.38b) and (8.38c), $(\partial T_{33}/\partial x_1)$ is independent of x_2 and x_3 and from Eq. (8.38a), we have

$$-\frac{dT_{12}}{dx_2} = \frac{\partial T_{33}}{\partial x_1}.$$

Since the left side of this equation depends only on x_2, therefore $\partial T_{33}/\partial x_1$ = constant. Let us denote $\partial T_{33}/\partial x_1$ by f. Now, $dT_{12}/dx_2 = -f$ from which

$$T_{12} = -fx_2 + C.$$

By the condition of symmetry, $T_{12} = 0$ at $x_2 = 0$. Therefore

$$C = 0$$

and

$$T_{12} = -fx_2.$$

Since $T_{12} = s(k)$ if we let λ (s) be the inverse of $s(k)$ [e.g., if $s(k) = \mu k$, then $\lambda(s) = s/\mu$]. We call $\lambda(s)$ the rate of shear function. From $s(k) = -fx_2$, we get

$$k = \lambda(-fx_2),$$

i.e.,

$$\frac{dv}{dx_2} = \lambda(-fx_2),$$

from which the velocity profile is obtained to be

$$v(x_2) = \int_{-h/2}^{x_2} \lambda(-fx)\,dx, \tag{8.39}$$

where $\lambda(s)$ is considered to be a known material function for a particular fluid, obtained from some other viscometric flow.

Let Q be the rate of volume discharge per unit length in x_3-direction across a cross-section of the channel, then

$$Q = \int_{-h/2}^{h/2} v(x)\,dx.$$

Integration by parts gives

$$Q = xv(x)\Big]_{-h/2}^{h/2} - \int_{-h/2}^{h/2} xv'(x)\,dx,$$

where

$$v'(x) \equiv k(x).$$

Noting that $v(\pm h/2) = 0$, and $dv/dx = \lambda(-fx)$, we have

$$Q = -\int_{-h/2}^{h/2} x\lambda(-fx)\,dx.$$

For a Newtonian fluid $s(k) = \mu k$, and $k = s/\mu \equiv \lambda(s)$. Thus $\lambda(-fx) = -fx/\mu$,

$$v(x_2) = \int_{-h/2}^{x_2} -\frac{fx}{\mu}\,dx = -\frac{f}{\mu}\frac{x^2}{2}\Big]_{-h/2}^{x_2} = -\frac{f}{\mu}\left[\frac{x_2{}^2}{2} - \frac{h^2}{8}\right] \tag{8.40}$$

and

$$Q = -\int_{-h/2}^{h/2} x\left(\frac{-fx}{\mu}\right)dx = \frac{f}{\mu}\frac{x^3}{3}\Big]_{-h/2}^{h/2} = \frac{fh^3}{12\mu}. \tag{8.41}$$

It is obvious that if $s(k)$ is not linearly proportional to k, the velocity profile will not be a parabola and the volume discharge will not be linearly proportional to the constant $f\,(= dT_{33}/dx_1)$.

PROBLEMS

8.1. For the following velocity field, (a) obtain the particle motion using the current time as the reference time, and (b) obtain the relative deformation tensor:

$$v_1 = 0, \quad v_2 = v(x_1), \quad v_3 = 0.$$

8.2. Do Problem 8.1 for the velocity field

$$v_1 = 0, \quad v_2 = 0, \quad v_3 = v(r), \quad r^2 = x_1{}^2 + x_2{}^2.$$

8.3. Do Problem 8.1 for the velocity field

$$v_1 = -kx_1, \quad v_2 = kx_2, \quad v_3 = 0.$$

8.4. Using $\mathbf{C}_t(\mathbf{x}, \tau)$ of Example 8.2, find the Rivlin–Ericksen tensors by computing

$$\left(\frac{\partial \mathbf{C}_t}{\partial \tau}\right)_{\tau=t} \quad \text{and} \quad \left(\frac{\partial^2 \mathbf{C}_t}{\partial \tau^2}\right)_{\tau=t}.$$

8.5. Do Problem 8.4 for the $\mathbf{C}_t(\mathbf{x}, \tau)$ of Example 8.4.

8.6. Show that (*see* Example 8.5)

$$\frac{D^3}{Dt^3}(ds^2) = d\mathbf{x} \cdot \mathbf{A}_3\, d\mathbf{x},$$

where

$$\mathbf{A}_3 = \frac{D}{Dt}\mathbf{A}_2 + \mathbf{A}_2(\nabla \mathbf{v}) + (\nabla \mathbf{v})^T \mathbf{A}_2.$$

8.7. Obtain the Rivlin–Ericksen tensors by using Eqs. (8.11a) and (8.11b) for the following velocity field:

$$v_1 = 0, \quad v_2 = v(x_1), \quad v_3 = 0.$$

8.8. Do Problem 8.7 for the velocity field

$$v_1 = -y\omega(r), \quad v_2 = x\omega(r), \quad v_z = 0, \quad r^2 = x^2 + y^2.$$

8.9. Do Problem 8.7 for the velocity field

$$v_1 = 0, \quad v_2 = 0, \quad v_3 = w(r), \quad r^2 = x^2 + y^2.$$

8.10. Obtain the first three Rivlin–Ericksen tensors for the velocity field, use Eqs. (8.11a) and (8.11b) and the results of Problem 8.6:

$$v_1 = -kx_1, \quad v_2 = kx_2, \quad v_3 = 0.$$

8.11. Do Problem 8.10 for the velocity field,

$$v_1 = kx_2, \quad v_2 = kx_1, \quad v_3 = 0.$$

8.12. Let $\mathbf{Q}$ be a tensor whose matrix with respect to the basis $\mathbf{n}_i$ is

$$[\mathbf{Q}]_{\mathbf{n}_i} = \begin{bmatrix} -1 & 0 & 0 \\ 0 & 1 & 0 \\ 0 & 0 & 1 \end{bmatrix}.$$

(a) Verify the following relations for the tensor $\mathbf{N}$ whose matrix with respect to $\mathbf{n}_i$ is given by Eq. (8.27):

$$\mathbf{Q}^T\mathbf{N}\mathbf{Q} = -\mathbf{N} \quad \text{and} \quad \mathbf{Q}^T\mathbf{N}^T\mathbf{N}\mathbf{Q} = \mathbf{N}^T\mathbf{N}.$$

(b) For $\mathbf{A}_1$ and $\mathbf{A}_2$ given by Eq. (8.26), verify the relations

$$\mathbf{Q}^T\mathbf{A}_1\mathbf{Q} = -\mathbf{A}_1 \quad \text{and} \quad \mathbf{Q}^T\mathbf{A}_2\mathbf{Q} = \mathbf{A}_2.$$

(c) From $\mathbf{T}(k) = -p\mathbf{I} + \mathbf{f}(\mathbf{A}_1, \mathbf{A}_2)$, where $\mathbf{A}_1$ and $\mathbf{A}_2$ are given by Eq. (8.26) and

$$\mathbf{Q}^T\mathbf{f}(\mathbf{A}_1, \mathbf{A}_2)\mathbf{Q} = \mathbf{f}(\mathbf{Q}^T\mathbf{A}_1\mathbf{Q}, \mathbf{Q}^T\mathbf{A}_2\mathbf{Q}).$$

Show that

$$\mathbf{Q}^T\mathbf{T}(k)\mathbf{Q} = \mathbf{T}(-k).$$

(d) From the results of part (c), show that the viscometric functions have the properties

$$s(k) = -s(-k),$$
$$\sigma_1(k) = \sigma(-k),$$
$$\sigma_2(k) = \sigma(-k).$$

8.13. Given that the only nonzero Rivlin–Ericksen tensors for the velocity field of Problem 8.8 are

$$[\mathbf{A}_1]_{\mathbf{e}_i} = k(r)\begin{bmatrix} -2\dfrac{xy}{r^2} & \dfrac{x^2-y^2}{r^2} & 0 \\ \dfrac{x^2-y^2}{r^2} & 2\dfrac{xy}{r^2} & 0 \\ 0 & 0 & 0 \end{bmatrix},$$

$$[\mathbf{A}_2]_{\mathbf{e}_i} = 2k^2(r)\begin{bmatrix} \dfrac{x^2}{r^2} & \dfrac{xy}{r^2} & 0 \\ \dfrac{xy}{r^2} & \dfrac{y^2}{r^2} & 0 \\ 0 & 0 & 0 \end{bmatrix},$$

where $k(r) = r(d\omega/dr)$. Verify that

$$\mathbf{A}_1 = k(\mathbf{N} + \mathbf{N}^T),$$
$$\mathbf{A}_2 = 2k^2\mathbf{N}^T\mathbf{N},$$

where

$$[\mathbf{N}]_{\mathbf{e}_i} = \begin{bmatrix} -\frac{xy}{r^2} & -\frac{y^2}{r^2} & 0 \\ \frac{x^2}{r^2} & \frac{xy}{r^2} & 0 \\ 0 & 0 & 0 \end{bmatrix}.$$

8.14. For the tensor **N** of the previous problem, (a) verify $\mathbf{NN} = 0$, $\operatorname{tr}\mathbf{N} = 0$, $\operatorname{tr}\mathbf{N}^T\mathbf{N} = 1$, and (b) find the basis $\mathbf{n}_i$ such that

$$[\mathbf{N}]_{\mathbf{n}_i} = \begin{bmatrix} 0 & 1 & 0 \\ 0 & 0 & 0 \\ 0 & 0 & 0 \end{bmatrix}.$$

8.15. Given that the only nonzero Rivlin–Ericksen tensors for the velocity field of Problem 8.9 are

$$[\mathbf{A}_1]_{\mathbf{e}_i} = k(r) \begin{bmatrix} 0 & 0 & \frac{x}{r} \\ 0 & 0 & \frac{y}{r} \\ \frac{x}{r} & \frac{y}{r} & 0 \end{bmatrix},$$

$$[\mathbf{A}_2]_{\mathbf{e}_i} = 2k^2(r) \begin{bmatrix} \frac{x^2}{r^2} & \frac{xy}{r^2} & 0 \\ \frac{xy}{r^2} & \frac{y^2}{r^2} & 0 \\ 0 & 0 & 0 \end{bmatrix},$$

(a) Find the tensor **N** such that

$$\mathbf{A}_1 = k(\mathbf{N} + \mathbf{N}^T)$$

and

$$\mathbf{A}_2 = 2k^2\mathbf{N}^T\mathbf{N}.$$

(b) Find the basis $\mathbf{n}_i$ such that

$$[\mathbf{N}]_{\mathbf{n}_i} = \begin{bmatrix} 0 & 1 & 0 \\ 0 & 0 & 0 \\ 0 & 0 & 0 \end{bmatrix}.$$

Appendix: Matrices

MATRIX

A matrix is an aggregate of elements arranged in rectangular array. The following is a matrix of m rows and n columns

$$T = \begin{bmatrix} T_{11}, & T_{12}, \ldots, & T_{1n} \\ T_{21}, & T_{22}, \ldots, & T_{2n} \\ T_{m1}, & T_{m2}, \ldots, & T_{mn} \end{bmatrix}. \tag{1}$$

There are $m \times n$ elements in the above matrix. The first subscript of each element refers to the row in which the element is located and the second subscript refers to the column. Thus T_{32} is located in the third row and second column of the matrix. In general, T_{ij} is the element in the ith row and j^{th} column. The matrix in Eq. (1) will be denoted by

$$T = [T_{ij}] \qquad \text{where} \qquad \begin{matrix} i = 1, 2, \ldots, m \\ j = 1, 2, \ldots, n \end{matrix} \tag{2}$$

Whenever necessary, we shall use the notation $T^{m \times n}$ to indicate that there are m rows and n columns in the matrix T.

TRANSPOSE OF A MATRIX

Let $S^{n \times m} = [S_{ij}]$ be a matrix of n rows and m columns. If $S_{ij} = T_{ji}$, then S is called the transpose of T, and will be denoted by S^T.

For example, if

$$T^{2\times 3} = \begin{bmatrix} 1 & 2 & 0 \\ 4 & 0 & 1 \end{bmatrix},$$

then,

$$S^{3\times 2} = \begin{bmatrix} 1 & 4 \\ 2 & 0 \\ 0 & 1 \end{bmatrix} = T^T.$$

Square Matrix

A matrix with equal number of rows and columns is called a square matrix. For example,

$$T = \begin{bmatrix} T_{11} & T_{12} & T_{13} \\ T_{21} & T_{22} & T_{23} \\ T_{31} & T_{32} & T_{33} \end{bmatrix}$$

is a square matrix of the third order. T_{11}, T_{22}, T_{33} are called the diagonal elements. The rest of the elements are called nondiagonal elements.

Symmetric Matrix

A square matrix with the property $T_{ij} = T_{ji}$ is called a symmetric matrix. Example:

$$T = \begin{bmatrix} T_{11} & 4 & -1 \\ 4 & T_{22} & 2 \\ -1 & 2 & T_{33} \end{bmatrix}.$$

Note that for a symmetric matrix

$$T = T^T.$$

Diagonal Matrix

A square matrix with the property that all nondiagonal elements are zero is called a diagonal matrix. Example:

$$T = \begin{bmatrix} 2 & 0 & 0 \\ 0 & -1 & 0 \\ 0 & 0 & 4 \end{bmatrix}.$$

Note that a diagonal matrix is a symmetric matrix.

Scalar Matrix

A diagonal matrix with the property that $T_{11} = T_{22} = T_{33} = \alpha$ is called a scalar matrix. Example:

$$T = \begin{bmatrix} 2 & 0 & 0 \\ 0 & 2 & 0 \\ 0 & 0 & 2 \end{bmatrix}.$$

Identity Matrix

A scalar matrix with the property that the diagonal elements equal unity is called the identity matrix. We shall denote the identity matrix by I. Thus

$$I = \begin{bmatrix} 1 & 0 & 0 \\ 0 & 1 & 0 \\ 0 & 0 & 1 \end{bmatrix}.$$

Row Matrix

A matrix with only one row is called a row matrix. For a row matrix, only one index is necessary to locate the position of an element. Thus

$$a = [a_1, a_2, a_3] = [a_i]_r,$$

where the subscript r indicates that a is a row matrix.

Column Matrix

A matrix with only one column is called a column matrix. Thus

$$a = \begin{bmatrix} a_1 \\ a_2 \\ a_3 \end{bmatrix} = [a_i]_c$$

is a column matrix. Note that

$$[a_r]^T = [a_c].$$

Matrix Operations

1. If $T^{m\times n} = [T_{ij}]$ and $S^{m\times n} = [S_{ij}]$, then $T = S$, if and only if $T_{ij} = S_{ij}$.
2. If α is a scalar, then

$$\alpha T = T\alpha \equiv [\alpha T_{ij}].$$

Example:

$$3\begin{bmatrix} 1 & 2 \\ 3 & 4 \end{bmatrix} = \begin{bmatrix} 3 & 6 \\ 9 & 12 \end{bmatrix}.$$

3. $T^{m\times n}+S^{m\times n} \equiv [T_{ij}+S_{ij}]$.

Example:

$$\begin{bmatrix}1 & 2\\3 & 4\end{bmatrix}+\begin{bmatrix}2 & 2\\0 & 2\end{bmatrix}=\begin{bmatrix}3 & 4\\3 & 6\end{bmatrix}.$$

The following rules follow from the operation rules of scalars

$$T+S = S+T, \qquad \text{(commutative law)}$$

$$T+[S+R] = [T+S]+R = T+S+R, \qquad \text{(associative law)}$$

$$[\alpha+\beta]T = \alpha T+\beta T,$$

$$\alpha[T+S] = \alpha T+\alpha S.$$

4. If $T^{m\times n} = [T_{ij}]$ is a matrix of m rows and n columns and $S^{n\times p} = [S_{ij}]$ is a matrix of n rows and p columns, then

$$T^{m\times n}S^{n\times p} \equiv R^{m\times p},$$

where the elements of R are given by

$$R_{ij} = \sum_{\alpha=1}^{n} T_{i\alpha}S_{\alpha j},$$

Example:

$$\begin{bmatrix}T_{11} & T_{12} & T_{13}\\T_{21} & T_{22} & T_{23}\end{bmatrix}\begin{bmatrix}S_{11} & S_{12}\\S_{21} & S_{22}\\S_{31} & S_{32}\end{bmatrix}$$

$$=\begin{bmatrix}T_{11}S_{11}+T_{12}S_{21}+T_{13}S_{31}, & T_{11}S_{12}+T_{12}S_{22}+T_{13}S_{32}\\T_{21}S_{11}+T_{22}S_{21}+T_{23}S_{31}, & T_{21}S_{12}+T_{22}S_{22}+T_{23}S_{32}\end{bmatrix},$$

$$[a_1, a_2, a_3]\begin{bmatrix}b_1\\b_2\\b_3\end{bmatrix} = a_1b_1+a_2b_2+a_3b_3.$$

It is important to note that only when the number of columns of the first matrix is the same as the number of rows of the second matrix is multiplication defined. Example:

$$\begin{bmatrix}1 & 2\\3 & 4\end{bmatrix}\begin{bmatrix}1\\2\end{bmatrix}=\begin{bmatrix}5\\11\end{bmatrix}$$

but

$$\begin{bmatrix}1\\2\end{bmatrix}\begin{bmatrix}1 & 2\\3 & 4\end{bmatrix}$$

is not defined.

The following examples show *even if both TS and ST are defined, they are in general not equal.* Example:

$$\begin{bmatrix} 1 & 2 \\ 3 & 4 \end{bmatrix}\begin{bmatrix} 1 & 1 \\ 2 & 3 \end{bmatrix} = \begin{bmatrix} 5 & 7 \\ 11 & 15 \end{bmatrix}$$

but

$$\begin{bmatrix} 1 & 1 \\ 2 & 3 \end{bmatrix}\begin{bmatrix} 1 & 2 \\ 3 & 4 \end{bmatrix} = \begin{bmatrix} 4 & 6 \\ 11 & 16 \end{bmatrix}.$$

However, if T is a scalar matrix, i.e.,

$$T = \begin{bmatrix} \alpha & 0 & 0 \\ 0 & \alpha & 0 \\ 0 & 0 & \alpha \end{bmatrix},$$

then

$$TS = ST,$$

provided both TS and ST are defined (that is, provided S is also a square matrix and the same order as that of T). In particular, if S is a square matrix of the same order as that of the unit matrix I, then

$$IS = SI = S.$$

It can be shown that the matrix product has the following properties:

$$T[SR] = [TS]R,$$

$$\alpha[TS] = [\alpha T]S = T[\alpha S],$$

$$[T+S]R = TR + SR,$$

$$R[T+S] = RT + RS.$$

The Reversed Rule for a Transposed Product

In the following we shall show that the transpose of a product of matrices is equal to the product of their transpose in reverse order, i.e.,

$$[TS]^T = S^T T^T.$$

Proof: Given

$$T^{m\times n} = [T_{ij}],$$

$$S^{n\times p} = [S_{ij}],$$

$$[T^{m\times n}S^{n\times p}] = \left(\sum_{\alpha=1}^{n} T_{i\alpha}S_{\alpha j}\right)^{m\times p},$$

$$[T^{m\times n}S^{n\times p}]^T = [R_{ij}]^{p\times m} \qquad \text{where } R_{ij} = \sum_{\alpha=1}^{n} T_{j\alpha}S_{\alpha i}.$$

Now

$$T^T = [T^T_{ij}]^{n\times m} \qquad \text{where } T^T_{ij} = T_{ji},$$

$$S^T = [S^T_{ij}]^{p\times n} \qquad \text{where } S^T_{ij} = T_{ji},$$

$$S^T T^T = \left[\sum_{\alpha=1}^{n} S^T_{i\alpha} T^T_{\alpha j}\right]^{p\times m} = \left[\sum_{\alpha=1}^{n} S_{\alpha i} T_{j\alpha}\right] = [R_{ij}]^{p\times m}.$$

That is,

$$S^T T^T = [TS]^T.$$

Inverse of a Matrix

A square matrix S is called the inverse of the square matrix T if

$$TS = I,$$

where I is a unit matrix.

We shall denote S by T^{-1}. Thus

$$TT^{-1} = I.$$

It can be proved that if the determinant of T, i.e., $|T_{ij}|$, is not equal to zero, then the inverse of T exists and that

$$TT^{-1} = T^{-1}T = I,$$

where T^{-1} is unique. The proof will not be given here.

The Reversed Rule of the Inverse of a Product of Matrices

In the following we shall show that the inverse of a product of matrices is equal to the product of their inverses in reversed order, i.e.,

$$[TS]^{-1} = S^{-1}T^{-1}.$$

Proof:

$$[TS][S^{-1}T^{-1}] = T[SS^{-1}]T^{-1} = TIT^{-1} = TT^{-1} = I.$$

Thus, $S^{-1}T^{-1}$ is the inverse of TS.

Differentiation of a Matrix

If $T = [T_{ij}]$, where T_{ij} are functions of x, y, z, then

$$\frac{\partial}{\partial x} T \equiv \left[\frac{\partial}{\partial x} T_{ij}\right],$$

$$\frac{\partial}{\partial y} T \equiv \left[\frac{\partial}{\partial y} T_{ij}\right], \text{etc.}$$

Answers to Problems

CHAPTER 2

A1. (a) 5 (b) 28 (c) 28 (d) 23

A2. (a) and (c)

A5. (a) $T_{11} = T_{22} = T_{33} = 0,\ T_{12} = -T_{21} = 0,\ T_{23} = -T_{32} = 1,\ T_{31} = -T_{13} = 2$
(b) $c_1 = 3,\ c_2 = 2,\ c_3 = 0$

B1. (a)
$$[\mathbf{T}]_{\mathbf{e}_i} = \begin{bmatrix} 0 & -1 & 0 \\ -1 & 0 & 0 \\ 0 & 0 & 1 \end{bmatrix}_{\mathbf{e}_i}$$
(b)
$$\begin{bmatrix} -2 \\ -1 \\ 0 \end{bmatrix}_{\mathbf{e}_i}$$

B3.
$$[\mathbf{mn}]_{\mathbf{e}_i} = \frac{1}{2}\begin{bmatrix} -1 & 0 & 1 \\ -1 & 0 & 1 \\ 0 & 0 & 0 \end{bmatrix}$$

B5. **T** is not a linear transformation

B6. (b)
$$\begin{bmatrix} 1 & 0 & 0 \\ 0 & -\dfrac{1}{\sqrt{2}} & -\dfrac{1}{\sqrt{2}} \\ 0 & -\dfrac{1}{\sqrt{2}} & \dfrac{1}{\sqrt{2}} \end{bmatrix}_{\mathbf{e}_i}$$
(c)
$$\begin{bmatrix} 1 & 0 & 0 \\ 0 & -\dfrac{1}{\sqrt{2}} & \dfrac{1}{\sqrt{2}} \\ 0 & \dfrac{1}{\sqrt{2}} & \dfrac{1}{\sqrt{2}} \end{bmatrix}_{\mathbf{e}_i}$$

B13. (a)
$$[\mathbf{Q}]_{\mathbf{e}_i} = \begin{bmatrix} \dfrac{\sqrt{3}}{2} & -\dfrac{1}{2} & 0 \\ \dfrac{1}{2} & \dfrac{\sqrt{3}}{2} & 0 \\ 0 & 0 & 1 \end{bmatrix}$$
(b)
$$[\mathbf{a}]_{\mathbf{e}'_i} = \begin{bmatrix} 2 \\ 0 \\ 0 \end{bmatrix}_{\mathbf{e}'_i}$$

B16. $T'_{11} = \frac{4}{5};\ T'_{12} = -3\sqrt{5};\ T'_{31} = \frac{2}{5}$

B17. (a)
$$[T'_{ij}] = \begin{bmatrix} 0 & -5 & 0 \\ -5 & 1 & 5 \\ 0 & 5 & 1 \end{bmatrix}$$
(b) $T_{ii} = T'_{ii} = 2$
$\det[T_{ij}] = \det[T'_{ij}] = -25$

B26. (a)
$$[\mathbf{T}^S] = \begin{bmatrix} 1 & 3 & 5 \\ 3 & 5 & 7 \\ 5 & 7 & 9 \end{bmatrix} \quad [\mathbf{T}^A] = \begin{bmatrix} 0 & -1 & -2 \\ 1 & 0 & -1 \\ 2 & 1 & 0 \end{bmatrix}$$
(b)
$$[\mathbf{t}^A] = \begin{bmatrix} 1 \\ -2 \\ 1 \end{bmatrix}$$

B28. Eigenvector of $\mathbf{T}$ is $\mathbf{n}$
Eigenvalue of $\mathbf{T}$ is $r_1 s_1$

B31. (a) $\mathbf{n} = \dfrac{\mathbf{a}}{|\mathbf{a}|}$ (b) $\mathbf{a}\cdot\mathbf{b} = a_i b_i$

B32. (a)
$$[\mathbf{R}]_{\mathbf{e}_i} = \begin{bmatrix} 0 & 0 & 1 \\ 1 & 0 & 0 \\ 0 & 1 & 0 \end{bmatrix}$$
(b) $\mathbf{n} = \pm\dfrac{1}{\sqrt{3}}(\mathbf{e}_1 + \mathbf{e}_2 + \mathbf{e}_3)$

B33. (b)
$$[\mathbf{R}^S]_{\mathbf{e}'_i} = \begin{bmatrix} \cos\theta & 0 & 0 \\ 0 & \cos\theta & 0 \\ 0 & 0 & 1 \end{bmatrix}; \quad [\mathbf{R}^A]_{\mathbf{e}'_i} = \begin{bmatrix} 0 & -\sin\theta & 0 \\ \sin\theta & 0 & 0 \\ 0 & 0 & 0 \end{bmatrix}$$

(c) $\lambda_1 = \cos\theta$; $\mathbf{n}_1 = \alpha\mathbf{e}'_1 + \beta\mathbf{e}'_2$;
$\alpha^2 + \beta^2 = 1$ otherwise arbitrary.
$\lambda_2 = \cos\theta$; $\mathbf{n}_2 = -\beta\mathbf{e}'_1 + \alpha\mathbf{e}'_2$;
$\lambda_3 = 1$; $\mathbf{n}_3 = \mathbf{e}'_3$

(d) $R'_{ii} = 2\cos\theta + 1$; (e) $\mathbf{t}^A = \sin\theta\,\mathbf{e}'_3$

(f) $\theta = \arccos\left(-\dfrac{1}{2}\right) = 120°$

$$\mathbf{e}'_3 = \frac{1}{2\sin\theta}(\mathbf{e}_1 + \mathbf{e}_2 + \mathbf{e}_3) = \frac{1}{\sqrt{3}}(\mathbf{e}_1 + \mathbf{e}_2 + \mathbf{e}_3)$$

B38. (a) $I_1 = 3$; $I_2 = -16$; $I_3 = -48$
$\lambda_1 = 3 \rightarrow \mathbf{n}_1 = \mathbf{e}_1$
$\lambda_2 = 4 \rightarrow \mathbf{n}_2 = \dfrac{1}{\sqrt{2}}(\mathbf{e}_2 + \mathbf{e}_3)$
$\lambda_3 = -4 \rightarrow \mathbf{n}_3 = \dfrac{1}{\sqrt{2}}(-\mathbf{e}_2 + \mathbf{e}_3)$

(b)
$$[\mathbf{T}]_{\mathbf{n}_i} = \begin{bmatrix} 3 & 0 & 0 \\ 0 & 4 & 0 \\ 0 & 0 & -4 \end{bmatrix}_{\mathbf{n}_i}$$

(c) No, the first invariants are not equal

B42. (a) $\mathbf{n} = \mathbf{e}_3$ (b) $|\nabla\varphi| = 2$ (c) $\dfrac{d\varphi}{dr} = \sqrt{2}$

B43. $\mathbf{n} = \left(\dfrac{x}{a^2}\mathbf{e}_1 + \dfrac{y}{b^2}\mathbf{e}_2 + \dfrac{z}{c^2}\mathbf{e}_3\right)\left(\dfrac{x^2}{a^4} + \dfrac{y^2}{b^4} + \dfrac{z^2}{c^4}\right)^{-\frac{1}{2}}$

B44. (a) $\mathbf{q} = -3k(\mathbf{e}_1 + \mathbf{e}_2)$
(b) $\mathbf{q} = -3k(\mathbf{e}_1 + 2\mathbf{e}_2)$

B45. (a) $\mathbf{E} = -\nabla\varphi = -\alpha(\cos\theta\mathbf{e}_1 + \sin\theta\mathbf{e}_2)$ (b) $\mathbf{D} = -\epsilon_1\alpha\cos\theta\mathbf{e}_1 - \epsilon_2\alpha\sin\theta\mathbf{e}_2$

(c) $|\mathbf{D}|^2 = \alpha^2(\epsilon_1{}^2\cos^2\theta + \epsilon_2{}^2\sin^2\theta)$

$|\mathbf{D}|_{max}$ for $\theta = 0; \pi/2; 3\pi/2 \ldots.$

B48. (a) $$[\nabla\mathbf{v}] = \begin{bmatrix} 2x & 0 & 0 \\ 0 & 0 & 2z \\ 0 & 2y & 0 \end{bmatrix}$$ (b) $$[(\nabla\mathbf{v})\mathbf{v}] = 2\begin{bmatrix} x^3 \\ y^2z \\ yz^2 \end{bmatrix}$$

(c) $div\ \mathbf{v} = 2x$ (d) $\mathbf{curl\ v} = 2(y-z)\mathbf{e}_1$ (e) $d\mathbf{v} = 2ds(x\mathbf{e}_1 + z\mathbf{e}_2 + y\mathbf{e}_3)$

B49. (b) $\left(\dfrac{dv}{dr}\right)_{max} = 12$

CHAPTER 3

3.1. (a) $\mathbf{v} = k\mathbf{e}_1, \mathbf{a} = 0$ (b) $\dfrac{D\theta}{Dt} = Ak$ (c) $\dfrac{D\theta}{Dt} = 0$

3.2. (b) $\mathbf{v} = 2kX_1{}^2t\mathbf{e}_2, \mathbf{a} = 2kX_1{}^2\mathbf{e}_2.$ (c) Use $X_1 = x_1$

3.3. (b) $v_1 = \dfrac{2ktx_2{}^2}{(1+kt)^2}, v_2 = \dfrac{kx_2}{1+kt}, v_3 = 0$ (c) $a_1 = \dfrac{2kx_2{}^2}{(1+kt)^2}, a_2 = 0, a_3 = 0$

3.4. (b) $\mathbf{v} = k\mathbf{e}_1, \mathbf{a} = 0$ (c) $\mathbf{v} = \left(\dfrac{k}{1+t}\right)\mathbf{e}_1, \mathbf{a} = 0$

3.5. (b) $\mathbf{v} = \left(\dfrac{x_1}{1+t}\right)\mathbf{e}_1$

3.9. (b) $\mathbf{a} = -\dfrac{1}{4}(\mathbf{e}_1 + \mathbf{e}_2)$

$\dfrac{D\theta}{Dt} = 2k$

3.13. (b) $E'_{11} = k/2$

3.14. (a) Unit elongations: $5k, 2k$

Decrease in angle: k

3.15. (a) $k/3$

3.16. (a) $\dfrac{58}{9}\times 10^{-4}$ (b) $\dfrac{32}{3\sqrt{5}}\times 10^{-4}$

3.19. $\left.\dfrac{\Delta l}{l}\right]_{max} = 3\times 10^{-6}$

3.22. $E_{11} = a, E_{22} = c, E_{12} = b - \frac{1}{2}(a+c)$

3.25. $E_{11} = a, E_{22} = \frac{1}{3}(2b + 2c - a), E_{12} = \dfrac{b-c}{\sqrt{3}}$

3.28. (b) $3k$

3.29. $\dfrac{(D/Dt)(ds_1)}{ds_1} = -(k+1), \dfrac{(D/Dt)(ds_2)}{ds_2} = -\dfrac{1}{2}(k+1)$

3.36. $k = 1$

3.38. $\upsilon_1 = \mathrm{v}_1(x_2, x_3), \upsilon_2 = 0$

3.39. (a) $\rho = \rho_0/(1+t)$ (b) $\rho = \alpha/x_1$

3.40. $\rho = \rho_0 e^{-t^2}$

CHAPTER 4

4.1. (a) On $\mathbf{e}_2$, $T_n = 4$ MPa (b) On $\mathbf{e}_3$, $T_s = 5.83$ MPa

4.2. (a) $\mathbf{t} = \dfrac{100}{3}(5\mathbf{e}_1 + 6\mathbf{e}_2 + 5\mathbf{e}_3)$ psi (b) $T_n = 300$ psi

$T_s = \dfrac{100\sqrt{5}}{3}$ psi

4.4. $\mathbf{t} = \dfrac{100}{4}(\sqrt{3}\mathbf{e}_1 + \mathbf{e}_2 - \sqrt{3}\mathbf{e}_3)$

4.5. $T'_{11} = -6.43$ Pa

$T'_{13} = 18.6$ Pa

4.6. $\mathbf{F} = 4\beta\mathbf{e}_2$

$\mathbf{M} = -(4\alpha/3)\mathbf{e}_3$

4.8. $\mathbf{F} = 4\alpha\mathbf{e}_1$

$\mathbf{M} = -\dfrac{4}{3}\alpha\mathbf{e}_1$

4.9. (a) $\mathbf{t} = 0$ (b) $\mathbf{F} = 0, \mathbf{M} = 8\pi\alpha\mathbf{e}_1$

4.16. $T_{11} = 10$ MPa

$T_{33} = 10$ MPa

4.18. (a) $T_1 = \tau, T_2 = 0, T_3 = -\tau$ (b) $T_s]_{\max} = \tau$

$\mathbf{n}_1 - \dfrac{1}{\sqrt{2}}(\mathbf{e}_1 + \mathbf{e}_2), \mathbf{n}_2 = \mathbf{e}_3, \mathbf{n}_3 = \dfrac{1}{\sqrt{2}}(-\mathbf{e}_1 + \mathbf{e}_2)$

$\mathbf{n} = \dfrac{\sqrt{2}}{2}(\mathbf{n}_1 \pm \mathbf{n}_3) = \mathbf{e}_1$ and $\mathbf{e}_2$

4.19. $T_s]_{\max} = \frac{1}{2}(\sigma_1 - \sigma_3)$

$\mathbf{n} = \dfrac{\sqrt{2}}{2}(\mathbf{e}_1 \pm \mathbf{e}_3)$

4.21. $T_{12} = 2x_1 - x_2 + 3$

4.22. $T_{33}=(1+\rho g)x_3+f(x_1,x_2)$

4.23. Yes

CHAPTER 5

5.4. E_Y = 207 GPa (30×10^6 psi), $\nu = 0.30$, k = 172 GPa (25×10^6 psi)

5.6. $\nu = 0.27$, λ = 91 GPa (13.2×10^6 psi), k = 141 GPa (20.5×10^6 psi)

5.8.
$$[\mathbf{T}] = \begin{bmatrix} 6.6 & -2.3 & 0 \\ -2.3 & 3.2 & 0 \\ 0 & 0 & 8.9 \end{bmatrix} \times 10^3 \text{ psi} = \begin{bmatrix} 45.5 & -15.9 & 0 \\ -15.9 & 22.1 & 0 \\ 0 & 0 & 61.4 \end{bmatrix} \text{MPa}$$

5.10. (a)
$$[\mathbf{E}] = \begin{bmatrix} 3.33 & 1.26 & 0 \\ 1.26 & -1.97 & 0 \\ 0 & 0 & -0.43 \end{bmatrix} \times 10^{-5}$$
(b) $\Delta V = 4.87 \times 10^{-3}$ cm^3

5.14. (a)
$$[\mathbf{T}] = k\mu \begin{bmatrix} 0 & 2X_3 & 2X_1+X_2 \\ 2X_3 & 0 & X_1-2X_2 \\ 2X_1+X_2 & X_1-2X_2 & 0 \end{bmatrix}$$
(b) Yes

5.17. c_L/c_T; 2; 7.15; 22.4

5.19. (a) Transverse wave in $\mathbf{e}_3$-direction (b) $c = c_T$ (c) $\alpha = -1$

(d) $\beta = \dfrac{n\pi}{l}$, $n = 1, 2, 3 \ldots$

5.20. (c) $\alpha = +1$ (d) $\beta = \dfrac{n\pi}{2l}$, $n = 1, 3, 5 \ldots$

5.21. (c) $\alpha = +1$ (d) $\beta = \dfrac{n\pi}{l}$, $n = 1, 2, 3 \ldots$

5.22. (a) Longitudinal in $\mathbf{e}_3$-direction (b) $c = c_L$ (c) $\alpha = -1$

(d) $\beta = \dfrac{n\pi}{l}$, $n = 1, 2, 3 \ldots$

5.27. (a) $\alpha_1 = \alpha_2 = \alpha_3 = 0$ (b) $\alpha_3 = 31.2°$, $\epsilon_2/\epsilon_1 = 0.74$, $\epsilon_3/\epsilon_1 = 0.50$

$\epsilon_2 = \epsilon_1$

$\epsilon_3 = 0$

5.28. $\alpha_1 = 30°$

5.34. (a) $u_1(X_1, t) = \alpha \cos \omega t \left[\cos \dfrac{\omega X_1}{c_L} + \left(\tan \dfrac{\omega l}{c_L}\right) \sin \dfrac{\omega X_1}{c_L}\right]$

(b) $\omega = \dfrac{n\pi c_L}{l}$ $n = 1, 3, 5 \ldots.$

5.37. (a) $\mathbf{u} = \alpha \left[\cos \dfrac{\omega}{c_T} X_1 - \cot \dfrac{\omega l}{c_L} \sin \dfrac{\omega X_1}{c_T}\right] [\cos \omega t \mathbf{e}_2 + \sin \omega t \mathbf{e}_3]$

(b) Circular

5.39. (a) $T_n]_{max} = 11{,}300$ psi (78 MPa) (b) $\Delta l = 3.62 \times 10^{-2}$ inch (9.2×10^{-2} cm)
$T_s]_{max} = 5{,}650$ psi (39 MPa) $\Delta d = -0.284 \times 10^{-3}$ inch (-0.721×10^{-3} cm

5.40. $\delta = 0.72$ cm

5.45. $P_1 = P/12, P_2 = P/3, P_3 = 7P/12$

5.48. $d = 6.12$ cm

5.49. $d = 3.38$ inch (8.59 cm)

5.51. $M_1 = \dfrac{M_t}{\dfrac{l_1}{l_2}\left(\dfrac{a_2}{a_1}\right)^4 + 1}, \quad M_2 = \dfrac{M_t}{1 + \dfrac{l_2}{l_1}\left(\dfrac{a_1}{a_2}\right)^4}$

5.53. $M_L = M \dfrac{(b+2c)}{(a+b+c)}$

5.54. (b) $T_n]_{max} = 21.4$ ksi (147 MPa)
$T_s]_{max} = 16.6$ ksi (114 MPa)

5.62. $d = 11.3$ cm

5.63. $b = 10$ inch (25.4 cm)

5.64. P = 0.21 MN

5.69. (a) Yes (b) $T_{11} = 0, T_{12} = -2\alpha X, T_{22} = 2\alpha, T_{33} = 2\nu\alpha$
(c) on $X_1 = a$: $\mathbf{t} = -2\alpha a \mathbf{e}_2$
$X_2 = b$: $\mathbf{t} = -2\alpha X_1 \mathbf{e}_1 + 2\alpha \mathbf{e}_2$

5.70. (a) If $\alpha = -\beta$

5.71. (a) Yes (b) $T_{11} = 2\alpha X_1, T_{22} = 0, T_{12} = -2\alpha X_2 - 3\beta X_2^2$
(c) $\alpha = -3/2\, \beta b$

CHAPTER 6

6.1. $B = 5.10 \times 10^4$ N

6.2. $p_A = p_0 + \rho (g + a) h$

6.3. $p_A = p_0 + \rho (h' + h)(g + a \sin \alpha)$

$h' = \dfrac{a r_0 \cos \alpha}{g + a \sin \alpha}$ ($h + h'$ = depth of water above A)

6.4. $z = \dfrac{\omega^2 r^2}{2g}$

6.5. $p = p_0 e^{-\rho_0 g (z - z_0)/p_0}$ for $n = 1$

6.6. (b) $T_n = -p + \mu, T_s = 0$

6.7. (*a*) -4μ, (*b*) 8μ

6.9. $v_1 = \dfrac{(\rho g \sin \theta) d^2}{\mu} \left[\left(\dfrac{x_2}{d}\right) - \dfrac{1}{2}\left(\dfrac{x_2}{d}\right)^2\right]$

6.12. (a) $v = A \ln \dfrac{r}{a} + v_a$ (b) $v = \dfrac{\beta}{4}(r^2 - a^2) + A \ln \dfrac{r}{a}$

$A = \dfrac{v_b - v_a}{\ln (b/a)}$ $\beta = \dfrac{1}{\mu}\dfrac{dp}{dx}$, $A = \dfrac{\beta(a^2 - b^2)}{4 \ln (b/a)}$

6.14. $A = \dfrac{\beta a^2 b^2}{2(a^2 + b^2)}$, $B = -A$

6.15. $A = -\dfrac{\beta\sqrt{3}}{6b}$, $B = 0$

6.17. $\theta = \left(\dfrac{\partial p}{\partial x}\right)^2 (b^4 - x_2{}^4)/(12\kappa\mu) + \dfrac{(\theta_2 - \theta_1)x_2}{2b} + \dfrac{\theta_1 + \theta_2}{2}$

6.21. (b) $T_{11} - p + 2\mu k$, $T_{22} = -p - 2\mu k$, $T_{33} = -p$, remaining $T_{ij} = 0$
(c) $\mathbf{a} = k^2(x_1\mathbf{e}_1 + x_2\mathbf{e}_2)$
(d) and (e) $p = -\dfrac{\rho k^2}{2}(x_1{}^2 + x_2{}^2) + p_0$ (f) $\Phi_{\text{inc}} = 4k^2\mu$ (g) No slip

6.22. (b) $T_{11} = -p + 4\mu k x_1$ (c) $a_1 = 2k^2(x_1{}^3 + x_1 x_2{}^2)$
$T_{22} = -p - 4\mu k x_1$ $a_2 = 2k^2(x_2{}^3 + x_1{}^2 x_2)$
$T_{12} = -4\mu k x_2$
$T_{33} = -p$
$T_{13} = T_{23} = 0$

6.27. $Q = 0.0762\ \text{m}^3/\text{sec}$

6.28. Q = 3.98 ft³/sec (0.113 m³/sec)

CHAPTER 7

7.7. (b) $3A\rho_0 e^{-t}, -3A\rho$ (c) $-3A\rho$

7.8. (b) $m = A\rho_0$ (c) $P = \frac{9}{2}\rho_0 A e^{t-t_0}\mathbf{e}_1$

7.11. (a) $\dfrac{4A\rho_0}{(1+t)^2}, -\dfrac{8A\rho_0}{(1+t)^3}$ (b) $\dfrac{8A\rho_0}{(1+t)^3}$ (c) 0

(d) $\dfrac{13}{3}\dfrac{\rho_0}{(1+t)^3}, -\dfrac{13\rho_0}{(1+t)^4}$ (e) $\dfrac{13\rho_0 A}{(1+t)^4}$

7.12. (a) $\frac{9}{2}\rho_0 A e^{-t}\mathbf{e}_1$, $-\frac{9}{2}\rho_0 A e^{-t}\mathbf{e}_1$ (b) $9\rho_0 A e^{-t}\mathbf{e}_1$ (c) $\frac{9}{2}\rho_0 A e^{-t}\mathbf{e}_1$

7.14. $\mathbf{F} = 8\rho(\mathbf{e}_1 + \mathbf{e}_2)$
$\mathbf{M} = 8\rho(-\mathbf{e}_1 + \mathbf{e}_2)$

7.18. $283\left(\dfrac{1}{2}\mathbf{e}_1 - \dfrac{\sqrt{3}}{2}\mathbf{e}_2\right) N$

7.20. $p v_0{}^2 A$

7.21. $y_2 = -\frac{y_1}{2} + \sqrt{\left(\frac{y_1}{2}\right)^2 + \frac{2Q^2}{gy_1}}$

7.24. 612 rad/sec

CHAPTER 8

8.1. $\xi_1 = x_1,\ \xi_2 = x_2 + v(x_1)(\tau - t),\ \xi_3 = x_3$

8.2. $\xi_1 = x_1,\ \xi_2 = x_2,\ \xi_3 = x_3 + v(r)(\tau - t)$

8.3. $\xi_1 = x_1 e^{-k(\tau-t)},\ \xi_2 = x_2 e^{k(\tau-t)},\ \xi_3 = x_3$

8.7. With $v' \equiv \frac{dv}{dx}$

$$[\mathbf{A}_1] = \begin{bmatrix} 0 & v' & 0 \\ v' & 0 & 0 \\ 0 & 0 & 0 \end{bmatrix},\quad [\mathbf{A}_2] = \begin{bmatrix} 2v'^2 & 0 & 0 \\ 0 & 0 & 0 \\ 0 & 0 & 0 \end{bmatrix}$$

8.8. With $\omega' = \frac{d\omega}{dr}$

$$[\mathbf{A}_1] = \frac{\omega'}{r}\begin{bmatrix} -2xy & x^2 - y^2 & 0 \\ x^2 - y^2 & 2xy & 0 \\ 0 & 0 & 0 \end{bmatrix},\quad [\mathbf{A}_2] = 2\omega'^2\begin{bmatrix} x^2 & xy & 0 \\ xy & y^2 & 0 \\ 0 & 0 & 0 \end{bmatrix}$$

$$\mathbf{A}_3 = \mathbf{A}_4 = \cdots = 0$$

8.9. With $w' = \frac{dw}{dr}$

$$[\mathbf{A}_1] = \frac{w'}{r}\begin{bmatrix} 0 & 0 & x \\ 0 & 0 & y \\ x & y & 0 \end{bmatrix},\quad [\mathbf{A}_2] = \frac{2w'^2}{r^2}\begin{bmatrix} x^2 & xy & 0 \\ xy & y^2 & 0 \\ 0 & 0 & 0 \end{bmatrix}$$

8.10.

$$[\mathbf{A}_1] = 2k\begin{bmatrix} -1 & 0 & 0 \\ 0 & 1 & 0 \\ 0 & 0 & 0 \end{bmatrix},\quad [\mathbf{A}_2] = 4k^2\begin{bmatrix} 1 & 0 & 0 \\ 0 & 1 & 0 \\ 0 & 0 & 0 \end{bmatrix},\quad [\mathbf{A}_3] = 8k^3\begin{bmatrix} -1 & 0 & 0 \\ 0 & 1 & 0 \\ 0 & 0 & 0 \end{bmatrix}$$

8.14. (b) $\mathbf{n}_1 = \mathbf{e}_\varphi = -\frac{y}{r}\mathbf{e}_1 + \frac{x}{r}\mathbf{e}_2$

$\mathbf{n}_2 = \mathbf{e}_r = \frac{x}{r}\mathbf{e}_1 + \frac{y}{r}\mathbf{e}_2$

$\mathbf{n}_3 = \mathbf{e}_z = \mathbf{e}_3$

8.15. (a)

$$[N] = \begin{bmatrix} 0 & 0 & 0 \\ 0 & 0 & 0 \\ x/r & y/r & 0 \end{bmatrix}$$

(b) $\mathbf{n}_1 = \mathbf{e}_z = \mathbf{e}_3$

$\mathbf{n}_2 = \mathbf{e}_r = (x/r)\,\mathbf{e}_1 + (y/r)\,\mathbf{e}_2$

$\mathbf{n}_3 = \mathbf{e}_\varphi = -(y/r)\,\mathbf{e}_1 + (x/r)\,\mathbf{e}_2$

Index